An Introduction to
COMPUTATIONAL
FLUID
MECHANICS

An Introduction to COMPUTATIONAL FLUID MECHANICS

Chuen-Yen Chow
Department of Aerospace Engineering Sciences
University of Colorado

JOHN WILEY & SONS
New York   Chichester   Brisbane   Toronto

Library of Congress Cataloging in Publication Data:

Chow, Chuen-Yen, 1932–
    An introduction to computational fluid mechanics.

    Bibliography: p.
    Includes index.
    1. Fluid mechanics.   2. Fluid mechanics--
Data processing.   I. Title.
TA357.C475   532′.05   78-27555
ISBN 0-471-15608-6

Printed in the United States of America

10 9 8 7 6 5 4 3 2 1

To my father
and
in memory of my mother

PREFACE

The motion of a realistic fluid flow is described by a set of nonlinear partial differential equations. Even for the simplest problem of a uniform flow past an aligned flat plate, many sophisticated mathematical techniques are required to obtain an analytical solution. In textbooks on introductory fluid mechanics, various assumptions are usually made to ignore viscosity or compressibility or to simplify the geometry so that closed-form solutions can be found for some idealized problems. Through these highly simplified or linearized problems, students learn the general behavior of the fluid flow; however, they cannot go too far beyond the material covered in a textbook because their analytical power is limited. Very quickly the analytically solvable problems run out, and the problems of practical importance, being either nonlinear or involving complex geometries, or both, are too difficult for beginning students.

This situation can be greatly improved with the help of numerical techniques, as I concluded from teaching computational fluid mechanics courses to seniors and first-year graduate students. In these courses, after a numerical method for solving a special type of ordinary or partial differential equation is introduced, its application to flow problems is immediately demonstrated through some representative computer programs. Equipped with these numerical techniques and having a previous basic knowledge of fluid mechanics, students can easily be led to the more advanced and realistic problems, including many that are otherwise impossible to tackle by using analytical methods.

This book was developed from the class notes that I have been using and revising for several years, and it is suitable for the senior-graduate level. Students are assumed to know the fundamentals of fluid mechanics and also to be familiar with the **FORTRAN IV** language. Numerical data are expressed in SI units (the International System of Units).

The topics are arranged so that ordinary differential equations appear before partial differential equations. For this reason, students will not be surprised to find that potential flows are considered after the data on dynamics of a body moving through a viscous fluid.

This is a book on fluid mechanics, not one on numerical computation. It deals with flow problems that either have to be solved numerically or that can be made much simpler with the help of a digital computer. Numerical techniques are dealt with merely as tools used to solve physical problems; detailed analyses and criticisms of these techniques are not attempted here. Usually, more than one numerical method exists for solving one type of differential equation. Because of the introductory nature of this book, the numerical methods adopted here are only the simpler or the commonly used ones and are by no means complete.

Thirty computer programs illustrate how to transform various finite-difference schemes into FORTRAN language and show a variety of techniques for printing or plotting the output data. Two plotting subroutines, PLOTN for printer plotting and EZPLOT for film plotting, have been used; they are available in the University of Colorado computer system. Before using these subroutines, students should check with the local computer library. Different versions of the subroutines written for the same purposes may be found at different institutions. If such subroutines are not available, students may write their own printer-plotting subroutines in comparison with the technique described in Section 2.13 just before Program 2.10.

I made the subject matter as broad as possible. The problems considered include those in classical aero- and hydrodynamics, and also those in ballistics, meteorology, hydraulics, gasdynamics, magnetohydrodynamics, and convection; however, many of the examples relate to aeronautical applications. In fact, Sections 2.4, 2.8, 3.3, and 3.4 complement the aerodynamics textbook by Kuethe and Chow*, in which the numerical methods are described without going into detail or the relevant problems are solved without using numerical methods.

This book is suitable for use in a two-semester course in computational fluid mechanics. It may be used as the text for a one-semester introductory course if some of the more advanced topics are omitted. It also may be used to supplement a traditional textbook for courses in fluid mechanics at various levels. Subprograms, constructed for special purposes, are used extensively. Those that may be helpful in solving other related problems are grouped according to their use and are listed in the appendix.

I am grateful to Professor Robert D. Richtmyer of the University of Colorado for reading part of the material and to Captain Thomas E. Miller and Dr. Wilbur Hankey, both of the Computational Aerodynamics Group, Wright-Patterson Air Force Base, Ohio, for reading the manuscript and providing valuable comments and suggestions. I thank the students in all my computational fluid mechanics classes for continuously helping me to improve the manuscript,

*A. M. Kuethe and C.-Y. Chow, Foundations of Aerodynamics: Bases of Aerodynamic Design, third edition. New York: Wiley, 1976.

and particularly Mr. Kurt M. Olender for his fruitful assistance. I especially thank my wife Julianna Huei Shek, whose helpful discussions, skillful programming techniques, and constant encouragement made this book possible.

The computer time and computing services provided by the University of Colorado Computing Center are greatly appreciated.

Boulder, Colorado   Chuen-Yen Chow

CONTENTS

An Introduction to
COMPUTATIONAL
FLUID
MECHANICS

chapter one

DYNAMICS OF A BODY MOVING THROUGH A FLUID MEDIUM

The numerical solution of initial-value problems that involve nonlinear ordinary differential equations is considered in this chapter. In Section 1.1 some numerical methods, especially the Runge-Kutta methods, are introduced for solving the first- and second-order equations. They are applied in Section 1.2 for finding the motion of a free-falling sphere through air and in Section 1.3 to simulate the motions of a simple pendulum and an aeroelastic system.

To extend the applications from one-dimensional to two-dimensional motions, Runge-Kutta formulas for solving simultaneous second-order equations are deduced in Section 1.4. After the motion of a spherical projectile in the presence of a fluid has been computed, the numerical integration procedure of Section 1.5 is combined with the half-interval method to find the maximum range of such a body. Section 1.6 deals with a meteorological problem concerning the collision of a raindrop with cloud droplets. Finally, while computing the trajectory of a glider in Section 1.7, a graphical method is employed to plot the result by the computer printer.

1.1 Numerical Solution of Ordinary Differential Equations—Initial-Value Problems

Consider the simplest case of a first-order ordinary differential equation having the general form

$$\frac{dx}{dt} = f(x, t) \tag{1.1.1}$$

where f is an analytic function. If, at a starting point $t = t_0$, the function x has a given value x_0, it is desired to find $x(t)$ for $t \geq t_0$ that satisfies both (1.1.1) and the prescribed initial condition. Such a problem is called an *initial-value problem*.

To solve the problem numerically, the axis of the independent variable is usually divided into evenly spaced small intervals of width h whose end points are situated at

$$t_i = t_0 + ih, \qquad i = 0, 1, 2, \cdots \tag{1.1.2}$$

The solution evaluated at the point t_i is denoted by x_i. Thus, by using a numerical method, the continuous function $x(t)$ is approximated by a set of discrete values x_i, $i = 0, 1, 2, \cdots$, as sketched in Fig. 1.1.1. Since h is small and f is an analytic function, the solution at any point can be obtained by means of a Taylor's series expansion about the previous point.

$$x_{i+1} \equiv x(t_{i+1}) = x(t_i + h)$$

$$= x_i + h \left(\frac{dx}{dt}\right)_i + \frac{h^2}{2!}\left(\frac{d^2x}{dt^2}\right)_i + \frac{h^3}{3!}\left(\frac{d^3x}{dt^3}\right)_i + \cdots$$

$$= x_i + hf_i + \frac{h^2}{2!}f'_i + \frac{h^3}{3!}f''_i + \cdots \tag{1.1.3}$$

where f_i^n denotes $d^n f / dt^n$ evaluated at (x_i, t_i). f is generally a function of both x and t, so that the first-order derivative is obtained according to the formula

$$\frac{df}{dt} = \frac{\partial f}{\partial t} + \frac{dx}{dt}\frac{\partial f}{\partial x}$$

Higher-order derivatives are obtained by using the same chain rule.

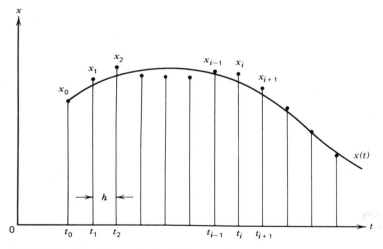

FIGURE 1.1.1 Numerical solution of an ordinary initial-value problem.

Dynamics of a Body Moving through a Fluid Medium

Alternatively (1.1.3) can be rewritten as

$$x_{i+1} = x_i + \Delta x_i \qquad (1.1.4)$$

where

$$\Delta x_i = hf_i + \frac{h^2}{2!} f_i' + \frac{h^3}{3!} f_i'' + \cdots \qquad (1.1.5)$$

Starting from $i = 0$ with x_0 given and Δx_0 computed based on any desired number of terms in (1.1.5), the value of x_1 is first calculated. Then, by letting $i = 1, 2$, etc., in (1.1.4), the values of x_2, x_3, etc., are obtained successively. Theoretically, if the number of terms retained in (1.1.5) increases indefinitely, the numerical result from this marching scheme approaches the exact solution. In reality, however, it is not permissible to do so, and the series has to be cut off after a certain finite number of terms. For example, if two terms are retained on the right-hand side of (1.1.5) in computing Δx_i, the value of x_{i+1} so obtained is smaller than the exact value by an amount $(h^3/3!)f_i'' + (h^4/4!)f_i''' + \cdots$. For small h the first term is dominant. We may say that the error involved in this numerical calculation is of the order of $h^3 f_i''$, or simply $O(h^3 f_i'')$. This is the *truncation error* that results from taking a finite number of terms in an infinite series.

It is termed *Euler's method* when only one term is used on the right-hand side of (1.1.5). The truncation error is $O(h^2 f_i')$, and the method should not be used if accuracy is demanded in the result.

It is impractical to use Taylor's series expansion method if f is a function that has complicated derivatives. Furthermore, because of the dependence of the series on the derivatives of f, a generalized computer program cannot be constructed for this method. The nth-order *Runge-Kutta method* is a commonly used alternative. Computations in this method require the evaluation of the function, f, instead of its derivatives, with properly chosen arguments; the accuracy is equivalent to that with n terms retained in the series expansion (1.1.5). The second-order Runge-Kutta formulas are

$$x_{i+1} = x_i + hf(x_i + \tfrac{1}{2}\Delta_1 x_i, t_i + \tfrac{1}{2}h) \qquad (1.1.6)$$

where

$$\Delta_1 x_i = hf(x_i, t_i) \qquad (1.1.7)$$

For better results the following fourth-order Runge-Kutta formulas are usually employed:

$$x_{i+1} = x_i + \tfrac{1}{6}(\Delta_1 x_i + 2\,\Delta_2 x_i + 2\,\Delta_3 x_i + \Delta_4 x_i) \qquad (1.1.8)$$

in which the increments are computed in the following order:

$$\Delta_1 x_i = hf(x_i, t_i)$$
$$\Delta_2 x_i = hf(x_i + \tfrac{1}{2}\Delta_1 x_i, t_i + \tfrac{1}{2}h)$$
$$\Delta_3 x_i = hf(x_i + \tfrac{1}{2}\Delta_2 x_i, t_i + \tfrac{1}{2}h)$$
$$\Delta_4 x_i = hf(x_i + \Delta_3 x_i, t_i + h) \tag{1.1.9}$$

The derivation of these formulas can be found, for instance, in Kuo (1972, p. 137). In the fourth-order method the dominant term in the truncation error is $(h^5/5!)f_i^{iv}$. Unless the slope of the solution is very steep, satisfactory results are usually obtained with reasonably small h. Stability and step-size control in Runge-Kutta methods are referred to in the book by Carnahan, Luther, and Wilkes (1969, p. 363) and are not discussed here.

Runge-Kutta methods can be extended to solving higher-order or simultaneous ordinary differential equations. Consider a second-order equation of the general form

$$\frac{d^2 x}{dt^2} = F\left(x, \frac{dx}{dt}, t\right) \tag{1.1.10}$$

accompanied by the initial conditions that $x = x_0$ and $dx/dt = p_0$ when $t = t_0$. By calling the first-order derivative a new variable p, the equation (1.1.10) can be written in the form of two simultaneous first-order equations

$$\frac{dx}{dt} = p$$

$$\frac{dp}{dt} = F(x, p, t) \tag{1.1.11}$$

with initial values $x(t_0) = x_0$ and $p(t_0) = p_0$. The fourth-order Runge-Kutta formulas for marching from t_i to t_{i+1} are

$$x_{i+1} = x_i + \tfrac{1}{6}(\Delta_1 x_i + 2\,\Delta_2 x_i + 2\,\Delta_3 x_i + \Delta_4 x_i)$$
$$p_{i+1} = p_i + \tfrac{1}{6}(\Delta_1 p_i + 2\,\Delta_2 p_i + 2\,\Delta_3 p_i + \Delta_4 p_i) \tag{1.1.12}$$

in which the individual terms are computed in the following order:

$$\Delta_1 x_i = hp_i$$
$$\Delta_1 p_i = hF(x_i, p_i, t_i)$$
$$\Delta_2 x_i = h(p_i + \tfrac{1}{2}\Delta_1 p_i)$$
$$\Delta_2 p_i = hF(x_i + \tfrac{1}{2}\Delta_1 x_i, p_i + \tfrac{1}{2}\Delta_1 p_i, t_i + \tfrac{1}{2}h)$$
$$\Delta_3 x_i = h(p_i + \tfrac{1}{2}\Delta_2 p_i)$$
$$\Delta_3 p_i = hF(x_i + \tfrac{1}{2}\Delta_2 x_i, p_i + \tfrac{1}{2}\Delta_2 p_i, t_i + \tfrac{1}{2}h)$$
$$\Delta_4 x_i = h(p_i + \Delta_3 p_i)$$
$$\Delta_4 p_i = hF(x_i + \Delta_3 x_i, p_i + \Delta_3 p_i, t_i + h) \tag{1.1.13}$$

1.2 Free Falling of a Spherical Body

As the first application of the Runge-Kutta methods, the motion of a free-falling body is to be studied. This problem is an example in which a solution cannot be obtained without using a numerical method.

It is said that Galileo released simultaneously two objects of different sizes from the Leaning Tower of Pisa and found that they touched the ground at the same instant. If Galileo did perform such an experiment, is the conclusion stated in the story correct? Certainly it is true in a vacuum. In the atmosphere, however, there are forces exerted on a body by the surrounding air that are determined by the size and motion of the body, and the conclusion seems doubtful. If it is not correct, which body should touch the ground first, the larger one or the smaller one?

To answer these questions, we will not repeat the experiment in a laboratory. Instead, we will first formulate the body motion, including the forces caused by the surrounding fluid, and then perform the experiment numerically on a digital computer. Because of its available drag data, a spherical body is preferred for our analysis.

The z-axis is chosen in the direction of gravitational acceleration g; its origin coincides with the center of the sphere at the initial instant $t = 0$, as shown in Fig. 1.2.1. At time $t > 0$, the sphere of diameter d and mass m is at a distance z from the origin and has a velocity v. It is surrounded by a fluid of

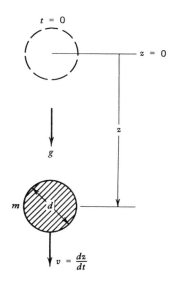

FIGURE 1.2.1 A free-falling spherical body.

Free Falling of a Spherical Body

density ρ_f and kinematic viscosity ν. In a vacuum the only external force acting on the body is the gravitational pull, mg, in the positive z direction. While moving through a fluid it is acted on by the following additional forces:

1. The buoyant force. According to Archimedes' principle, the buoyant force is equal to the weight of the fluid displaced by the body. It has the expression $-m_f g$, where

$$m_f = \tfrac{1}{6}\pi \, d^3 \rho_f \tag{1.2.1}$$

and the negative sign means that the force is in the direction of the negative z-axis.

2. The force on an accelerating body. When a body immersed in a stationary fluid is suddenly set in motion, a flow field is induced in the fluid. The kinetic energy associated with the fluid motion is generated by "doing work," or moving the body against a drag force. This drag exists even if the fluid is frictionless, and it has the value $-\tfrac{1}{2} m_f \, dv/dt$, derived from an inviscid theory (Lamb, 1932, p. 124).

3. The forces caused by viscosity. Around a body moving through a real fluid, a region of rapid velocity change exists adjacent to the surface. The velocity gradient causes a shear stress on the surface; the force resulting from integrating the shear stresses throughout the body is termed the *skin friction*. In addition, in a viscous fluid the pressure at the rear of the body becomes lower than that at the front. The pressure difference gives rise to a *pressure* or *form drag* on the body. The total viscous force, or the sum of skin friction and form drag, is sometimes expressed in a dimensionless form called a *drag coefficient*, which is a function of the body shape and the *Reynolds number*. Except for a few simple shapes and at very low Reynolds numbers, this function is difficult to find analytically, and it is usually determined through experiment. Figure 1.2.2 shows a typical experimental curve for smooth spheres (Goldstein, 1938, p. 16) in which the drag coefficient c_d, defined as the total viscous force divided by $\tfrac{1}{2}\rho_f v^2 \tfrac{1}{4}\pi \, d^2$, is plotted against the Reynolds number Re, defined as $v \, d/\nu$.

4. The wave drag. When the body speed is comparable to the speed of sound in a fluid medium, shock waves may develop on or ahead of the body, causing a *wave drag*.

If we consider only low subsonic speeds, the wave drag can be safely omitted. Including all the other forces discussed, Newton's law of motion, when applied to a spherical body, has the form

$$m \frac{dv}{dt} = mg - m_f g - \frac{1}{2} m_f \frac{dv}{dt} - \frac{1}{2} \rho_f v |v| \frac{\pi}{4} d^2 c_d(v)$$

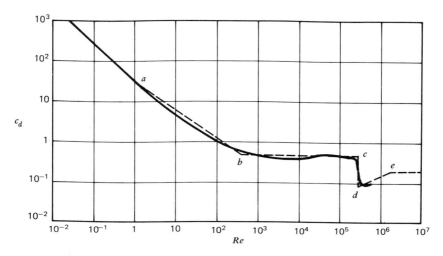

FIGURE 1.2.2 Drag coefficient for smooth spheres.

$v|v|$ is used instead of v^2, so that the direction of viscous drag is always opposite to the direction of v. After rearranging, it can be written as

$$\left(m + \frac{1}{2}m_f\right)\frac{dv}{dt} = (m - m_f)g - \frac{\pi}{8}\rho_f v|v|d^2 c_d(v) \tag{1.2.2}$$

The left-hand side indicates that in an accelerating or decelerating motion through a fluid, the body behaves as if its mass were increased. The term $\frac{1}{2}m_f$ is sometimes referred to as the *added mass*. Upon substitution from (1.2.1) and from $m = \frac{1}{6}\pi d^3\rho$, ρ being the density of the body, (1.2.2) becomes

$$\frac{dv}{dt} = \frac{1}{A}[B - Cv|v|c_d(v)] \tag{1.2.3}$$

with

$$\frac{dz}{dt} = v \tag{1.2.4}$$

where $A = 1 + \frac{1}{2}\bar{\rho}$, $B = (1 - \bar{\rho})g$, and $C = 3\bar{\rho}/4d$; $\bar{\rho}$ stands for the density ratio ρ_f/ρ. These equations fit the general form (1.1.11) and can be solved by applying Runge-Kutta methods.

Special care is needed for the empirical function $c_d(v)$ in numerical computation. To present it in a form suitable for the computer, the curve is replaced by several broken lines, as shown in Fig. 1.2.2. To the left of the point a, where $Re \leqslant 1$, the *Stokes formula*

$$c_d = \frac{24}{Re} \tag{1.2.5}$$

Free Falling of a Spherical Body

is used, which coincides well with the experimental curve for Reynolds numbers that are much less than unity and deviates only slightly from it in the neighborhood of the point a (where $Re = 1$, $c_d = 24$). On the log-log plot a straight line is drawn between points a and b (where $Re = 400$, $c_d = 0.5$) to approximate the actual curve, which gives

$$c_d = 24/Re^{0.646} \qquad \text{for } 1 < Re \leqslant 400 \qquad (1.2.6)$$

Between b and c (where $Re = 3 \times 10^5$), we assume that the drag coefficient has a constant value of 0.5. The abrupt drop of drag coefficient around c is an indication of transition from laminar to turbulent boundary layer before flow separates from the body surface (see Kuethe and Chow, 1976, Section 17.11). Another straight line is drawn between d (where $Re = 3 \times 10^5$, $c_d = 0.08$) and e (where $Re = 2 \times 10^6$, $c_d = 0.18$). Thus

$$c_d = 0.000366 \times Re^{0.4275} \qquad \text{for } 3 \times 10^5 < Re \leqslant 2 \times 10^6 \qquad (1.2.7)$$

Finally, in the high-Reynolds number region beyond e, a constant value of 0.18 is assumed for the drag coefficient.

Here we have chosen a simple way to approximate a complicated function. Better results can be expected by dividing the curve into a larger number of broken lines. Alternatively, the drag coefficient can be fed into the computer as a tabulated function from which the value at any Reynolds number can be interpolated or extrapolated.

Now let us return to the numerical experiment. Suppose steel spheres are dropped in air under standard atmospheric conditions at sea level; we have $\rho = 8000$ kg/m^3, $\rho_f = 1.22$ kg/m^3, $\nu = 1.49 \times 10^{-5}$ m^2/s, and $g = 9.8$ m/s^2. Because of the low value of $\bar{\rho}$ in this particular example, the effects of buoyancy and added mass are negligible, as can be seen in (1.2.3). They become significant if the experiment is repeated in water or in any fluid whose density is comparable to that of the body. However, for generality these terms are still kept in our computer program.

The result for a sphere 0.01 m in diameter will be shown. Its motion is to be compared with that of the same body when dropped in a vacuum. In the latter case $\bar{\rho} = 0$, (1.2.3) and (1.2.4) can be integrated directly to give the solution in the vacuum:

$$v_v = v_0 + gt$$
$$z_v = z_0 + v_0 t + \tfrac{1}{2}gt^2$$

where z_0 and v_0 are, respectively, the initial position and velocity of the sphere at $t_0 = 0$. We will choose the initial values $z_0 = 0$ and $v_0 = 0$ in the computation.

Since the right-hand side of (1.2.3) does not explicitly contain z or t, let us call it $F(v)$, which has a form simpler than the general expression shown in

(1.1.11). We compute the position and velocity at time t_{i+1} based on those at t_i according to the following formulas simplified from (1.1.12) and (1.1.13).

The eight increments are first calculated.

$$\Delta_1 z_i = h v_i$$
$$\Delta_1 v_i = h F(v_i)$$
$$\Delta_2 z_i = h(v_i + \tfrac{1}{2}\Delta_1 v_i)$$
$$\Delta_2 v_i = h F(v_i + \tfrac{1}{2}\Delta_1 v_i)$$
$$\Delta_3 z_i = h(v_i + \tfrac{1}{2}\Delta_2 v_i)$$
$$\Delta_3 v_i = h F(v_i + \tfrac{1}{2}\Delta_2 v_i)$$
$$\Delta_4 z_i = h(v_i + \Delta_3 v_i)$$
$$\Delta_4 v_i = h F(v_i + \Delta_3 v_i)$$

They enable us to compute the new values

$$z_{i+1} = z_i + \tfrac{1}{6}(\Delta_1 z_i + 2\,\Delta_2 z_i + 2\,\Delta_3 z_i + \Delta_4 z_i)$$
$$v_{i+1} = v_i + \tfrac{1}{6}(\Delta_1 v_i + 2\,\Delta_2 v_i + 2\,\Delta_3 v_i + \Delta_4 v_i)$$

According to the above formulas, the variables t_i, z_i, and v_i are one-dimensional arrays whose elements consist of their individual values at $t_0, t_1, t_2, \cdots$. However, if instead of printing the complete set of data at the last time step we print out the time, position, and velocity at every time step t_i, the same variable names may be used for the quantities evaluated at t_{i+1}. In this way the subscripts i and $i+1$ can be dropped from the variables to simplify the appearance of the program.

In Program 1.1 the instantaneous Reynolds number is shown in addition to the position and velocity for our reference. A time limit TMAX is introduced in the program. The numerical integration is stopped when t exceeds this value.

In Program 1.1 a time step of 0.1 s is used. To check the accuracy, the same program has been repeated with step sizes of 0.02 and 0.2 s. Since no appreciable difference can be found between the three sets of data, it can be concluded that the result is reliable. Keeping $h = 0.1$ s, more data are obtained by varying the diameter while doubling the maximum time of integration. They are plotted in Figs. 1.2.3 and 1.2.4.

We have thus obtained some results from the numerical experiment of a free-falling steel sphere. The displacement curves in Fig. 1.2.3 indicate that to travel through a given distance, a larger body takes less time than a smaller one. For two spheres of comparable sizes, the difference in arrival times at a distance of 56 m (the height of the Tower of Pisa) is only a small fraction of a second. Such a time difference is difficult to detect without the help of some instruments.

Free Falling of a Spherical Body

List of Principal Variables in Program 1.1

Program Symbol	Definition
(Main Program)	
A	$1 + \bar{\rho}/2$
B	$(1 - \bar{\rho})g$
C	$3\bar{\rho}/4d$
D	Diameter of sphere, d, m
D1V, D2V, etc.	Velocity increments $\Delta_1 v_i$, $\Delta_2 v_i$, etc.
D1Z, D2Z, etc.	Displacement increments $\Delta_1 z_i$, $\Delta_2 z_i$, etc.
G	Gravitational acceleration, g, m/s^2
H	Time increment, h, s
NU	Kinematic viscosity of fluid, ν, m^2/s
PI	π
RE	Reynolds number vd/ν
RHO	Density of body, ρ, kg/m^3
RHOBAR	Density ratio, $\bar{\rho}$, or ρ_f/ρ
RHOF	Density of fluid, ρ_f, kg/m^3
T, T0	Time, t_i, and initial time, t_0, respectively, s
TMAX	Maximum time of integration, s
V, V0	Velocity, v_i, and initial velocity, v_0, respectively, of the sphere, m/s
VV	Velocity of sphere in vacuum, v_v, m/s
Z, Z0	Position, z, and initial position, z_0, respectively, of the sphere, m
ZV	Position of sphere in vacuum, z_v, m
(Function F)	
CD	Drag coefficient, c_d
F	$(B - CW|W|c_d)/A$
R	Dummy name for Reynolds number
W	Dummy name for velocity

```
C                    ***** PROGRAM 1.1 *****

C        COMPARISON OF THE MOTION OF A STEEL SPHERE DROPPED IN AIR WITH
C        THAT IN VACUUM, BASED ON FOURTH-ORDER RUNGE-KUTTA METHOD

         COMMON A,B,C,D,NU
         REAL NU
```

```
C      ..... SPECIFY DATA AND PRINT HEADINGS .....
       DATA T0,Z0,V0 / 0.0, 0.0, 0.0 /
       DATA RHO,RHOF,G,H,TMAX / 8000.0, 1.22, 9.8, 0.1, 5.0 /
       RHOBAR = RHOF/RHO
       NU = 0.0000149
       A = 1.+RHOBAR/2.
       B = (1.-RHOBAR)*G
       D = 0.01
       C = 3.*RHOBAR/(4.*D)
       WRITE(6,51)
C      ..... INITIALIZE THE PROBLEM .....
       T = T0
       Z = Z0
       V = V0
       RE = V*D/NU
C      ..... COMPUTE POSITION AND VELOCITY IN VACUUM, AND PRINT DATA ....
     1 ZV = Z0 + V0*T + G*T*T/2.
       VV = V0 + G*T
       WRITE(6,52) T,ZV,Z,VV,V,RE
C      ..... RUNGE-KUTTA METHOD .....
       D1Z = H*V
       D1V = H*F(V)
       D2Z = H*(V + D1V/2.)
       D2V = H*F(V+D1V/2.)
       D3Z = H*(V + D2V/2.)
       D3V = H*F(V+D2V/2.)
       D4Z = H*(V + D3V)
       D4V = H*F(V+D3V)
       T = T + H
       Z = Z + (D1Z + 2.*D2Z + 2.*D3Z + D4Z)/6.
       V = V + (D1V + 2.*D2V + 2.*D3V + D4V)/6.
       RE = V*D/NU
C      ..... REPEAT THE COMPUTATION IF T≤TMAX, OTHERWISE STOP .....
       IF (T-TMAX) 1,1,2
     2 STOP

    51 FORMAT(1H1, 3X1HT, 6X12HZ(IN VACUUM), 6X1HZ, 8X12HV(IN VACUUM),
     A        6X1HV, 13X2HRE / 2X5H(SEC), 9X3H(M), 9X3H(M),
     B        10X7H(M/SEC), 5X7H(M/SEC) /)
    52 FORMAT( F7.2, 3X, 1P2E12.3, 3X, 1P2E12.3, 1PE15.3 )

       END


       FUNCTION F(W)
C      ..... F REPRESENTS THE FUNCTION ON THE RIGHT-HAND SIDE OF EQUATION
C            (1.2.3). W STANDS FOR VELOCITY AND R FOR REYNOLDS NUMBER ...

       COMMON A,B,C,D,NU
       REAL NU
       R = ABS(W)*D/NU

C      ..... STOKES FORMULA CANNOT BE USED WHEN R=0. IN
C            THIS CASE THE VALUE ZERO IS ASSIGNED TO CD .....
       IF (R.EQ.0.) CD=0.
       IF (R.GT.0. .AND. R.LE.1.) CD=24./R
       IF (R.GT.1. .AND. R.LE.400.) CD=24./R**0.646
       IF (R.GT.400. .AND. R.LE.3.0E+5) CD=0.5
       IF (R.GT.3.0E+5 .AND. R.LE.2.0E+6) CD=3.66E-4*R**0.4275
       IF (R.GT.2.0E+6) CD=0.18
       F = (B - C*W*ABS(W)*CD)/A
       RETURN
       END
```

T (SEC)	Z(IN VACUUM) (M)	Z (M)	V(IN VACUUM) (M/SEC)	V (M/SEC)	RE
0.00	0.	0.	0.	0.	0.
.10	4.900E-02	4.898E-02	9.800E-01	9.796E-01	6.574E+02
.20	1.960E-01	1.959E-01	1.960E+00	1.958E+00	1.314E+03
.30	4.410E-01	4.405E-01	2.940E+00	2.934E+00	1.969E+03
.40	7.840E-01	7.826E-01	3.920E+00	3.907E+00	2.622E+03
.50	1.225E+00	1.222E+00	4.900E+00	4.876E+00	3.273E+03
.60	1.764E+00	1.758E+00	5.880E+00	5.839E+00	3.919E+03
.70	2.401E+00	2.390E+00	6.860E+00	6.796E+00	4.561E+03
.80	3.136E+00	3.117E+00	7.840E+00	7.746E+00	5.199E+03
.90	3.969E+00	3.938E+00	8.820E+00	8.687E+00	5.830E+03
1.00	4.900E+00	4.854E+00	9.800E+00	9.619E+00	6.456E+03
1.10	5.929E+00	5.862E+00	1.078E+01	1.054E+01	7.074E+03
1.20	7.056E+00	6.962E+00	1.176E+01	1.145E+01	7.685E+03
1.30	8.281E+00	8.152E+00	1.274E+01	1.235E+01	8.288E+03
1.40	9.604E+00	9.431E+00	1.372E+01	1.324E+01	8.883E+03
1.50	1.103E+01	1.080E+01	1.470E+01	1.411E+01	9.469E+03
1.60	1.254E+01	1.225E+01	1.568E+01	1.497E+01	1.005E+04
1.70	1.416E+01	1.379E+01	1.666E+01	1.581E+01	1.061E+04
1.80	1.588E+01	1.541E+01	1.764E+01	1.664E+01	1.117E+04
1.90	1.769E+01	1.712E+01	1.862E+01	1.745E+01	1.171E+04
2.00	1.960E+01	1.890E+01	1.960E+01	1.825E+01	1.225E+04
2.10	2.161E+01	2.077E+01	2.058E+01	1.903E+01	1.277E+04
2.20	2.372E+01	2.271E+01	2.156E+01	1.980E+01	1.329E+04
2.30	2.592E+01	2.473E+01	2.254E+01	2.054E+01	1.379E+04
2.40	2.822E+01	2.682E+01	2.352E+01	2.127E+01	1.428E+04
2.50	3.062E+01	2.898E+01	2.450E+01	2.199E+01	1.476E+04
2.60	3.312E+01	3.122E+01	2.548E+01	2.268E+01	1.522E+04
2.70	3.572E+01	3.352E+01	2.646E+01	2.336E+01	1.568E+04
2.80	3.842E+01	3.589E+01	2.744E+01	2.402E+01	1.612E+04
2.90	4.121E+01	3.832E+01	2.842E+01	2.466E+01	1.655E+04
3.00	4.410E+01	4.082E+01	2.940E+01	2.528E+01	1.697E+04
3.10	4.709E+01	4.338E+01	3.038E+01	2.589E+01	1.737E+04
3.20	5.018E+01	4.599E+01	3.136E+01	2.647E+01	1.777E+04
3.30	5.336E+01	4.867E+01	3.234E+01	2.704E+01	1.815E+04
3.40	5.664E+01	5.140E+01	3.332E+01	2.760E+01	1.852E+04
3.50	6.002E+01	5.419E+01	3.430E+01	2.813E+01	1.888E+04
3.60	6.350E+01	5.703E+01	3.528E+01	2.865E+01	1.923E+04
3.70	6.708E+01	5.992E+01	3.626E+01	2.915E+01	1.957E+04
3.80	7.076E+01	6.286E+01	3.724E+01	2.964E+01	1.989E+04
3.90	7.453E+01	6.585E+01	3.822E+01	3.011E+01	2.021E+04
4.00	7.840E+01	6.888E+01	3.920E+01	3.056E+01	2.051E+04
4.10	8.237E+01	7.196E+01	4.018E+01	3.100E+01	2.081E+04
4.20	8.644E+01	7.508E+01	4.116E+01	3.142E+01	2.109E+04
4.30	9.060E+01	7.824E+01	4.214E+01	3.183E+01	2.136E+04
4.40	9.486E+01	8.144E+01	4.312E+01	3.222E+01	2.163E+04
4.50	9.922E+01	8.469E+01	4.410E+01	3.260E+01	2.188E+04
4.60	1.037E+02	8.796E+01	4.508E+01	3.297E+01	2.213E+04
4.70	1.082E+02	9.128E+01	4.606E+01	3.332E+01	2.236E+04
4.80	1.129E+02	9.463E+01	4.704E+01	3.366E+01	2.259E+04
4.90	1.176E+02	9.801E+01	4.802E+01	3.398E+01	2.281E+04
5.00	1.225E+02	1.014E+02	4.900E+01	3.430E+01	2.302E+04

A body falling through a fluid always reaches a constant velocity, called the *terminal velocity*, which increases with the diameter of a spherical body, as shown in Fig. 1.2.4. For a small sphere the Reynolds number is relatively low and the boundary layer always remains laminar before the point of separation. At a sufficiently high Reynolds number, when the transition from a laminar to

Dynamics of a Body Moving through a Fluid Medium

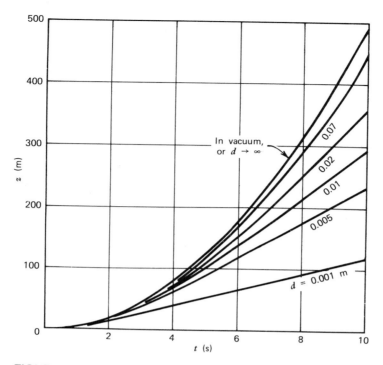

FIGURE 1.2.3 Displacement of steel spheres falling in air.

a turbulent boundary layer occurs on a larger sphere, the abrupt decrease in drag shown in Fig. 1.2.2 will cause a sudden acceleration of the body. This phenomenon can be observed in Fig. 1.2.4 on the curve for a sphere with $d = 0.07$ m around $t = 7.5$ s. For an extremely large sphere the effect of the surrounding fluid becomes negligible in comparison with body inertia, so that the sphere behaves as if it were moving in a vacuum. In this case the velocity increases indefinitely with time, and a terminal velocity can never be reached.

When traveling at the terminal velocity, the gravitational force is balanced by the sum of buoyancy and viscous drag, so that the acceleration is zero. From (1.2.2) an expression is derived for the terminal velocity.

$$v_t = \sqrt{\frac{4(1 - \bar{\rho})g\,d}{3\bar{\rho}c_d(v_t)}} \tag{1.2.8}$$

Based on the numerical result for the steel sphere of 0.01 m diameter, the terminal Reynolds number is estimated to be not much higher than 2.73×10^4. According to the approximation shown in Fig. 1.2.2, the drag coefficient at the

Free Falling of a Spherical Body

13

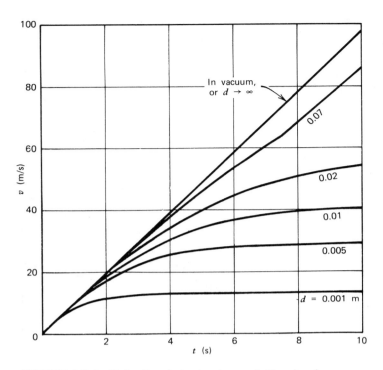

FIGURE 1.2.4 Velocity of steel spheres falling in air.

terminal velocity is $c_d(v_t) = 0.5$. Substitution of this together with other given values into (1.2.8) gives $v_t = 41.4$ m/s.

In general, $c_d(v_t)$ is not known a priori, and the terminal velocity cannot be obtained unless the time history of the motion is computed first. However, from the data computed for steel spheres falling through air, it is concluded that the terminal Reynolds number exceeds 2×10^6 if the diameter of a sphere is greater than 0.12 m. $c_d(v_t)$ has a constant value of 0.18 for such a sphere according to our approximation, and v_t can readily be calculated. On the other hand, for tiny particles whose terminal Reynolds numbers are in the Stokes flow regime, the Stokes formula (1.2.5) is used in (1.2.8) to obtain

$$v_t = \frac{(1 - \bar{\rho})g\, d^2}{18\bar{\rho}\nu} \qquad \text{for } Re < 1 \qquad (1.2.9)$$

Let us now estimate the minimum diameter of a steel sphere whose terminal velocity in air can exceed the speed of sound, 340 m/s at sea level. The terminal Reynolds number of such a body is well beyond 2×10^6, so that the value 0.18 is used for $c_d(v_t)$. By assigning the sound speed to v_t, (1.2.8)

Dynamics of a Body Moving through a Fluid Medium

gives $d = 0.243$ m. Remember that before reaching sonic speed, a transonic region appears on the body, causing a wave drag that should also be included in the drag coefficient. Furthermore, because of the limited drag data available for high Reynolds numbers, the constant c_d assumption is doubtful. The minimum diameter estimated here is expected to be too low.

In the atmosphere or in an ocean, if the displacement of a body is so large that the variations in fluid density and kinematic viscosity are significant, the program must be modified by specifying the dependence of these quantities on the height z. In this case the function $F(v)$ in Program 1.1 should be changed to $F(v, z)$.

Problem 1.1 Find the motion of a ping-pong ball 0.036 m in diameter released at the bottom of a water tank. The density of water is 1000 kg/m^3 and the kinematic viscosity is 1×10^{-6} m^2/s. Assume that the shell of the ball is so thin that the density of the body is the same as that of the filling air, or 1.22 kg/m^3.

Hint: Because of the large buoyant force, the stationary ping-pong ball initially experiences a large acceleration. You may verify that the numerical solution diverges if $h = 0.1$ s is still used. The value $h = 0.01$ s is recommended in this problem. Since the ball reaches an upward steady motion within a fraction of a second, a maximum time of 0.5 s is sufficient for the computation.

Problem 1.2 For a given body falling in a specified fluid, the terminal velocity is fixed independent of the initial velocity of the body. If the initial velocity is slower than or in a direction opposite to the terminal velocity, the body will accelerate toward the terminal value. On the other hand, if the initial velocity is faster than the terminal velocity, it will decelerate and finally reach the same velocity. Verify this phenomenon by assigning several values to v_0 in Program 1.1 with the addition of a DO statement.

On a rainy day it can be observed that the larger raindrops come down faster than the smaller ones. If a raindrop is considered to be a rigid ball of water, its motion can be computed from Program 1.1 by assigning the numerical value of water density to the variable name RHO. The resulting terminal velocities for drops of various diameters are plotted in Fig. 1.2.5 in comparison with the measured data taken from Blanchard (1967, p. 12). Good agreements between computed and measured values are obtained for drops whose diameters are less than 3.5 mm.

The difference in terminal velocity between a liquid drop and a rigid sphere occurs because the binding surface of the liquid drop moves with the

Free Falling of a Spherical Body

15

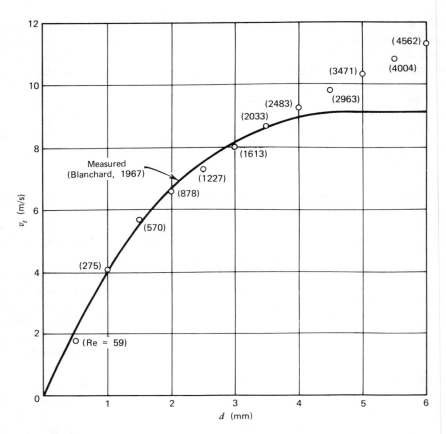

FIGURE 1.2.5 Measured terminal velocity of a raindrop in comparison with that computed for a rigid sphere. Numbers in parentheses are Reynolds numbers at the points indicated.

surrounding fluid and, at the same time, it is deformed. At small Reynolds numbers a liquid drop is approximately spherical in shape. The tangential motion of the interior fluid at the interface reduces the skin friction and delays flow separation from the surface. Thus a small spherical liquid drop has a smaller drag than a rigid sphere of the same size and density and, therefore, has a faster terminal velocity. When the Reynolds number is large, however, the drop is deformed by the external flow into a shape that is flattened in the vertical direction (see Blanchard, 1967, Plate III, for a photograph of a water drop in an airflow). This shape causes an early flow separation and results in a tremendous increase in form drag. This explains

Dynamics of a Body Moving through a Fluid Medium

the phenomenon that the terminal velocity of a large liquid drop is slower than that computed for a rigid sphere, as shown by the flattening portion of the measured curve. Certain instabilities grow in very large liquid drops, causing them to break into smaller droplets. Because of this, raindrops having diameters greater than 6 mm are rarely observed.

For Reynolds numbers that are much less than unity, the nonlinear inertia terms in the equations of motion can be ignored, and both the internal and external flow fields can be found analytically. The drag of a liquid sphere so derived has the expression (Happel and Brenner, 1965, p. 127)

$$\text{Drag} = 3\pi\mu_e \, dv \, \frac{1 + 2\mu_e/3\mu_i}{1 + \mu_e/\mu_i} \tag{1.2.10}$$

where μ_i and μ_e are, respectively, the coefficients of viscosity of the internal and external fluid media. For a raindrop falling in air, $\mu_i = 1 \times 10^{-3}$ and $\mu_e = 1.818 \times 10^{-5}$ kg/ms, the drag of a liquid sphere is only slightly lower than $3\pi\mu_e \, dv$, which is the drag of a rigid sphere in the Stokes flow regime.

The expression (1.2.10) has been modified by Taylor and Acrivos (1964) to include the inertial effect under the restriction that $Re \ll 1$. Because of the nonlinearity of the governing equations, a closed-form solution cannot be found for a liquid drop at a Reynolds number much higher than unity.

When a raindrop freezes, its diameter increases because of the lower density. You may verify that at a low Reynolds number a freezing raindrop experiences a deceleration while falling.

1.3 Computer Simulation of Some Restrained Motions

In the previous example of a free-falling body the physical system was first replaced by a system of mathematical equations that describes approximately the body motion. The algorithm for solving this system of equations was then programmed in Program 1.1 under a set of specified conditions. By varying the input data in the same program, one can find the motion of a spherical body of an arbitrary material falling through a given fluid starting from any desired initial conditions. Thus, instead of measuring the real motion, which is usually tedious and in some cases extremely difficult, one can perform the experiment numerically on a computer, which is a simple process once the physical laws have been correctly formulated. In this respect the motion of a free-falling sphere is said to have been simulated numerically by Program 1.1.

Discrepancies certainly exist between the real and the simulated motions because of the approximations used in deriving the governing equations and the errors involved in numerical computations. However, numerical simulation gives an approximate outcome before an experiment is actually conducted, and the result can be used as a guide in designing the experiment and in choosing the right instruments for measurements. For example, such a practice has been used in predicting the orbit of a spacecraft or the landing site of a reentry vehicle.

In this section some restrained body motions in a fluid are simulated. We first consider the oscillatory motion of a simple pendulum whose small-amplitude motion in vacuum is well known. The weight of the pendulum is a spherical body of diameter d and mass m suspended on a weightless thin cord of length $(l - \frac{1}{2}d)$. At time t the displacement angle is θ, the tangential velocity is v, and the body is acted on by forces shown in Fig. 1.3.1. Fluid dynamic forces consist of the viscous drag and the force caused by acceleration. For the motion in tangential direction, the governing equations are

$$v = l \frac{d\theta}{dt}$$

and

$$m \frac{dv}{dt} = -(m - m_f)g \sin \theta - \frac{1}{2} m_f \frac{dv}{dt} - \frac{1}{2} \rho_f v |v| \frac{\pi}{4} d^2 c_d(v)$$

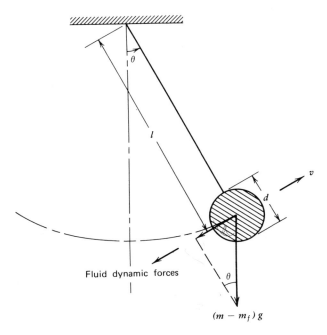

FIGURE 1.3.1   A simple pendulum.

Dynamics of a Body Moving through a Fluid Medium

where m_f, ρ_f, g, and c_d have the same meanings as defined in the previous section. By letting $y = l\theta$ and defining A, B, and C in the same manner as shown in (1.2.3), the above equations become

$$\frac{dy}{dt} = v \tag{1.3.1}$$

$$\frac{dv}{dt} = \frac{1}{A}\left[-B \sin\frac{y}{l} - Cv|v|c_d(v)\right] \tag{1.3.2}$$

The general initial conditions are that $y = y_0$ (or $l\theta_0$) and $v = v_0$ at time $t = t_0$. For the motion in a vacuum the equations reduce to

$$\frac{dy_v}{dt} = v_v \tag{1.3.3}$$

$$\frac{dv_v}{dt} = -g \sin\left(\frac{y_v}{l}\right) \tag{1.3.4}$$

obtained by setting $\rho_f = 0$ in (1.3.2). Even in this simplified case, analytical solution is not straightforward if the amplitude of oscillation is large. The fourth-order Runge-Kutta formulas (1.1.12) and (1.1.13) will be used in solving the preceding two systems of equations.

For numerical computations a glass sphere ($\rho = 2500$ kg/m^3) with $d = 0.01$ m and $l = 2$ m is considered. At $t = 0$ the sphere is released from a stationary position of the cord with 45° angular displacement. Its motions in the vacuum and in the air are computed and compared in Program 1.2.

Instead of putting the Runge-Kutta formulas in the main program, as shown in Program 1.1, we group them together in a subroutine named RUNGE. This subroutine is written according to the general form (1.1.11) for a second-order ordinary differential equation. In this way the formulas need not be repeated in the program when motions in several different fluid media are to be computed. Furthermore, once such a subroutine is written, it can be saved and attached to any program in which a second-order equation is to be solved through fourth-order Runge-Kutta method. Although the right-hand sides of (1.3.2) and (1.3.4) do not contain all three variables (i.e., displacement, velocity, and time), they are still defined as FN and FV, respectively, in Program 1.2 as functions of all three variables in order to conform with the general function F which appears in the subroutine RUNGE.

Another modification from Program 1.1 is that the drag coefficient of a sphere is now defined in a function subprogram named CD. Both subprograms RUNGE and CD will be used later in other programs.

In the above formulation the angular displacement θ is expressed in radians, but we prefer degree as the unit of θ in Program 1.2. It has to be multiplied by the factor $\pi/180$ in order to convert back to radians.

Computer Simulation of Some Restrained Motions **19**

Program Symbol	Definition
(Main Program)	
A	$1 + \bar{\rho}/2$
B	$(1 - \bar{\rho})g$
C	$3\bar{\rho}/4\,d$
D	Diameter of sphere, d, m
DT	Time increment, s
G	Gravitational acceleration, g, m/s^2
L	Distance between the pivot axis and the center of sphere, l, m
NU	Kinematic viscosity of fluid, ν, m^2/s
PI	π
RHO	Density of spherical body, ρ, kg/m^3
RHOBAR	Density ratio, $\bar{\rho}$, or ρ_f/ρ
RHOF	Density of fluid, ρ_f, kg/m^3
T, T0	Time, t_i, and initial time, t_0, respectively, s
THETA, THETA0	Angular displacement of pendulum, θ, and its initial value, θ_0, respectively, degrees
THETAV	Angular displacement in vacuum, θ_v, degrees
TMAX	Maximum time of integration, s
V, V0	Tangential velocity, v, and its initial value, v_0, respectively, m/s
VV	Tangential velocity in vacuum, v_v, m/s
Y, Y0	Circumferential displacement, y or $l\theta$, and its initial value, y_0, respectively, m
YV	Circumferential displacement in vacuum, y_v, m
(Subroutine RUNGE)	
D1P, D2P, etc.	Increments $\Delta_1 p_i$, $\Delta_2 p_i$, etc. in (1.1.13)
D1X, D2X, etc.	Increments $\Delta_1 x_i$, $\Delta_2 x_i$, etc. in (1.1.13)
(Function CD)	
RE	Reynolds number, Re

Because of the length of the output of Program 1.2, only the result for $t \leqslant 5$ s is shown. From the plot of angular displacement in Fig. 1.3.2 based on the computer output, we can see that the pendulum swings undamped in a vacuum with a constant amplitude of 45°. The period is approximately 2.95 s compared to 2.838 s, calculated from the well-known expression $2\pi\sqrt{l/g}$ for

Dynamics of a Body Moving through a Fluid Medium

```
C                    ***** PROGRAM 1.2 *****

C          MOTION OF A SIMPLE PENDULUM IN VACUUM AND THAT IN A FLUID
         EXTERNAL FV,FN
         COMMON A,B,C,D,G,L,NU
         REAL L,NU

C          ..... ASSIGN INPUT DATA AND PRINT HEADINGS .....
         DATA RHO,RHOF,PI,DT,TMAX / 2500.,  1.22, 3.14159, 0.1, 20. /
         D = 0.01
         G = 9.8
         L = 2.0
         NU = 1.49E-5
         RHOBAR = RHOF/RHO
         A = 1.0 + RHOBAR/2.0
         B = (1.0 - RHOBAR)*G
         C = 3.0*RHOBAR/(4.*D)
         WRITE(6,100)

C          ..... ASSIGN INITIAL CONDITIONS AND PRINT VALUES .....
         TO = 0.0
         VO = 0.0
         THETA0 = 45.0
         YO = L*THETA0*PI/180.0
         T   = TO
         YV  = YO
         VV  = VO
         THETAV = THETA0
         Y  = YO
         V  = VO
         THETA = THETA0
       1 WRITE(6,200)  T, THETAV, VV, THETA, V

C          ..... INTEGRATE THE EQUATIONS OF MOTION BY USING FOURTH-ORDER
C             RUNGE-KUTTA METHOD. STOP THE COMPUTATION WHEN T>TMAX .....
         CALL RUNGE( YV, VV, T, DT, FV )
         THETAV = YV/L * 180.0/PI
C          ..... T HAS BEEN CHANGED TO T+DT AFTER CALLING RUNGE.  BEFORE
C             CALLING RUNGE TO COMPUTE THE MOTION IN A FLUID AT THE SAME
C             TIME, T HAS TO BE RETURNED TO ITS PREVIOUS VALUE .....
         T = T - DT
         CALL RUNGE( Y, V, T, DT, FN )
         THETA = Y/L * 180.0/PI
         IF (T-TMAX) 1,1,2
       2 STOP

     100 FORMAT( 1H1, 16X9HIN VACUUM, 9X10HIN A FLUID / 5X1HT, 9X5HTHETA,
        A         5X1HV, 8X5HTHETA, 5X1HV / 3X5H(SEC), 7X5H(DEG),
        B         2X7H(M/SEC), 5X5H(DEG), 2X7H(M/SEC) /)
     200 FORMAT( F8.2, 5X,2F7.2, 5X,2F7.2 )

         END
```

the period, assuming small amplitudes. The time history of angular displacement is no longer a cosine curve, as predicted by the linearized theory. In the presence of air the amplitude is slowly damped by air resistance and, in the meantime, the period is shortened.

Parameters can be varied in Program 1.2 to simulate the motions of a pendulum that has any combination of size and material in different fluid

```fortran
      SUBROUTINE RUNGE( X, P, T, H, F )
C     ..... FOURTH-ORDER RUNGE-KUTTA FORMULAE (1.1.12) AND (1.1.13) ARE
C           PROGRAMMED IN THIS SUBROUTINE FOR SOLVING A SECOND-ORDER
C           DIFFERENTIAL EQUATION REWRITTEN IN THE FORM :
C               DX/DT = P     AND    DP/DT = F(X,P,T).
C           H IS THE SIZE OF INCREMENT IN THE INDEPENDENT VARIABLE T,
C           AND F IS A FUNCTION DEFINED IN ANOTHER SUBPROGRAM .....

      D1X = H*P
      D1P = H*F( X, P, T )
      D2X = H*(P + D1P/2.)
      D2P = H*F(X+D1X/2., P+D1P/2., T+H/2.)
      D3X = H*(P + D2P/2.)
      D3P = H*F(X+D2X/2., P+D2P/2., T+H/2.)
      D4X = H*(P + D3P)
      D4P = H*F(X+D3X, P+D3P, T+H)
      T = T + H
      X = X + (D1X + 2.*D2X + 2.*D3X + D4X) / 6.0
      P = P + (D1P + 2.*D2P + 2.*D3P + D4P) / 6.0
      RETURN
      END



      FUNCTION FN( Y, V, T )
C     ..... THIS REPRESENTS THE FUNCTION ON THE RIGHT-HAND SIDE OF
C           EQUATION (1.3.2) .....

      COMMON A,B,C,D,G,L,NU
      REAL L
      FN = (-B*SIN(Y/L) - C*V*ABS(V)*CD(V)) / A
      RETURN
      END



      FUNCTION FV( YV, VV, T )
C     ..... THIS REPRESENTS THE FUNCTION ON THE RIGHT-HAND SIDE OF
C           EQUATION (1.3.4) .....

      COMMON A,B,C,D,G,L,NU
      REAL L
      FV = -G*SIN(YV/L)
      RETURN
      END



      FUNCTION CD(V)
C     ..... CD IS THE APPROXIMATE DRAG COEFFICIENT OF A SPHERE .....

      COMMON A,B,C,D,G,L,NU
      REAL NU
      RE = ABS(V)*D/NU
      IF( RE.EQ.0. ) CD=0.
      IF( RE.GT.0. .AND. RE.LE.1. ) CD=24./RE
      IF( RE.GT.1. .AND. RE.LE.400. ) CD=24./RE**0.646
      IF( RE.GT.400. .AND. RE.LE.3.0E+5 ) CD=0.5
      IF( RE.GT.3.0E+5 .AND. RE.LE.2.0E+6 ) CD=3.66E-4*RE**0.4275
      IF( RE.GT.2.0E+6 ) CD=0.18
      RETURN
      END
```

T (SEC)	IN VACUUM THETA (DEG)	IN VACUUM V (M/SEC)	IN A FLUID THETA (DEG)	IN A FLUID V (M/SEC)
0.00	45.00	0.00	45.00	0.00
.10	44.01	-.69	44.01	-.69
.20	41.08	-1.35	41.08	-1.35
.30	36.30	-1.97	36.33	-1.96
.40	29.88	-2.50	29.93	-2.49
.50	22.06	-2.93	22.18	-2.90
.60	13.19	-3.23	13.42	-3.19
.70	3.68	-3.38	4.06	-3.32
.80	-6.00	-3.36	-5.43	-3.29
.90	-15.39	-3.17	-14.61	-3.10
1.00	-24.04	-2.84	-23.03	-2.76
1.10	-31.56	-2.38	-30.32	-2.31
1.20	-37.61	-1.83	-36.18	-1.76
1.30	-41.95	-1.20	-40.37	-1.15
1.40	-44.43	-.52	-42.74	-.50
1.50	-44.94	.17	-43.21	.17
1.60	-43.48	.85	-41.77	.83
1.70	-40.09	1.51	-38.46	1.46
1.80	-34.90	2.10	-33.43	2.04
1.90	-28.11	2.62	-26.89	2.52
2.00	-20.01	3.02	-19.11	2.89
2.10	-10.95	3.28	-10.46	3.12
2.20	-1.36	3.39	-1.36	3.21
2.30	8.30	3.33	7.75	3.13
2.40	17.55	3.11	16.44	2.91
2.50	25.97	2.74	24.30	2.55
2.60	33.15	2.26	30.97	2.09
2.70	38.82	1.68	36.18	1.54
2.80	42.72	1.04	39.73	.93
2.90	44.73	.36	41.49	.29
3.00	44.77	-.33	41.39	-.36
3.10	42.84	-1.01	39.45	-.99
3.20	39.00	-1.66	35.73	-1.59
3.30	33.41	-2.24	30.40	-2.12
3.40	26.27	-2.73	23.68	-2.56
3.50	17.90	-3.10	15.87	-2.87
3.60	8.67	-3.32	7.34	-3.06
3.70	-.98	-3.39	-1.50	-3.09
3.80	-10.58	-3.29	-10.22	-2.97
3.90	-19.67	-3.03	-18.39	-2.71
4.00	-27.82	-2.64	-25.65	-2.33
4.10	-34.66	-2.13	-31.67	-1.85
4.20	-39.92	-1.53	-36.20	-1.30
4.30	-43.38	-.88	-39.08	-.70
4.40	-44.92	-.19	-40.19	-.07
4.50	-44.48	.50	-39.49	.56
4.60	-42.09	1.17	-37.02	1.16
4.70	-37.81	1.80	-32.87	1.72
4.80	-31.82	2.36	-27.22	2.21
4.90	-24.36	2.83	-20.32	2.59
5.00	-15.75	3.16	-12.49	2.86

media. For example, the motion of the same pendulum in water is obtained by changing the input values of ρ_f and ν and is plotted in Fig. 1.3.2 for comparison. In this case the density of water is comparable to that of the glass sphere. The influence of fluid is so large that the displacement curve is far different from that in a vacuum. Starting from the same inclination, it

Computer Simulation of Some Restrained Motions

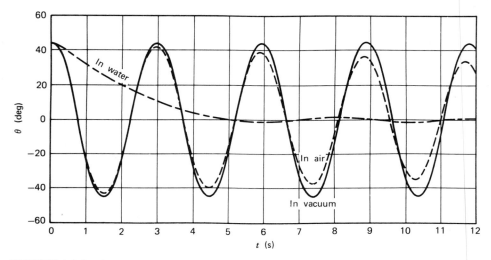

FIGURE 1.3.2 Angular displacements of a suspended glass sphere in different fluid media.

takes 5 s to reach the vertical position in water compared with 0.74 s in a vacuum or in air. After that it oscillates at a small but decreasing amplitude with a period approximately equal to 4 s.

Problem 1.3 Find the variation of period with amplitude for a simple pendulum swinging in a vacuum.

Problem 1.4 If the body in Program 1.2 is replaced by a thin spherical shell containing air, and the cord is rigid but weightless and frictionless, find the angular motion of the shell in water, starting from a stationary position of 5° inclination. Assume that the shell is so thin that its weight is negligible, and that the rigid arm can rotate freely about the hinge.

An elastically restrained wing vibrates under certain flight conditions. Instead of studying the problem analytically, which is usually done by the use of linearized theories, we will simulate a simplified aeroelastic system and study on the computer the response of a wing to various wind conditions. By analytical methods such a study is difficult and, in some cases, it is impossible if a high degree of accuracy is required.

Figure 1.3.3 shows schematically a wing installed in a wind tunnel. The weight of the wing, mg, is supported by a spring of spring constant k, which is equivalent to the stiffness of a wing on an airplane. The cross section of the

Dynamics of a Body Moving through a Fluid Medium

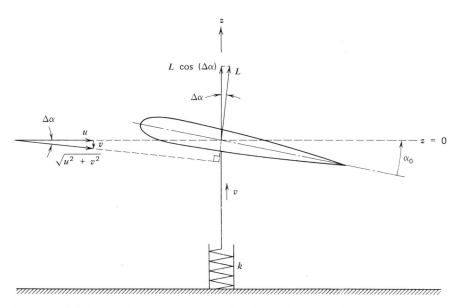

FIGURE 1.3.3 Vertical motion of a wing in a wind tunnel.

wing is a symmetric airfoil. In the absence of a wind, the airfoil makes an angle α_0 with the horizontal and its center of mass under a state of equilibrium is at a height $z = 0$. Assume that the wing is installed so that it can move only vertically. The displacement of the center of mass is z, considered positive when going upward. The wind tunnel supplies a uniform horizontal wind of speed u at the test section. When the wing moves upward at a speed v, the surrounding air moves downward relative to the wing at the same speed, as shown in the velocity diagram. To the oncoming air flow, the angle of attack of the wing is

$$\alpha = \alpha_0 - \Delta\alpha$$
$$= \alpha_0 - \tan^{-1}(v/u) \tag{1.3.5}$$

The lift of the wing, in the direction normal to the flow, has the expression

$$L = \tfrac{1}{2}\rho_f(u^2 + v^2)Sc_l \tag{1.3.6}$$

where ρ_f is air density, S the projected wing area, and c_l the *lift coefficient* of the wing. According to the *thin-airfoil theory* for a wing of large aspect ratio, the lift coefficient is a linear function of the angle of attack, which has the following form for a symmetric airfoil (Kuethe and Chow, 1976, Section 5.5).

$$c_l = 2\pi\alpha \tag{1.3.7}$$

Computer Simulation of Some Restrained Motions

The theory agrees with the experiment if α is within a certain limit. Beyond the limit the wing no longer behaves like a thin body, and the flow separates from the surface. The sudden decrease of lift associated with this phenomenon causes the wing to stall. For simplicity let us assume that the formula (1.3.7) holds for the wing if $-18° \le \alpha \le +18°$, and that beyond this range the lift is zero.

The equations of motion for the center of mass are

$$\frac{dz}{dt} = v \quad \text{and} \quad m\frac{dv}{dt} = -kz + L\cos(\Delta\alpha) \tag{1.3.8}$$

Upon substitution of the value $\cos(\Delta\alpha) = u/\sqrt{u^2 + v^2}$ and L from (1.3.6), the second equation becomes

$$m\frac{dv}{dt} = -kz + \tfrac{1}{2}\rho_f Sc_l u\sqrt{u^2 + v^2} \tag{1.3.9}$$

As in previous examples, this problem also contains several parameters, and each of them may vary. Instead of solving the equations in the present form for each set of parameters, it is usually more economical to solve them in a nondimensionalized form. In this way we need only to vary the values of a reduced number of dimensionless parameters, so that a class of problems can be solved through a single computation.

The procedure in nondimensionalizing the governing equations is first to find the basic reference quantities for the problem. For instance, in kinematics they are the reference time, length, and velocity.

In the present case of a spring-supported wing, we may use the period of free oscillation of the system, $2\pi\sqrt{m/k}$, as the reference time; the deformation of spring due to the weight of wing, mg/k, as the reference length; and the ratio of the two as the reference velocity. Based on these reference quantities, the following dimensionless variables are defined.

$$T = t \Big/ 2\pi\sqrt{\frac{m}{k}} \qquad Z = z \Big/ \frac{mg}{k}$$
$$U = u \Big/ \frac{g}{2\pi}\sqrt{\frac{m}{k}} \qquad V = v \Big/ \frac{g}{2\pi}\sqrt{\frac{m}{k}} \tag{1.3.10}$$

After being expressed in terms of these new variables, (1.3.8) and (1.3.9) become

$$\frac{dZ}{dT} = V \tag{1.3.11}$$

$$\frac{dV}{dT} = -(2\pi)^2 Z + \beta c_l U\sqrt{U^2 + V^2} \tag{1.3.12}$$

where $\beta = \rho_f g S/2k$ is a dimensionless parameter. Thus the six parameters, ρ_f,

g, m, S, k, and u, are reduced to only two dimensionless groups β and U. Each set of values of β and U represents a large number of combinations of the dimensional parameters. For our numerical computations, the following conditions are assumed.

$$\rho_f = 1.22 \text{ kg/m}^3 \qquad g = 9.8 \text{ m/s}^2$$
$$m = 3 \text{ kg} \qquad S = 0.3 \text{ m}^2$$
$$k = 980 \text{ kg/s}^2 \qquad \alpha_0 = 10°$$

Corresponding to these assumptions, $\beta = 0.00183$, the reference time is 0.348 s, the reference length is 0.03 m, and the reference velocity is 0.0862 m/s. Thus $U = 100$ means that the horizontal wind speed is 8.62 m/s. The numerical result in dimensionless form applies to other problems under a variety of conditions, for example, if m, S, and k are multiplied by the same factor.

The subroutine RUNGE is borrowed from Program 1.2 in solving (1.3.11) and (1.3.12). The lift coefficient is defined in the function subprogram CL(V) according to (1.3.7) under the assumed limitations, and the angle of attack that appears in the function is expressed by (1.3.5). Program 1.3 computes the

---

List of Principal Variables in Program 1.3

Program Symbol	Definition
(Main Program)	
ALPHA, ALPHAD	Angle of attack, α, in radians and that in degrees, respectively
ALPHA0	Angle of attack of a stationary wing, α_0, rad.
BETA	β, the dimensionless parameter $\rho_f g S / 2k$
DT	Dimensionless time increment
DU	Increment in dimensionless horizontal velocity
PI	π
T, T0	Dimensionless time, T, and its initial value, respectively
TMAX	Maximum time of integration, dimensionless
U	Dimensionless horizontal wind velocity, U
V, V0	Dimensionless vertical velocity of wing, V, and its initial value, respectively
Z, Z0	Dimensionless displacement of the wing, Z, and its initial value, respectively
(Function CL)	
CL	Lift coefficient of the wing, c_l

Computer Simulation of Some Restrained Motions

```
C                        ***** PROGRAM 1.3 *****

C          SIMULATION OF THE VERTICAL MOTION OF AN AIRFOIL IN WINDTUNNEL

      EXTERNAL F
      COMMON ALPHA0,BETA,PI,U

C      ..... ASSIGN VALUES TO ALPHA0, BETA, DT, AND TMAX. SPECIFY INITIAL
C            POSITION AND VELOCITY OF THE AIRFOIL, THEN COMPUTE THE
C            MOTION AT VARIOUS WIND SPEEDS. HEADINGS ARE PRINTED FOR EACH
C            WIND SPEED. .....
      PI = 3.14159
      ALPHA0 = 10.*PI/180.
      BETA = 0.00183
      DT = 0.1
      TMAX = 20.0
      T0 = 0.0
      Z0 = 0.0
      V0 = 0.0
      U = 0.0
      DU = 25.0
      DO 2 I=1,5
      U = U + DU
      WRITE(6,100) U
      T = T0
      Z = Z0
      V = V0
    1 ALPHA = ALPHA0 - ATAN(V/U)
      ALPHAD = ALPHA*180./PI
      WRITE(6,150)  T, Z, V, ALPHAD

C      ..... COMPUTE THE MOTION AND PRINT UNTIL T>TMAX .....
      CALL RUNGE( Z, V, T, DT, F )
      IF (T-TMAX) 1,1,2
    2 CONTINUE

  100 FORMAT( 1H1, 13X14HWIND SPEED U =,F7.1 //
     A         7X1HT, 10X1HZ, 10X1HV, 5X10HALPHA(DEG) / )
  150 FORMAT( F10.2, 3F11.3 )

      END

      FUNCTION F( Z, V, T )
C      ..... IT REPRESENTS THE RIGHT-HAND SIDE OF EQUATION (1.3.12) .....

      COMMON ALPHA0,BETA,PI,U
      F = -(2.*PI)**2*Z + BETA*U*SQRT(U*U+V*V)*CL(V)
      RETURN
      END

      FUNCTION CL(V)
C      ..... LIFT COEFFICIENT OF THE AIRFOIL EXPRESSED AS
C            A FUNCTION OF VERTICAL VELOCITY OF THE WING .....

      COMMON ALPHA0,BETA,PI,U
      ALPHA = ALPHA0 - ATAN(V/U)
      CL = 0.0
      IF(ABS(ALPHA) .LE. PI/10.) CL=2.*PI*ALPHA
      RETURN
      END

      SUBROUTINE RUNGE( X, P, T, H, F )
C      ..... TAKEN FROM PROGRAM 1.2 .....
```

T	Z	V	ALPHA(DEG)
0.00	0.000	0.000	10.000
.10	.006	.116	9.735
.20	.022	.184	9.578
.30	.040	.182	9.583
.40	.056	.111	9.745
.50	.061	.001	9.998
.60	.056	-.107	10.244
.70	.041	-.171	10.392
.80	.024	-.169	10.388
.90	.010	-.104	10.238
1.00	.004	-.002	10.004
1.10	.009	.098	9.774
1.20	.023	.159	9.636
1.30	.039	.157	9.639
1.40	.052	.097	9.777
1.50	.057	.002	9.995
1.60	.053	-.091	10.208
1.70	.040	-.147	10.337
1.80	.025	-.147	10.336
1.90	.013	-.091	10.208
2.00	.008	-.003	10.006
2.10	.012	.084	9.808
2.20	.024	.136	9.687
2.30	.038	.136	9.688
2.40	.049	.085	9.805
2.50	.054	.003	9.993
2.60	.050	-.077	10.177
2.70	.039	-.127	10.290
2.80	.026	-.127	10.291
2.90	.016	-.079	10.182
3.00	.011	-.004	10.008
3.10	.015	.071	9.836
3.20	.025	.117	9.731
3.30	.037	.118	9.730
3.40	.047	.074	9.830
3.50	.051	.004	9.991
3.60	.048	-.066	10.151
3.70	.038	-.109	10.250
3.80	.027	-.110	10.252
3.90	.018	-.069	10.159
4.00	.014	-.004	10.009
4.10	.017	.061	9.861
4.20	.025	.101	9.768
4.30	.036	.102	9.766
4.40	.045	.065	9.851
4.50	.048	.004	9.990
4.60	.045	-.056	10.129
4.70	.038	-.094	10.215
4.80	.028	-.095	10.218
4.90	.020	-.061	10.139
5.00	.017	-.004	10.010

displacement, velocity, and angle of attack of the wing, initially stationary at its equilibrium position, after an airflow of various speeds is passed through the wind tunnel.

Only a part of the numerical output is shown. The computed displacement of the wing is plotted in Fig. 1.3.4 for five different wind speeds. After the

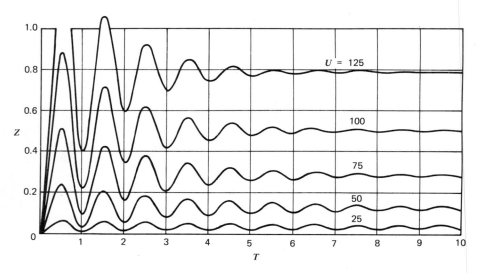

FIGURE 1.3.4 Displacements of a wing under five different wind speeds.

wind is turned on, the aerodynamic force raises the wing upward from its static equilibrium position and the wing starts to oscillate; meanwhile, the motion of the wing causes the lift to change. At a low wind speed the wing oscillates at the natural frequency of the mass-spring system, and the amplitude of oscillation damps with time at a slow rate. As the wind speed increases there is a slow increase in period in contrast with a faster damping in amplitude. In each case the wing will finally settle at a new state of equilibrium under which the lift is just balanced by the force from the stretched spring.

You may verify that in a much stronger wind the wing may approach the equilibrium position without going through any oscillatory motion.

In Program 1.3 the lift coefficient is calculated only approximately from the steady-state formula (1.3.7), with α defined as the instantaneous angle between the resultant wind velocity and the chordline. Finding the actual lift on a wing in an unsteady motion is quite involved (Theodorsen, 1935; Bisplinghoff and Ashley, 1962, Chapter 4) and will not be discussed here. Furthermore, the angular displacement of the wing plays an important role in causing the wing to flutter. With the rotational degree of freedom omitted in the present formulation, the wing-spring system becomes stable; that is, the amplitude of oscillation cannot grow indefinitely with time.

Before an airplane is built, the designer would like to know the response of the wing to a gusty wind. Such motions of the wing may also be simulated by using our simplified model based on a quasisteady approximation. Suppose

Dynamics of a Body Moving through a Fluid Medium

that superimposed on the uniform flow there is a fluctuating horizontal velocity component, so that the total velocity can be expressed in dimensionless form as $U(1 + a \sin \omega T)$. The fluctuation has an amplitude aU, and its frequency is ω times the natural frequency of the wing.

Computations have been performed after replacing U by the present form and changing the function CL(V) to CL(V, T) in Program 1.3. The results for $U = 100$, $a = 0.2$, and $\omega = 0.5, 1, 2$ are presented in Fig. 1.3.5. In response to a gust of any frequency, the wing initially tries to vibrate at its natural frequency, but finally sets into a periodical (*not* sinusoidal) motion whose frequency is smaller than both the natural frequency and the frequency of the fluctuating wind.

It is interesting to examine also the response to a fluctuating wind in the vertical direction. Figure 1.3.6 shows the results for $U = 100$ and $a = 0.2$, under the assumption that the dimensionless vertical velocity is described by $-aU \sin \omega T$. Data were obtained from Program 1.3 after replacing V by $V + aU \sin \omega T$ in the function F. In all three cases computed for $\omega = 0.5, 1$, and 2, the wing finally vibrates approximately at its natural frequency with an

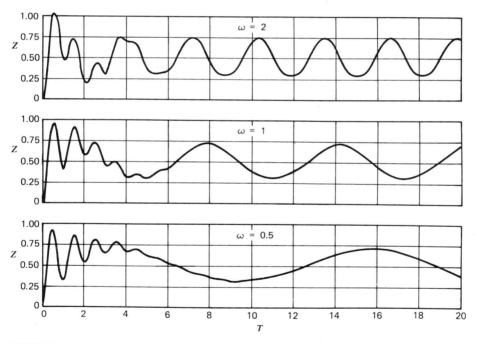

FIGURE 1.3.5 Response of a wing to an unsteady wind of speed $U(1 + a \sin \omega T)$, where $U = 100$ and $a = 0.2$.

Computer Simulation of Some Restrained Motions

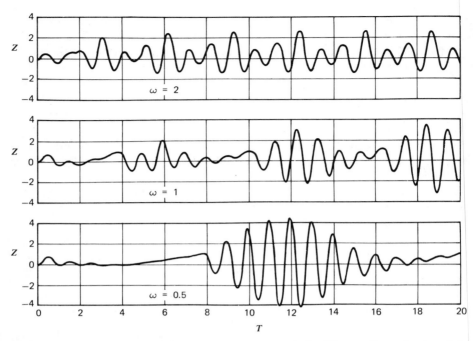

FIGURE 1.3.6 Response of a wing to a gusty wind with a uniform horizontal speed $U = 100$ and an unsteady vertical speed $0.2U \sin \omega T$.

amplitude varying in time. After some initial adjustment the seemingly irregular motion actually repeats the pattern every definite period of time. This period increases with decreasing ω, and it is equal to the period of the final oscillatory motion shown in Fig. 1.3.5 for the same value of ω.

In both examples resonance does not occur when the frequency of the gust coincides with the natural frequency of the wing. In a wind with horizontal fluctuations, the final amplitude of the wing motion increases with ω, while the opposite is true in a wind with vertical fluctuations. You may experiment with the program to find the value of ω that causes a maximum amplitude of oscillation for each case.

Program 1.3 can easily be generalized for computing the motion of a wing in a turbulent atmosphere with fluctuating velocities in both horizontal and vertical directions. The fluctuations may be expressed as functions of time in the form of Fourier series.

Problem 1.5 If a horizontal oscillatory wind is added and the sphere is replaced by a circular cylinder in Fig. 1.3.1, the arrangement

Dynamics of a Body Moving through a Fluid Medium

may be used as a crude model for studying the motion of a power transmission line in time-varying winds.

Consider a copper wire ($\rho = 8950$ kg/m^3) of diameter $d = 0.005$ m, initially stationary at its lowest sagging position with $l = 0.3$ m. Simulate numerically its motion in a wind of speed

$$u = 30(1 + 0.2 \sin \omega t) \text{ m/s}$$

with $\omega = 0$, 0.5, 1, and 2.

The drag coefficient of a circular cylinder is defined by

$$c_d = (\text{drag per unit length})/\tfrac{1}{2}\rho_f v_r^2 \, d,$$

where v_r is the *total* velocity of the fluid *relative* to the body. The drag coefficient of a long circular cylinder varies with Reynolds number, as does that of a sphere. For simplicity let us assume that $c_d = 1.2$, which is the approximate value for Reynolds numbers between 10^4 and 1.5×10^5 (Lindsey, 1938).

Hint: Since the density of copper is much higher than that of air, the buoyancy and added mass can be neglected in the equations of motion. It can be shown that the component of viscous force in the direction of v is

$$\tfrac{1}{2}\rho_f \, dc_d(u \cos \theta - v)\sqrt{(u \sin \theta)^2 + (u \cos \theta - v)^2}$$

where θ and v are indicated in Fig. 1.3.1. The direction of this force component is controlled by the sign of $(u \cos \theta - v)$.

1.4 Fourth-Order Runge-Kutta Method for Computing Two-Dimensional Motions of a Body Through a Fluid

Consider the translation motion of a body through a fluid in the x-y plane, where y is in the direction opposite to that of the gravitational acceleration, as sketched in Fig. 1.4.1. The velocity vector of the body relative to the stationary coordinate system is $\mathbf{w}$, which has components u and v. In general the fluid is in motion and, at the location of the body, its velocity, $\mathbf{w}_f$, has components u_f and v_f. The fluid dynamic forces are determined by the velocity $\mathbf{w}_r(= \mathbf{w}_f - \mathbf{w})$ of the fluid relative to that of the body.

If the resultant fluid dynamic force acting on the body is $\mathbf{f}$, with components f_x and f_y, which excludes the force associated with the added mass m', the

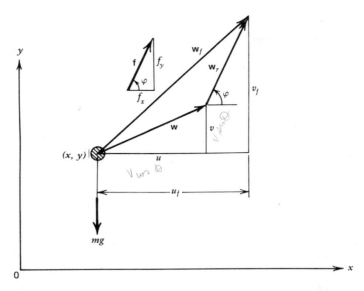

FIGURE 1.4.1 Two-dimensional motion of a body through a fluid.

equations of motion of the body of mass m are

$$(m + m') \frac{d^2x}{dt^2} = f_x$$

$$(m + m') \frac{d^2y}{dt^2} = -(m - m_f)g + f_y \tag{1.4.1}$$

where x, y are the coordinates of the projectile and m_f is the mass of fluid displaced by the body.

Since f is generally a function of position, velocity, and time, the simultaneous ordinary differential equations (1.4.1) can be expressed in the following functional form, with each second-order equation replaced by two simultaneous equations of the first order.

$$\frac{dx}{dt} = u \qquad \frac{du}{dt} = F_1(x, y, u, v, t)$$

$$\frac{dy}{dt} = v \qquad \frac{dv}{dt} = F_2(x, y, u, v, t) \tag{1.4.2}$$

The forms of the functions F_1 and F_2 vary in different problems. Suppose at an initial instant t_0 the position (x_0, y_0) and velocity (u_0, v_0) are given, the trajectory and motion of the body for $t > t_0$ are to be sought as functions of time.

The initial-value problem can again be solved numerically by use of Runge-Kutta methods. Similar to (1.1.12) and (1.1.13), with h representing the size of increments in time and subscripts i and $i+1$ respectively denoting the values evaluated at time steps t_i and t_{i+1} $(= t_i + h)$, the fourth-order formulas are:

$$\Delta_1 x_i = h u_i$$
$$\Delta_1 y_i = h v_i$$
$$\Delta_1 u_i = h F_1(x_i, y_i, u_i, v_i, t_i)$$
$$\Delta_1 v_i = h F_2(x_i, y_i, u_i, v_i, t_i)$$
$$\Delta_2 x_i = h(u_i + \tfrac{1}{2}\Delta_1 u_i)$$
$$\Delta_2 y_i = h(v_i + \tfrac{1}{2}\Delta_1 v_i)$$
$$\Delta_2 u_i = h F_1(x_i + \tfrac{1}{2}\Delta_1 x_i, y_i + \tfrac{1}{2}\Delta_1 y_i, u_i + \tfrac{1}{2}\Delta_1 u_i, v_i + \tfrac{1}{2}\Delta_1 v_i, t_i + \tfrac{1}{2}h)$$
$$\Delta_2 v_i = h F_2(x_i + \tfrac{1}{2}\Delta_1 x_i, y_i + \tfrac{1}{2}\Delta_1 y_i, u_i + \tfrac{1}{2}\Delta_1 u_i, v_i + \tfrac{1}{2}\Delta_1 v_i, t_i + \tfrac{1}{2}h)$$
$$\Delta_3 x_i = h(u_i + \tfrac{1}{2}\Delta_2 u_i)$$
$$\Delta_3 y_i = h(v_i + \tfrac{1}{2}\Delta_2 v_i)$$
$$\Delta_3 u_i = h F_1(x_i + \tfrac{1}{2}\Delta_2 x_i, y_i + \tfrac{1}{2}\Delta_2 y_i, u_i + \tfrac{1}{2}\Delta_2 u_i, v_i + \tfrac{1}{2}\Delta_2 v_i, t_i + \tfrac{1}{2}h)$$
$$\Delta_3 v_i = h F_2(x_i + \tfrac{1}{2}\Delta_2 x_i, y_i + \tfrac{1}{2}\Delta_2 y_i, u_i + \tfrac{1}{2}\Delta_2 u_i, v_i + \tfrac{1}{2}\Delta_2 v_i, t_i + \tfrac{1}{2}h)$$
$$\Delta_4 x_i = h(u_i + \Delta_3 u_i)$$
$$\Delta_4 y_i = h(v_i + \Delta_3 v_i)$$
$$\Delta_4 u_i = h F_1(x_i + \Delta_3 x_i, y_i + \Delta_3 y_i, u_i + \Delta_3 u_i, v_i + \Delta_3 v_i, t + h)$$
$$\Delta_4 v_i = h F_2(x_i + \Delta_3 x_i, y_i + \Delta_3 y_i, u_i + \Delta_3 u_i, v_i + \Delta_3 v_i, t + h)$$
$$x_{i+1} = x_i + \tfrac{1}{6}(\Delta_1 x_i + 2\Delta_2 x_i + 2\Delta_3 x_i + \Delta_4 x_i)$$
$$y_{i+1} = y_i + \tfrac{1}{6}(\Delta_1 y_i + 2\Delta_2 y_i + 2\Delta_3 y_i + \Delta_4 y_i)$$
$$u_{i+1} = u_i + \tfrac{1}{6}(\Delta_1 u_i + 2\Delta_2 u_i + 2\Delta_3 u_i + \Delta_4 u_i)$$
$$v_{i+1} = v_i + \tfrac{1}{6}(\Delta_1 v_i + 2\Delta_2 v_i + 2\Delta_3 v_i + \Delta_4 v_i) \qquad (1.4.3)$$

Based on these formulas, a subroutine named **KUTTA** will be constructed in Program 1.4 in the next section. To use the subroutine, one needs only to attach it to the main program and define in two separate subprograms the functions F_1 and F_2. The usage will be demonstrated in some of the following programs.

1.5 Ballistics of a Spherical Projectile

It is well known that in small-scale motions a projectile traces out a parabola when shooting upward in a direction not perpendicular to the earth's surface. But this conclusion is derived by considering trajectories in a vacuum. To study the effect of the surrounding fluid on the motion of a spherical projectile, let us go back to the equations of motion (1.4.1). If the sphere does not rotate, $\mathbf{f}$ becomes the viscous drag in the direction of the relative velocity $\mathbf{w}_r$, which makes an angle φ with the x-axis. From Fig. 1.4.1 we have $\sin \varphi = (v_f - v)/w_r$ and $\cos \varphi = (u_f - u)/w_r$, where $w_r = \sqrt{(u_f - u)^2 + (v_f - v)^2}$. Thus

$$f_x = |\mathbf{f}| \cos \varphi = \tfrac{1}{8}\pi \rho_f d^2 c_d(u_f - u)w_r$$

and

$$f_y = |\mathbf{f}| \sin \varphi = \tfrac{1}{8}\pi \rho_f d^2 c_d(v_f - v)w_r$$

The directions of f_x and f_y are determined by the signs of $(u_f - u)$ and $(v_f - v)$, respectively, and c_d is the drag coefficient of the sphere moving at the speed w_r, described in Fig. 1.2.2.

As discussed in Section 1.2, the added mass m' for a sphere is $m_f/2$. With m and m_f expressed in terms of densities ρ and ρ_f, and f_x, f_y replaced by the above expressions, (1.4.1) becomes, after some rearrangement,

$$\frac{d^2x}{dt^2} = \frac{3\bar{\rho}}{4d} c_d(u_f - u)w_r/(1 + \tfrac{1}{2}\bar{\rho}) \tag{1.5.1}$$

and

$$\frac{d^2y}{dt^2} = \left[-(1 - \bar{\rho})g + \frac{3\bar{\rho}}{4d} c_d(v_f - v)w_r \right] \Big/ (1 + \tfrac{1}{2}\bar{\rho}) \tag{1.5.2}$$

Since the fluid velocity components are generally functions of location and time, the right-hand sides of (1.5.1) and (1.5.2) will be called FX(x, y, u, v, t) and FY(x, y, u, v, t), respectively. They are the special forms of the functions F_1 and F_2 in (1.4.2).

We consider a steel sphere 0.05 m in diameter moving in air. Initially, when $t = 0$, the sphere shoots from the origin of the coordinate system with a speed w_0 at an elevation of $\theta_0°$. Elevation is the angle between the initial velocity of a projectile and the x-axis. The motions of the body for a fixed initial speed $w_0 = 50$ m/s are to be computed under the conditions that $\theta_0 = 30°$, $45°$, and $60°$. For numerical computations in Program 1.4, the Runge-Kutta formulas (1.4.3) are programmed in the subroutine KUTTA, which will also be used in some later problems.

To avoid printing a large amount of data, two integer variables I and IFRPRT are introduced in Program 1.4. I is the count of the number of times the subroutine KUTTA has been called in the main program; in fact, it is the subscript i in (1.4.3). IFRPRT is the frequency of printing; that is, the calculated data will be printed out when I is an integral multiple of IFRPRT. This can be achieved by comparing the value of I with the product (I/IFRPRT)*IFRPRT, obtained by using the rule for integer division. To make the ratio I/IFRPRT an integer, the quotient is truncated after the decimal

List of Principal Variables in Program 1.4

Program Symbol	Definition
(Main Program)	
A	$1 + \bar{\rho}/2$
B	$(1 - \bar{\rho})g$
C	$3\bar{\rho}/4d$
D	Diameter of sphere, d, m
DT	Size of time steps, s
G	Gravitational acceleration, g, m/s^2
I	Step counter
IFRPRT	Frequency of printing
J	Counter for different cases
NU	Kinematic viscosity of fluid, ν, m^2/s
PI	π
RHO	Density of body, ρ, kg/m^3
RHOBAR	Density ratio, $\bar{\rho}$, or ρ_f/ρ
RHOF	Density of fluid, ρ_f, kg/m^3
T, T0	Time, t, and initial time, t_0, respectively, s
THETA0	Elevation of the projectile, θ_0, deg
U	Horizontal component of body velocity, u, m/s
V	Vertical component of body velocity, v, m/s
W0	Initial speed of body, w_0, m/s
X, X0	Horizontal position of body, x, and its initial value, x_0, respectively, m
Y, Y0	Vertical position of body, y, and its initial value, y_0, respectively, m
(Subroutine KUTTA)	
D1U, D2U, etc.	Increments $\Delta_1 u_i$, $\Delta_2 u_i$, etc.
D1V, D2V, etc.	Increments $\Delta_1 v_i$, $\Delta_2 v_i$, etc.
D1X, D2X, etc.	Increments $\Delta_1 x_i$, $\Delta_2 x_i$, etc.
D1Y, D2Y, etc.	Increments $\Delta_1 y_i$, $\Delta_2 y_i$, etc.
H	Size of increment in T
(Functions FX, FY)	
UF	Horizontal component of fluid velocity, u_f, m/s
VF	Vertical component of fluid velocity, v_f, m/s
WR	Speed of fluid relative to body, w_r, m/s
CD	Drag coefficient of a sphere

```
C                    ***** PROGRAM 1.4 *****

C       MOTION AND TRAJECTORY OF A SPHERICAL PROJECTILE IN A FLUID

        EXTERNAL FX,FY
        COMMON A,B,C,D,NU,PI
        REAL NU

C       ..... ASSIGN INPUT DATA AND START THE OUTPUT ON A NEW PAGE .....
        DATA RHO,RHOF,G,DT,IFRPRT / 8000.0, 1.22, 9.8, 0.1, 5 /
        DATA TO,XO,YO,WO / 3*0.0, 50.0 /
        RHOBAR = RHOF/RHO
        NU = 1.49E-5
        PI = 3.14159
        D = 0.05
        A = 1. + RHOBAR/2.
        B = (1. - RHOBAR)*G
        C = 3.*RHOBAR/(4.*D)
        WRITE(6,10)

C       ..... PRINT HEADINGS AND ASSIGN INITIAL CONDITIONS FOR EACH OF THE
C             3 CASES WITH THETA0 = 30, 45, AND 60 DEGREES, RESPECTIVELY.
C             THEN PRINT THE INITIAL VALUES .....
        THETA0 = 15.0
        DO 4 J=1,3
        THETA0 = THETA0 + 15.0
        WRITE(6,20)  J, WO, THETA0
        T = TO
        X = XO
        Y = YO
        U = WO * COS(THETA0*PI/180.)
        V = WO * SIN(THETA0*PI/180.)
        I = 0
      1 WRITE(6,30)  T, X, Y, U, V

C       ..... FOURTH-ORDER RUNGE-KUTTA METHOD IS USED TO INTEGRATE THE
C             EQUATIONS OF MOTIN. STOP THE COMPUTATION FOR ONE CASE IF
C             THE PROJECTILE DROPS BACK TO ITS INITIAL LEVEL OR BELOW.
C             PRINT DATA WHEN THE LAST TIME STEP IS REACHED OR WHEN THE
C             COUNT I IS AN INTEGRAL MULTIPLE OF IFRPRT .....
      2 CALL KUTTA( X, Y, U, V, T, DT, FX, FY )
        IF (Y.LE.0. .AND. V.LT.0.) GO TO 3
        I = I+1
        IF ( (I/IFRPRT)*IFRPRT .EQ. I )  GO TO 1
        GO TO 2
      3 WRITE(6,30)  T, X, Y, U, V
      4 CONTINUE

     10 FORMAT( 1H1 )
     20 FORMAT(// 1X, 5HCASE(, I1, 26H) ......   INITIAL SPEED =, F3.0,
      A        19H M/SEC,     THETA0 =, F3.0, 8H DEGREES /
      B        4X,1HT, 12X,1HX, 12X1HY, 15X,1HU, 12X,1HV / 2X,5H(SEC),
      C        9X,3H(M), 10X,3H(M), 11X,7H(M/SEC), 6X,7H(M/SEC) )
     30 FORMAT( F7.2, 2X, 1P2E13.4, 3X, 1P2E13.4 )

        END
```

38

```
      SUBROUTINE KUTTA( X, Y, U, V, T, H, F1, F2 )
C     ..... FOURTH-ORDER RUNGE-KUTTA FORMULAE (1.4.3) FOR SOLVING TWO
C           SIMULTANEOUS SECOND-ORDER DIFFERENTIAL EQUATIONS WRITTEN AS
C                  DX/DT = U ,      DU/DT = F1(X,Y,U,V,T) ,
C                  DY/DT = V ,      DV/DT = F2(X,Y,U,V,T) .
C           H IS STEP SIZE IN T. F1,F2 ARE DEFINED IN OTHER SUBPROGRAMS.
      D1X = H * U
      D1Y = H * V
      D1U = H * F1( X, Y, U, V, T )
      D1V = H * F2( X, Y, U, V, T )

      D2X = H * (U+D1U/2.)
      D2Y = H * (V+D1V/2.)
      D2U = H * F1(X+D1X/2., Y+D1Y/2., U+D1U/2., V+D1V/2., T+H/2.)
      D2V = H * F2(X+D1X/2., Y+D1Y/2., U+D1U/2., V+D1V/2., T+H/2.)

      D3X = H * (U+D2U/2.)
      D3Y = H * (V+D2V/2.)
      D3U = H * F1(X+D2X/2., Y+D2Y/2., U+D2U/2., V+D2V/2., T+H/2.)
      D3V = H * F2(X+D2X/2., Y+D2Y/2., U+D2U/2., V+D2V/2., T+H/2.)

      D4X = H * (U+D3U)
      D4Y = H * (V+D3V)
      D4U = H * F1(X+D3X, Y+D3Y, U+D3U, V+D3V, T+H)
      D4V = H * F2(X+D3X, Y+D3Y, U+D3U, V+D3V, T+H)

      T = T + H
      X = X + (D1X + 2.*D2X + 2.*D3X + D4X) / 6.
      Y = Y + (D1Y + 2.*D2Y + 2.*D3Y + D4Y) / 6.
      U = U + (D1U + 2.*D2U + 2.*D3U + D4U) / 6.
      V = V + (D1V + 2.*D2V + 2.*D3V + D4V) / 6.
      RETURN
      END


      FUNCTION FX( X, Y, U, V, T )
C     ..... IT REPRESENTS THE RIGHT-HAND SIDE OF EQUATION (1.5.1) .....
C           ( IN THE PRESENT PROBLEM THE FLUID IS STATIONARY )

      COMMON A,B,C,D,NU,PI
      UF = 0.0
      VF = 0.0
      WR = SQRT( (UF-U)**2 + (VF-V)**2 )
      FX = C*CD(WR)*(UF-U)*WR/A
      RETURN
      END


      FUNCTION FY( X, Y, U, V, T )
C     ..... IT REPRESENTS THE RIGHT-HAND SIDE OF EQUATION (1.5.2) .....
C           ( IN THE PRESENT PROBLEM THE FLUID IS STATIONARY )

      COMMON A,B,C,D,NU,PI
      UF = 0.0
      VF = 0.0
      WR = SQRT( (UF-U)**2 + (VF-V)**2 )
      FY = ( -B + C*CD(WR)*(VF-V)*WR )/A
      RETURN
      END


      FUNCTION CD(V)
C     ..... TAKEN FROM PROGRAM 1.2 WITH THE COMMON STATEMENT
C           REPLACED BY THE ONE SHOWN BELOW .....
      COMMON A,B,C,D,NU,PI
```

```
CASE(1) ......   INITIAL SPEED =50. M/SEC.    THETA0 =30. DEGREES
   T            X              Y              U              V
 (SEC)         (M)            (M)          (M/SEC)        (M/SEC)
  0.00    0.             0.              4.3301E+01     2.5000E+01
   .50    2.1352E+01     1.1114E+01      4.2125E+01     1.9488E+01
  1.00    4.2143E+01     1.9518E+01      4.1058E+01     1.4157E+01
  1.50    6.2424E+01     2.5296E+01      4.0079E+01     8.9788E+00
  2.00    8.2234E+01     2.8519E+01      3.9169E+01     3.9315E+00
  2.50    1.0160E+02     2.9247E+01      3.8310E+01    -9.9994E-01
  3.00    1.2055E+02     2.7537E+01      3.7485E+01    -5.8244E+00
  3.50    1.3909E+02     2.3440E+01      3.6679E+01    -1.0545E+01
  4.00    1.5723E+02     1.7009E+01      3.5880E+01    -1.5161E+01
  4.50    1.7497E+02     8.2977E+00      3.5078E+01    -1.9665E+01
  4.90    1.8887E+02    -2.7471E-01      3.4428E+01    -2.3184E+01


CASE(2) ......   INITIAL SPEED =50. M/SEC.    THETA0 =45. DEGREES
   T            X              Y              U              V
 (SEC)         (M)            (M)          (M/SEC)        (M/SEC)
  0.00    0.             0.              3.5355E+01     3.5355E+01
   .50    1.7435E+01     1.6222E+01      3.4405E+01     2.9571E+01
  1.00    3.4423E+01     2.9608E+01      3.3564E+01     2.4009E+01
  1.50    5.1015E+01     4.0261E+01      3.2816E+01     1.8628E+01
  2.00    6.7252E+01     4.8262E+01      3.2143E+01     1.3397E+01
  2.50    8.3168E+01     5.3679E+01      3.1530E+01     8.2895E+00
  3.00    9.8789E+01     5.6568E+01      3.0962E+01     3.2852E+00
  3.50    1.1413E+02     5.6979E+01      3.0423E+01    -1.6283E+00
  4.00    1.2921E+02     5.4954E+01      2.9897E+01    -6.4566E+00
  4.50    1.4403E+02     5.0537E+01      2.9373E+01    -1.1199E+01
  5.00    1.5859E+02     4.3771E+01      2.8838E+01    -1.5849E+01
  5.50    1.7287E+02     3.4705E+01      2.8285E+01    -2.0396E+01
  6.00    1.8687E+02     2.3393E+01      2.7709E+01    -2.4829E+01
  6.50    2.0057E+02     9.8973E+00      2.7106E+01    -2.9133E+01
  6.90    2.1131E+02    -2.4276E+00      2.6603E+01    -3.2476E+01


CASE(3) ......   INITIAL SPEED =50. M/SEC.    THETA0 =60. DEGREES
   T            X              Y              U              V
 (SEC)         (M)            (M)          (M/SEC)        (M/SEC)
  0.00    0.             0.              2.5000E+01     4.3301E+01
   .50    1.2330E+01     2.0142E+01      2.4334E+01     3.7313E+01
  1.00    2.4349E+01     3.7356E+01      2.3756E+01     3.1586E+01
  1.50    3.6099E+01     5.1763E+01      2.3256E+01     2.6073E+01
  2.00    4.7616E+01     6.3458E+01      2.2824E+01     2.0734E+01
  2.50    5.8932E+01     7.2519E+01      2.2448E+01     1.5533E+01
  3.00    7.0072E+01     7.9009E+01      2.2119E+01     1.0442E+01
  3.50    8.1057E+01     8.2976E+01      2.1825E+01     5.4364E+00
  4.00    9.1901E+01     8.4457E+01      2.1553E+01     5.0016E-01
  4.50    1.0261E+02     8.3486E+01      2.1289E+01    -4.3747E+00
  5.00    1.1319E+02     8.0093E+01      2.1020E+01    -9.1870E+00
  5.50    1.2363E+02     7.4311E+01      2.0735E+01    -1.3927E+01
  6.00    1.3392E+02     6.6180E+01      2.0426E+01    -1.8581E+01
  6.50    1.4405E+02     5.5747E+01      2.0089E+01    -2.3132E+01
  7.00    1.5400E+02     4.3067E+01      1.9722E+01    -2.7563E+01
  7.50    1.6377E+02     2.8206E+01      1.9324E+01    -3.1857E+01
  8.00    1.7332E+02     1.1236E+01      1.8899E+01    -3.5999E+01
  8.40    1.8081E+02    -3.8064E+00      1.8539E+01    -3.9195E+01
```

40

point. Unless I is an integral multiple of IFRPRT, the product will not return to the value I. An IF statement is used in the program to compare these two integer values before making the decision for printing.

Instead of setting a time limit to terminate the computation, we stop it when the projectile falls back to or below its initial height, in other words, when $y \leq 0$ for a negative v. The drag coefficient of the spherical body is still approximated by the function described already in the subprogram CD attached to Program 1.2.

The program is constructed so that the fluid velocity components u_f and v_f are specified in the function subprograms FX and FY instead of in the main program. In the present problem both components are zero. The subprograms can easily be modified to describe any steady or unsteady wind fields.

The motions in a vacuum are obtained when the value zero is assigned to ρ_f in Program 1.4, and the resultant trajectories are plotted in Fig. 1.5.1 for comparison. It shows that air resistance reduces both the range and the maximum height of a projectile. The data also reveal that the flight time is also reduced in the presence of air. It is interesting to note that the ranges are the same for $\theta_0 = 30°$ and $60°$ in a vacuum; while in air, because of the longer flight time for the trajectory with $\theta_0 = 60°$, its range becomes shorter due to the longer action of air resistance.

Problem 1.6 Find the effect of a 20 m/s wind on the trajectory of a sphere described in Program 1.4. The wind may either be in the direction of the x-axis or against it.

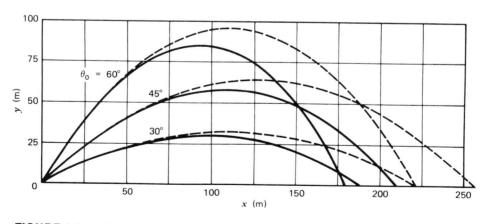

FIGURE 1.5.1 Trajectories of a sphere in air (solid lines) and those in vacuum (dashed lines).

Ballistics of a Spherical Projectile

By examining Fig. 1.5.1 a question naturally arises. It is known that in a vacuum the range of a projectile becomes maximum when $\theta_0 = 45°$. Is it still true in the presence of a fluid?

Modifications on Program 1.4 are needed before we can do some calculations to find an answer to this question. Position and velocity of the body are computed in Program 1.4 according to a constant time increment; therefore, the range is not specifically shown in the output. Furthermore, an algorithm has to be found so that the maximum of a function can be located automatically by the computer.

The range can be found approximately by using Fig. 1.5.2. Suppose that in the numerical integration the point Q is the *first* computed point at which the body falls back on or below the horizon. Tracing back to the previous time step, the body was at the point P. The path connecting P and Q is in general a curve, but can be approximated by a straight line if the time interval is small. The range x_r is the distance between the origin and the point where the line PQ intersects the x-axis. Similarity of the two shaded triangles gives

$$\frac{y_P}{x_r - x_P} = \frac{-y_Q}{x_Q - x_r}$$

or, after solving for x_r,

$$x_r = \frac{x_Q y_P - x_P y_Q}{y_P - y_Q} \qquad (1.5.3)$$

In this equation y_Q is either negative or zero.

Q is the last point on a trajectory at which numerical computation is performed in Program 1.4. When this point is reached, the data associated with the previous points have been erased from the computer memory according to the way the Runge-Kutta formulas are programmed. The position of P, which is one time step ahead of Q, can be obtained at this stage by

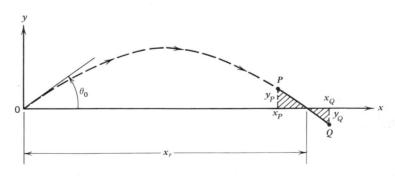

FIGURE 1.5.2 Finding the range of a projectile.

Dynamics of a Body Moving through a Fluid Medium

calling the subroutine **KUTTA** with the argument DT replaced by − DT. By doing so it is equivalent to integrating the equations of motion backward through one time step.

If one keeps integrating backward from the point P until time returns to its initial value, the deviations between the computed and the assumed conditions at $t = 0$ will show the accuracy of the numerical method. A test on Program 1.4 reveals that the error in position is only of the order of 10^{-9} m and that in velocity cannot be detected when printed according to the present field specification.

The range computed from (1.5.3) is a function of the elevation of a projectile. If the variation is described by the curve sketched in Fig. 1.5.3, we would like to locate the angle θ_0 at which the range is maximum.

Let us choose an arbitrary point a on the curve and locate a second point b according to the relation

$$(\theta_0)_b = (\theta_0)_a + \delta_1$$

where δ_1 is an arbitrary incremental quantity. If $(x_r)_a < (x_r)_b$ as shown in the sketch, the maximum is to the right of this interval, and a third point c is located that is a distance δ_1 to the right of b. If $(x_r)_b < (x_r)_c$, we repeat the process by locating points successively at a constant pace δ_1 to the right. However, if $(x_r)_b \geq (x_r)_c$ as shown, the maximum should appear to the left of c. This time we change the increment to $\delta_2 (= -\delta_1/2)$ and find the next point d according to

$$(\theta_0)_d = (\theta_0)_c + \delta_2$$

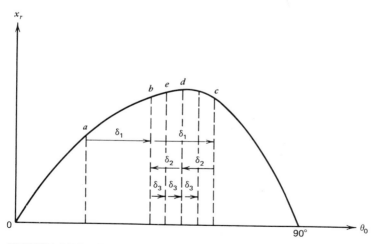

FIGURE 1.5.3 Finding maximum range by the use of a half-interval method.

Ballistics of a Spherical Projectile

Now $(x_r)_c < (x_r)_d$; that is, the range at the "old" point c is less than that at the "new" point d. Following the previous rule, we should go to a point whose abscissa is $(\theta_0)_d + \delta_2$, which is essentially the point b, although the maximum lies on the other side of d in the present case. But the condition that $(x_r)_d \geq (x_r)_b$, or that the range at the old point is greater than or equal to that at the new point, demands the change of increment to $\delta_3 = -\delta_2/2$. This increment brings the next point e back to the right of b. Two steps later we end up at a point between d and c Repeating the same process, we can arrive at a point that is as close as desired to the point where the range is maximum.

To make it convenient for computer calculation, the process for finding maximum range is generalized as follows. At any step in this method we have a point on the curve whose coordinates are called $(\theta_0)_{\text{old}}$ and $(x_r)_{\text{old}}$, and an increment called δ. A new point is found by using the relation

$$(\theta_0)_{\text{new}} = (\theta_0)_{\text{old}} + \delta \tag{1.5.4}$$

at which the range is $(x_r)_{\text{new}}$. The ranges at these two points are then compared. The increment keeps the same value if $(x_r)_{\text{new}} > (x_r)_{\text{old}}$; otherwise δ is replaced by $-\delta/2$. At this stage the old point is no longer needed, so we call the new point an old one and, in the meantime, rename its coordinates as $(\theta_0)_{\text{old}}$ and $(x_r)_{\text{old}}$. A new point is then located according to (1.5.4). The process is repeated until $|\delta|$ becomes less than a specified small quantity ϵ. At this final step the two ranges $(x_r)_{\text{new}}$ and $(x_r)_{\text{old}}$ are again compared. The greater one is the approximated maximum range, and the corresponding θ_0 is the optimum elevation. The accuracy of the result is controlled by the magnitude of ϵ.

Because the increment δ is consecutively reduced by half in the preceding method, it is called the *half-interval* or *interval halving method*. With some modifications the method can be used to find the minimum or the zeros of a function.

The formula (1.5.3) and the half-interval method are included in Program 1.5 to find the optimum shooting angle of a spherical projectile causing a maximum range. To emphasize the effect of air resistance, a smaller steel sphere of 1 cm diameter is considered instead of the 5-cm diameter projectile examined in Program 1.4. The conclusion that the surrounding fluid has less influence on the motion of larger spheres has been drawn from the result of Program 1.1 for free-falling bodies. The initial speed is still kept at 50 m/s.

Three wind conditions are assumed in Program 1.5 with horizontal wind speeds of 20, 0, and -20 m/s, respectively. Because of the variable wind speed, u_f can no longer be defined in the subprograms FX and FY as in Program 1.4. In the present case, the wind components are defined in the main program and are transmitted into the subprograms by using COMMON statements.

Most symbols to appear in Program 1.5 can be found in Program 1.4 except those listed below.

Program Symbol	Definition
DELTA	Increment δ in half-interval method, deg
EPSLON	ϵ, used to control the accuracy of the optimum shooting angle, deg
M	A counter for different wind conditions
N	Number of steps used in the half-interval method
THENEW, THEOLD	$(\theta_0)_{new}$ and $(\theta_0)_{old}$, respectively, deg
THETA	Optimum shooting angle at which range is maximum, deg
XP, XQ, YP, YQ	x_P, x_Q, y_P, and y_Q, respectively, m
XRMAX	Maximum range, m
XRNEW, XROLD	$(x_r)_{new}$ and $(x_r)_{old}$, respectively, m

```
C                   ***** PROGRAM 1.5 *****
C          FINDING THE MAXIMUM RANGE OF A SPHERICAL PROJECTILE

      EXTERNAL FX,FY
      COMMON A,B,C,D,NU,PI,UF,VF
      REAL NU

C      ..... ASSIGN INPUT DATA AND PRINT HEADINGS ON A NEW PAGE .....
      DATA RHO,RHOF,G,DT,EPSLON / 8000.0, 1.22, 9.8, 0.1, 0.001 /
      DATA T0,X0,Y0,W0 / 3*0.0, 50.0 /
      RHOBAR = RHOF/RHO
      NU = 1.49E-5
      PI = 3.14159
      D = 0.01
      A = 1. + RHOBAR/2.
      B = (1.-RHOBAR)*G
      C = 3.*RHOBAR/(4.*D)
      WRITE(6,50)  D, W0, EPSLON

C      ..... ASSIGN THREE DIFFERENT WIND CONDITIONS .....
      VF = 0.
      UF = 40.
      DO 10 M=1,3
      UF = UF-20.

C      ..... START WITH A HORIZONTAL SHOOTING POSITION AND WITH A
C              10-DEGREE INCREMENT .....
      THEOLD = 0.
      XROLD = 0.
      DELTA = 10.
      N = 0
```

Ballistics of a Spherical Projectile

```
C      ..... ACCORDING TO HALF-INTERVAL METHOD, THE ABSCISSA OF THE NEXT
C            POINT ON THE SHOOTING ANGLE-RANGE CURVE IS .....
     1 THENEW = THEOLD + DELTA
       N = N + 1

C      ..... TO FIND THE ORDINATE OF THAT POINT, WE FIRST USE RUNGE-KUTTA
C            METHOD TO DETERMINE THE POINTS P AND Q ON THE TRAJECTORY,
C            THEN USE EQUATION (1.5.3) TO COMPUTE THE RANGE .....
       T = T0
       X = X0
       Y = Y0
       U = W0 * COS(THENEW*PI/180.)
       V = W0 * SIN(THENEW*PI/180.)
     2 CALL KUTTA( X, Y, U, V, T, DT, FX, FY )
       IF( Y.LE.0. .AND. V.LT.0. ) GO TO 3
       GO TO 2
     3 XQ = X
       YQ = Y
       CALL KUTTA( X, Y, U, V, T, -DT, FX, FY )
       XP = X
       YP = Y
       XRNEW = (XQ*YP-XP*YQ)/(YP-YQ)

C      ..... CHANGE THE VALUE OF DELTA IF XROLD≥XRNEW.
C            STOP HALF-INTERVAL METHOD IF DELTA<EPSLON .....
       IF (XRNEW-XROLD) 4,4,5
     4 DELTA = -DELTA/2.
       IF (ABS(DELTA) .LT. EPSLON) GO TO 6
     5 CONTINUE

C      ..... CALL THE NEW POINT AN OLD ONE,
C            THEN LOCATE THE NEXT NEW POINT .....
       THEOLD = THENEW
       XROLD = XRNEW
       GO TO 1

C      ..... THE MAXIMUM RANGE, XRMAX, AND THE OPTIMUM SHOOTIMG
C            ANGLE, THETA, ARE DETERMINED .....
     6 IF (XRNEW-XROLD) 7,7,8
     7 XRMAX = XROLD
       THETA = THEOLD
       GO TO 9
     8 XRMAX = XRNEW
       THETA = THENEW
     9 WRITE(6,60)  UF, THETA, XRMAX, N
    10 CONTINUE

    50 FORMAT(1H1,////5X,40HCONSIDER A STEEL SPHERICAL PROJECTILE OF,
      A      F5.3, 11H M DIAMETER / 5X,
      B      40HSHOOTING IN AIR WITH AN INITIAL SPEED OF, F6.2,
      C      7H M/SEC. // 8X,
      D      41HSYMBOLS USED IN THE FOLLOWING TABLE ARE :/ 14X,
      E      26HUF = HORIZONTAL WIND SPEED / 11X,
      F      34HTHETA = THE OPTIMUM SHOOTING ANGLE / 11X,
      G      25HXRMAX = THE MAXIMUM RANGE / 15X,
      H      41HN = NUMBER OF STEPS USED IN HALF-INTERVAL / 19X,
      I      35HMETHOD TO OBTAIN THE PRINTED RESULT / 8X,
      J      37HTHE ERROR IN THETA IS OF THE ORDER OF, F5.3,
      K      9H DEGREES. //// 13X,2HUF, 9X,5HTHETA, 8X,5HXRMAX, 8X,1HN
      L      / 11X,7H(M/SEC), 6X,5H(DEG), 9X,3H(M) /
      M      11X,7H-------, 6X,5H-----, 8X,5H-----, 7X,3H--- / )
    60 FORMAT( F17.2, F13.3, F14.4, I8 )

       END
```

```
      FUNCTION FX( X, Y, U, V, T )
C     ..... IT REPRESENTS THE RIGHT-HAND SIDE OF EQUATION (1.5.1) .....

      COMMON A,B,C,D,NU,PI,UF,VF
      WR = SQRT( (UF-U)**2 + (VF-V)**2 )
      FX = C*CD(WR)*(UF-U)*WR / A
      RETURN
      END




      FUNCTION FY( X, Y, U, V, T )
C     ..... IT REPRESENTS THE RIGHT-HAND SIDE OF EQUATION (1.5.2) .....

      COMMON A,B,C,D,NU,PI,UF,VF
      WR = SQRT( (UF-U)**2 + (VF-V)**2 )
      FY = ( -B + C*CD(WR)*(VF-V)*WR ) / A
      RETURN
      END




      FUNCTION CD(V)
C     ..... TAKEN FROM PROGRAM 1.2 WITH THE COMMON STATEMENT
C           REPLACED BY THE ONE SHOWN BELOW .....
      COMMON A,B,C,D,NU,PI,UF,VF




      SUBROUTINE KUTTA( X, Y, U, V, T, H, F1, F2 )
C     ..... TAKEN FROM PROGRAM 1.4 .....
```

CONSIDER A STEEL SPHERICAL PROJECTILE OF .010 M DIAMETER
SHOOTING IN AIR WITH AN INITIAL SPEED OF 50.00 M/SEC.

    SYMBOLS USED IN THE FOLLOWING TABLE ARE :
        UF = HORIZONTAL WIND SPEED
     THETA = THE OPTIMUM SHOOTING ANGLE
     XRMAX = THE MAXIMUM RANGE
         N = NUMBER OF STEPS USED IN HALF-INTERVAL
             METHOD TO OBTAIN THE PRINTED RESULT
    THE ERROR IN THETA IS OF THE ORDER OF .001 DEGREES.

UF (M/SEC)	THETA (DEG)	XRMAX (M)	N
20.00	44.611	189.7606	47
0.00	39.745	130.2014	40
-20.00	27.372	75.6149	42

The computer output shows that for the specified projectile the optimum shooting angles are all below 45°. The optimum angle in a favorable wind is the highest, and that in an adverse wind is the lowest among the three. The result can be explained as follows. Shooting a projectile in a vacuum at a 45° angle gives a range longer than the one that resulted from a lower shooting angle, and the flight time of the projectile in the former is longer than that in the latter case. In the presence of air without a wind, because of the shorter action of air resistance on the body, less kinetic energy is dissipated from the projectile shooting at a lower angle and, under appropriate conditions, the loss in horizontal distance because of air friction can be less. The plot of trajectories for $\theta_0 = 30°$ and 60° in Fig. 1.5.1 is such an example. By shooting the body at a properly chosen angle below 45°, the frictional loss can be minimized to give a maximum range. A wind blowing in the direction of the body motion carries the body with it. To aim the projectile higher increases the contact time with air and therefore increases the range. On the other hand, in an adverse wind, the optimum angle should be lower than that in a

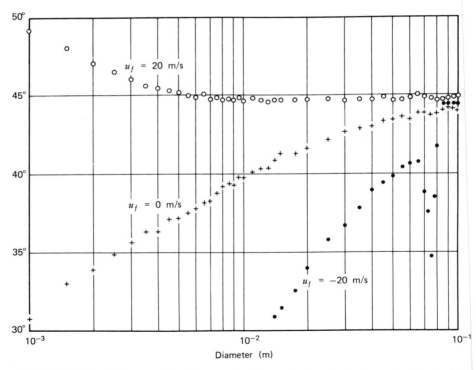

FIGURE 1.5.4 Optimum shooting angle for a steel spherical projectile having an initial speed of 50m/s.

Dynamics of a Body Moving through a Fluid Medium

quiet atmosphere in order to reduce the retarding effect of the wind. The computed results are in agreement with the experiences of a golfer.

The optimum angles are not always below 45°, however, if the size of the projectile is changed. The variations of the optimum angle with diameter under three wind conditions are plotted in Fig. 1.5.4. The data are obtained by varying the value of D in Program 1.5. Figure 1.5.4 shows that for small projectiles in a favorable wind, the optimum angle can be higher than 45°. The influence of air on the motion of a projectile is diminishing with increasing diameter, and the optimum angle finally approaches 45°.

On the curve for the adverse-wind case a sharp dip appears in a region where the Reynolds number starts to exceed the value 3×10^5. The phenomenon is caused by the abrupt decrease of drag at that particular Reynolds number (see Fig. 1.2.2).

Problem 1.7 Contrary to common sense, under certain conditions the maximum range of a projectile can be made longer when it is thrown against the wind instead of in the wind direction. To prove that this is possible, run Program 1.5 for a steel sphere 0.09 m in diameter while keeping the other conditions the same. The result will show that among the three cases the maximum range is the longest for $u_f = -20$ m/s. Print out the Reynolds numbers and give an explanation of this phenomenon.

Problem 1.8 It was discovered accidentally during World War I that aiming a cannon at an angle higher than what had previously been believed to give maximum range resulted in a great increase in the range of the shell. The reason is that the projectile reaches a higher altitude by aiming the gun higher, and the smaller air resistance there may cause a longer range for certain angles.

To verify this phenomenon, let us consider a cannon shell whose muzzle velocity is 800 m/s. At a supersonic speed the drag coefficient is a function of Mach number *and* of Reynolds number, and it varies with the shape of the projectile. For the sake of simplicity we assume that the shell is equivalent to a steel sphere of 0.3 m diameter having a drag coefficient of a constant value of 0.4 throughout the flight. Find the optimum shooting angle and the maximum range of the shell in an atmosphere of constant density of 1.22 kg/m³; then find the angle and range in an atmosphere whose density is described by the exponential law

$$\rho_f = 1.22 \exp(-0.000118y) \text{ kg/m}^3$$

Ballistics of a Spherical Projectile

where y is the height above sea level measured in meters. The result will show a higher optimum shooting angle and a longer range in the case of the variable-density atmosphere.

Problem 1.9 Consider the cannon shell and the stratified atmosphere described in Problem 1.8. Write a computer program that computes the angle at which the cannon should be aimed in order to hit a target a distance x_t away. To make the program more general, the wind components are to be included. Test the program for $x_t = 8000$ m and 15,000 m in the absence of a wind.

Hint: The half-interval method can again be applied to this problem. Let the curve in Fig. 1.5.5 represent the variation of range with shooting angle. There are in general two angles at which the range is x_t, if x_t is within the maximum range. Let the lower one be θ_t. Similar to what we did in finding the maximum range, two arbitrary points whose ordinates are called $(x_r)_{old}$ and $(x_r)_{new}$, respectively, are chosen on the curve at a distance δ apart. When the abscissas of both points are on the left side of θ_t, we will choose a new point a distance δ to the right of the second point. If they end up on two sides of θ_t, as shown in Fig. 1.5.5, the new point has just passed the location we are looking for, and the next new point will be located midway between the

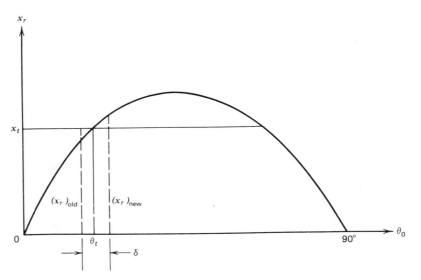

FIGURE 1.5.5 Finding the aiming angles of a cannon for hitting a target at a distance x_t away.

Dynamics of a Body Moving through a Fluid Medium

present new and old points. Thus an algorithm has been obtained. We start from the left end of the curve with two points and a positive value of δ. At each step the product of $[x_t - (x_r)_{old}]$ and $[x_t - (x_r)_{new}]$ is examined. δ keeps the same value if the product is positive; it is replaced by $-\delta/2$ otherwise. By adding δ to $(x_r)_{new}$, the next point is located. Repeat the process until the absolute value of δ is within a specified small quantity; the approximate value of θ_t is then obtained. Proceed to the right from there with the initially assumed value of δ; the second value of the aiming angle is then found. If no value of θ_t can be found within 90°, the searching should be stopped, because the target is out of the maximum range of the cannon. Print a statement in the output for such a case.

1.6   Collision of a Raindrop with Cloud Droplets

The formation and growth of raindrops is an important topic in cloud physics. Tiny water droplets of a spectrum of sizes have been observed in the clouds. Collisions between droplets take place because of the different terminal velocities for drops of various sizes (see Fig. 1.2.5). As a result of coalescence during successive collisions, the radius of a droplet grows and its terminal velocity increases with time. The larger size and the faster velocity of the droplet cause an increase in the number of collisions per unit time, so that it grows at a constantly increasing rate. After traveling a long distance, a droplet may finally fall out of the cloud as a raindrop.

Consider the collisions between a water drop of radius r_1 with terminal velocity u_{t1} and smaller cloud droplets of radius r_2 with terminal velocity u_{t2}. To the large drop the small droplets far away are moving toward it at a speed $u_t = u_{t1} - u_{t2}$, as sketched in Fig. 1.6.1. If there were no air, all the droplets contained within a cylinder of radius r_1 ahead of the drop would collide with it. In reality the airflow around the drop has a tendency to carry the droplets away from the central axis, so that only those contained within a smaller cylinder of radius r_c can make the collision. The ratio of the cross-sectional areas of these two cylinders, or $(r_c/r_1)^2$, is called the *collision efficiency*, E, which is determined by the radii r_1 and r_2 of the colliding spheres. Based on the collision efficiency, the growth rate of a raindrop in a cloud of given water content can be calculated if the fraction of collisions leading to coalescence is known or assumed.

In this section we will construct a program for finding the collision efficiency of a raindrop falling through smaller cloud droplets. We assume that

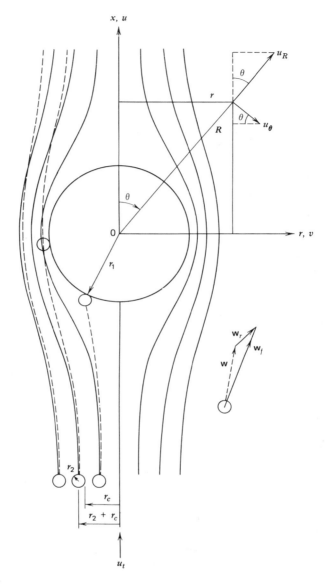

FIGURE 1.6.1 Collision of a raindrop with smaller droplets. Solid curves are the streamlines of airflow around the drop, and dashed curves are the trajectories of droplets.

r_2 is so small in comparison with r_1 that the presence of the small droplets does not alter the flow around the large drop, and that the changes in both radius and velocity of the drop are negligible within a period when a droplet has traveled through a distance of the order of r_1. The formulation is derived in a cylindrical coordinate system attached to the center of the raindrop.

Suppose at time t the position of a cloud droplet is at (x, r) with a velocity $\mathbf{w}(u, v)$ and, at the same location, the airflow has a velocity $\mathbf{w}_f(u_f, v_f)$. The force acting on the droplet is determined by the relative velocity $\mathbf{w}_r = \mathbf{w}_f - \mathbf{w}$, shown in Fig. 1.6.1. The quantity $\bar{\rho}$, or ρ_f/ρ, is the ratio of the density of air to that of water in the present case. The equations of motion for a droplet are almost the same as (1.5.1) and (1.5.2) derived for a projectile, except that the gravitational acceleration does not appear in the moving coordinate system. Thus

$$\frac{d^2x}{dt^2} \doteq \frac{3\bar{\rho}}{8r_2} \frac{c_d(u_f - u)w_r}{1 + \frac{1}{2}\bar{\rho}} \qquad (1.6.1)$$

$$\frac{d^2r}{dt^2} = \frac{3\bar{\rho}}{8r_2} \frac{c_d(v_f - v)w_r}{1 + \frac{1}{2}\bar{\rho}} \qquad (1.6.2)$$

in which $u = dx/dt$ and $v = dr/dt$. The Reynolds numbers computed in Fig. 1.2.5 for falling drops are generally much higher than unity, indicating that viscous effects are confined within a thin boundary layer around the drop, if there is no flow separation. Even if separation takes place, the flow ahead of the drop, on which the collision efficiency is based, is only slightly affected. It is therefore reasonable to assume a potential flow around the drop.

The velocity components of a potential flow of uniform speed u_t past a sphere of radius r_1 are, in spherical coordinates,

$$u_R = u_t\left(1 - \frac{r_1^3}{R^3}\right)\cos\theta \qquad (1.6.3)$$

$$u_\theta = -u_t\left(1 + \frac{r_1^3}{2R^3}\right)\sin\theta \qquad (1.6.4)$$

which can be found in many textbooks in fluid mechanics. To transform them into cylindrical coordinates, the diagram in Fig. 1.6.1 is used. It shows that

$$u_f = u_R \cos\theta - u_\theta \sin\theta$$

and

$$v_f = u_R \sin\theta + u_\theta \cos\theta$$

where the subscript f is added to denote the fluid velocity components. Substituting from (1.6.3) and (1.6.4) and using geometrical relations between

Collision of a Raindrop with Cloud Droplets

(R, θ) and (x, r), we obtain

$$u_f = u_t \left[1 + \frac{r_1^3 (\frac{1}{2} r^2 - x^2)}{(x^2 + r^2)^{5/2}} \right] \tag{1.6.5}$$

$$v_f = - u_t \frac{\frac{3}{2} r_1^3 xr}{(x^2 + r^2)^{5/2}} \tag{1.6.6}$$

The system of equations can be nondimensionalized by using the following dimensionless variables.

$$X = \frac{x}{r_1}, \qquad R = \frac{r}{r_1}, \qquad T = t \frac{u_t}{r_1}$$

$$U = \frac{u}{u_t}, \qquad V = \frac{v}{u_t}, \qquad U_f = \frac{u_f}{u_t}, \qquad V_f = \frac{v_f}{u_t}$$

$$W_r = \frac{w_r}{u_t} = \sqrt{(U_f - U)^2 + (V_f - V)^2} \tag{1.6.7}$$

Thus the differential equations become

$$\frac{d^2 X}{dT^2} = \frac{3 r_1}{8 r_2} c_d (U_f - U) W_r \frac{\bar{\rho}}{1 + \frac{1}{2} \bar{\rho}} \tag{1.6.8}$$

$$\frac{d^2 R}{dT^2} = \frac{3 r_1}{8 r_2} c_d (V_f - V) W_r \frac{\bar{\rho}}{1 + \frac{1}{2} \bar{\rho}} \tag{1.6.9}$$

where the dimensionless fluid velocity components are

$$U_f = 1 + \frac{\frac{1}{2} R^2 - X^2}{(X^2 + R^2)^{5/2}} \tag{1.6.10}$$

$$V_f = - \frac{\frac{3}{2} XR}{(X^2 + R^2)^{5/2}} \tag{1.6.11}$$

The drag coefficient c_d is to be computed based on the dimensional relative velocity w_r. On tiny droplets the drag is most likely in the Stokes flow range. But, in order to cover droplets of various sizes, we assume that the drag coefficient of a cloud droplet is approximated by that of a rigid sphere, described by the function subprogram CD in Program 1.2.

The collision efficiency will be computed according to the following procedure. We assume that at an initial instant $T = 0$ the motion of a droplet at a position (X_0, R_0), where $X_0 = -10$, say, moves at the local fluid velocity. Consider a first droplet at (X_0, R_{01}), $R_{01} = 0$, which can move along the centerline and will collide with the drop. An integer variable I1 associated with this droplet is defined as follows. When collision takes place, I1 is assigned the value $+1$; otherwise it takes the value -1. Next consider a second droplet initially at (X_0, R_{02}), $R_{02} = R_{01} + \Delta R_0$ where ΔR_0 is a properly chosen increment. The trajectory of this second droplet can be found by solving the simultaneous nonlinear equations (1.6.8) and (1.6.9) by use of the fourth-order Runge-Kutta method. Let another integer variable I2 be associated with this droplet. If at a later time the dimensional distance between the center of the second droplet and that of the large drop is equal to or less than $(r_1 + r_2)$, it collides with the drop and we let I2 = +1. If the condition for collision has not yet been satisfied at a position just past the drop, it cannot make the collision and we let I2 = −1 instead. At this stage the product of I1 and I2 is examined. If it is positive, either both the first and second droplets collide with the large drop, as in the present case that has one droplet along the centerline, or neither can collide with it. Under this condition ΔR_0 keeps the same value. When the product is negative, indicating that only one of the two droplets collides with the drop, we let ΔR_0 assume a new value $-\Delta R_0/2$. In the next step we call the second droplet our first and, at the same time, assign the previous value of R_{02} to the variable R_{01}. A new second droplet situated at (X_0, R_{02}), $R_{02} = R_{01} + \Delta R_0$, is then located. The process is repeated until $|\Delta R_0|$ is less than a specified small quantity ϵ. This is again a half-interval method.

Program 1.6 computes the collision efficiency of a raindrop 100 microns (or 0.0001 m) in radius colliding with cloud droplets of radii ranging from 5 to 20 microns. To find the terminal velocity of a water drop of any size, a function subprogram TRMVEL(RADIUS) is constructed in which the argument is the radius of the drop measured in meters. In this subprogram, written by modifying Program 1.1, the free-falling motion of an initially stationary drop is computed by using the fourth-order Runge-Kutta method. When the difference between the velocities at two consecutive time steps becomes less than a small value, say 0.00001 m/s, the velocity at the last time step is returned to the main program as the terminal velocity of the drop.

For solving two simultaneous ordinary differential equations, the subroutine KUTTA is borrowed from Program 1.4. The symbols in subprograms TRMVEL, F, CD, and KUTTA can be found in Programs 1.1, 1.2, and 1.4, respectively, and will not be listed in here.

The result shows that the collision efficiency increases with the size of cloud drop'ets. The values of E are higher than those computed by Langmuir (1948, Table 4), who pioneered such a quantitative analysis. The discrepancy occurs because Langmuir calculated terminal velocities by using the Stokes law of drag and thus obtained a much higher terminal velocity for the raindrop of a 100-micron radius.

In the presence of a droplet of radius comparable to that of the raindrop, the distortions on the flow around the two bodies are not negligible and must be included in the analysis. Such computations were performed for small drops in the Stokes flow regime by Hocking (1959). The result reveals that as the radius of the droplet becomes greater than a certain value, a repulsive

Collision of a Raindrop with Cloud Droplets

List of Principal Variables in Program 1.6

Program Symbol	Definition
(Main Program)	
DR0	Dimensionless radial increment, ΔR_0
DT	Dimensionless time increment
E	Collision efficiency
EPSLON	ϵ, used to control the accuracy of E
I1, I2	Integer variables indicating whether the first and second droplets collide with the raindrop
M	A counter for different droplet sizes
NU	Kinematic viscosity of air, ν, m²/s
PI	π
R	Dimensionless radial coordinate
RHO, RHOF	Densities of water and air, ρ and ρ_f, respectively, kg/m³
RHOBAR	Density ratio, $\bar{\rho}$, or ρ_f/ρ
R01, R02	Initial dimensionless radial coordinates of the first and the second droplets, R_{0_1} and R_{0_2}, respectively
R1, R2	Radius of the raindrop, r_1, and that of the cloud droplet, r_2, respectively, m
T, T0	Dimensionless time and its initial value, respectively
U, U0	Dimensionless axial velocity and its initial value, respectively
UT	Velocity of a faraway droplet relative to the raindrop, u_t, m/s
UT1, UT2	Terminal velocities of the raindrop and of the droplet, u_{t1} and u_{t2}, respectively, m/s
V, V0	Dimensionless radial velocity and its initial value, respectively
X, X0	Dimensionless axial coordinate and its initial value, respectively
(Functions FX, FY)	
UF	Dimensionless air velocity in the axial direction, U_f
VF	Dimensionless air velocity in the radial direction, V_f
WR	Dimensionless air velocity relative to the droplet, W_r

```
C                        ***** PROGRAM 1.6 *****

C          FINDING THE COLLISION EFFICIENCY OF A RAINDROP OF RADIUS R1
C          FALLING THROUGH CLOUD DROPLETS OF UNIFORM RADIUS R2
       EXTERNAL FX,FR
       COMMON D,NU,PI,R1,R2,RHOBAR,UT
       REAL NU

C       ..... ASSIGN RADIUS OF THE RAINDROP AND OTHER INPUT DATA.
C             FIND THE TERMINAL VELOCITY, UT1, OF THE DROP,
C             AND PRINT HEADINGS .....
       R1 = 0.0001
       RHO = 1000.0
       RHOF = 1.22
       RHOBAR = RHOF/RHO
       NU = 1.49E-5
       PI = 3.14159
       DT = 0.05
       EPSLON = 0.005
       UT1 = TRMVEL(R1)
       WRITE(6,10) R1

C       ..... CONSIDER SEVEN DIFFERENT DROPLET SIZES .....
       R2 = 2.5E-6
       DO 7  M = 1,7
       R2 = R2 + 2.5E-6

C       ..... FIND THE TERMINAL VELOCITY, UT2, OF THE DROPLET.
C             THE RELATIVE VELOCITY IS CALLED UT .....
       UT2 = TRMVEL(R2)
       UT = UT1-UT2

C       ..... SPECIFY THE LOCATION OF THE FIRST DROPLET ON THE
C             CENTERLINE. IT DEFINITELY COLLIDES WITH THE RAINDROP .....
       X0 = -10.
       R01 = 0.
       I1 = +1

C       ..... SPECIFY THE LOCATION OF THE SECOND DROPLET. ITS INITIAL VEL-
C             OCITY IS ASSUMED TO BE THE SAME AS THE LOCAL FLUID VELOCITY.
       DR0 = 0.2
       R02 = R01 + DR0
       T0 = 0.
     1 U0 = 1.+(R02**2/2.-X0**2)/(X0**2+R02**2)**2.5
       V0 = -1.5*X0*R02/(X0**2+R02**2)**2.5

C       ..... FIND THE TIME HISTORY OF THE POSITION AND VELOCITY FOR THE
C             SECOND DROPLET BY USING FOURTH-ORDER RUNGE-KUTTA METHOD. THE
C             VALUE OF I2 IS DETERMINED BY WHETHER OR NOT IT COLLIDES
C             WITH THE RAINDROP .....
       T = T0
       X = X0
       R = R02
       U = U0
       V = V0
     2 CALL KUTTA( X, R, U, V, T, DT, FX, FR )
       IF( SQRT(X*X+R*R) .LE. (1.+R2/R1) ) GO TO 3
       IF( X .GT. 1. ) GO TO 4
       GO TO 2
     3 I2 = +1
       GO TO 5
     4 I2 = -1
     5 CONTINUE
```

```
C       ..... DETERMINE THE VALUE OF DRO FOR THE NEXT STEP BY CHECKING THE
C             PRODUCT OF I1 AND I2. THEN CALL THE SECOND DROPLET THE
C             FIRST. LOCATE THE NEW SECOND DROPLET, AND REPEAT THE SAME
C             PROCESS.   IF ABS(DRO) IS LESS THAN EPSILON, COMPUTE THE
C             APPROXIMATED COLLISION EFFICIENCY AND PRINT THE RESULT .....
        IF( (I1*I2) .LT. 0 )  DRO=-DRO/2.
        IF( ABS(DRO) .LT. EPSILON )  GO TO 6
        RO1 = RO2
        I1 = I2
        RO2 = RO1 + DRO
        GO TO 1
      6 E = ( RO2-R2/R1 )**2
        WRITE(6,11)  R2, E, UT
      7 CONTINUE

     10 FORMAT(1H1////7X, 43HTHE COLLISION EFFICIENCY E OF A RAINDROP OF,
     A          F6.4, 9H M RADIUS, / 7X,
     B          50HCOLLIDING WITH DROPLETS OF RADIUS R2. THE RELATIVE / 7X,
     C          56HVELOCITY BETWEEN RAINDROP AND A DROPLET FAR AHEAD IS UT.
     D          // 20X,2HR2, 14X,1HE. 13X,2HUT / 19X3H(M), 25X,7H(M/SEC) /
     E          17X,8H--------, 9X,4H----, 9X,7H------- )
     11 FORMAT( / F25.7, F13.3, F15.4 )

        END


        FUNCTION TRMVEL( RADIUS )
C       ..... FINDING THE TERMINAL VELOCITY IN METERS PER SECOND OF A
C             WATER DROP OF GIVEN RADIUS IN METERS.  THIS FUNCTION
C             SUBPROGRAM IS WRITTEN BY MODIFYING PROGRAM 1.1 .....

        COMMON D,NU,PI,R1,R2,RHOBAR,UT

C       ..... CALCULATE THE DIAMETER OF THE DROP, AND SPECIFY THE TIME
C             STEP SIZE AND THE INITIAL CONDITIONS .....
        D = 2.*RADIUS
        H = 0.0005
        T = 0.
        Z = 0.
        V = 0.

C       ..... USE RUNGE-KUTTA METHOD TO COMPUTE VELOCITY VT AT TIME
C             INTERVAL H LATER. IF (VT-V) IS LESS THAN 0.00001, VT IS
C             CONSIDERED THE TERMINAL VELOCITY, OTHERWISE ASSIGN THE VALUE
C             OF VT TO THE VARIABLE NAME V AND COMPUTE VELOCITY AT THE
C             NEXT TIME STEP .....
      1 D1Z = H*V
        D1V = H*F(V)
        D2Z = H*(V + D1V/2.)
        D2V = H*F(V+D1V/2.)
        D3Z = H*(V + D2V/2.)
        D3V = H*F(V+D2V/2.)
        D4Z = H*(V + D3V)
        D4V = H*F(V+D3V)
        T = T + H
        Z = Z + (D1Z + 2.*D2Z + 2.*D3Z + D4Z)/6.
        VT = V + (D1V + 2.*D2V + 2.*D3V + D4V)/6.
        IF( ABS(VT-V) .LT. 0.00001 )  GO TO 2
        V = VT
        GO TO 1
      2 TRMVEL = VT
        RETURN
        END
```

```fortran
      FUNCTION F( V )
C     ..... THIS IS THE FUNCTION F APPEARING IN
C           THE FUNCTION SUBPROGRAM TRMVEL .....

      COMMON D,NU,PI,R1,R2,RHOBAR,UT
      G = 9.8
      A = 1. + RHOBAR/2.
      B = (1. - RHOBAR) * G
      C = 3.*RHOBAR / (4.*D)
      F = ( B - C*V*ABS(V)*CD(V) ) / A
      RETURN
      END




      FUNCTION FX( X, R, U, V, T )
C     ..... IT REPRESENTS THE RIGHT-HAND SIDE OF EQUATION (1.6.8) .....

      COMMON D,NU,PI,R1,R2,RHOBAR,UT
      UF = 1. + (R*R/2.-X*X)/(X*X+R*R)**2.5
      VF = -1.5*X*R/(X*X+R*R)**2.5
      WR = SQRT( (UF-U)**2+(VF-V)**2 )

C     ..... CD IS COMPUTED BASED ON DIMENSIONAL RELATIVE VELOCITY .....
      FX = 0.375*R1/R2*CD(UT*WR)*(UF-U)*WR*RHOBAR/(1.+RHOBAR/2.)
      RETURN
      END




      FUNCTION FR( X, R, U, V, T )
C     ..... IT REPRESENTS THE RIGHT-HAND SIDE OF EQUATION (1.6.9) .....

      COMMON D,NU,PI,R1,R2,RHOBAR,UT
      UF = 1. + (R*R/2.-X*X)/(X*X+R*R)**2.5
      VF = -1.5*X*R/(X*X+R*R)**2.5
      WR = SQRT( (UF-U)**2+(VF-V)**2 )

C     ..... CD IS COMPUTED BASED ON DIMENSIONAL RELATIVE VELOCITY .....
      FR = 0.375*R1/R2*CD(UT*WR)*(VF-V)*WR*RHOBAR/(1.+RHOBAR/2.)
      RETURN
      END




      FUNCTION CD(V)
C     ..... TAKEN FROM PROGRAM 1.2 WITH THE COMMON STATEMENT
C           REPLACED BY THE ONE SHOWN BELOW .....
      COMMON D,NU,PI,R1,R2,RHOBAR,UT




      SUBROUTINE KUTTA( X, Y, U, V, T, H, F1, F2 )
C     ..... TAKEN FROM PROGRAM 1.4 .....
```

R2 (M)	E	UT (M/SEC)
.0000050	.640	.5752
.0000075	.799	.5714
.0000100	.879	.5662
.0000125	.914	.5595
.0000150	.938	.5513
.0000175	.951	.5416
.0000200	.963	.5304

action exists between the two spheres and the collision efficiency starts to decrease. Droplets with radii exceeding a limiting value cannot collide with the raindrop unless they are along the centerline.

Problem 1.10 It can be observed when driving a car on a snowy day that the number of snowflakes hitting the windshield decreases as the car speed increases. At high speeds the windshield wipers are hardly needed.

To simulate the motion of snowflakes in a region influenced by a moving vehicle, let us assume that a snowflake is equivalent to a sphere 1.5 mm in diameter having a density of 100 kg/m³. The terminal velocity of such a sphere, given by (1.2.8), is approximately 1.21 m/s. The flow around the car moving at a speed u_0 is to be approximated by the upper half of a potential flow past a circular cylinder of radius R, as shown in Fig. 1.6.2. In a coordinate system attached to the center of the cylinder, the air velocity components in the x and y directions are, respectively,

$$u_f = u_0\left[1 + \frac{R^2(y^2 - x^2)}{(x^2 + y^2)^2}\right]$$

and

$$v_f = -u_0\frac{2R^2xy}{(x^2 + y^2)^2}$$

Suppose that at a distance $10R$ ahead the snowflakes are not influenced by the car, so that they move at a velocity with

Dynamics of a Body Moving through a Fluid Medium

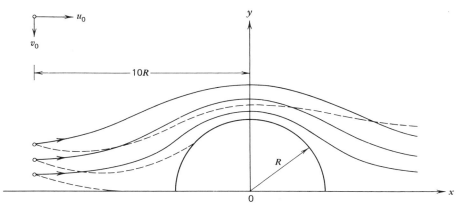

FIGURE 1.6.2 Paths of snowflakes (dashed curves) ahead of a moving vehicle. Streamlines of air are represented by solid curves.

components $(u_0, -v_0)$, where v_0 is the terminal velocity. At that distance, find the vertical range within which all the snowflakes contained will collide with the car. With $R = 1.5$ m, repeat the computation for u_0 ranging from 10 to 30 m/s.

Problem 1.11 The wavy pattern of dirt on the bottom of a water tank suggests that the sediment movement in the tank is influenced by water waves. The phenomenon may be studied numerically as follows.

Consider a layer of water contained between a rigid horizontal surface at $y = 0$ and a free surface at $y = h$. In the presence of a standing wave of wave number k and frequency $\omega/2\pi$, the height of the deformed free surface is described by $h + a \cos kx \sin \omega t$, where a is the amplitude of deformation. The frequency is dependent on k, h, and the gravitational acceleration g. With surface tension ignored and for small deformations, it has been found that $\omega = \sqrt{gk \tanh kh}$. The corresponding fluid motion has the following components (see Lamb, 1932, p. 364).

$$u_f = \frac{gka \cosh ky \sin kx \cos \omega t}{\omega \cosh kh}$$

$$v_f = -\frac{gka \sinh ky \cos kx \cos \omega t}{\omega \cosh kh}$$

Spherical sand particles of diameter d having a specific gravity of 2.5 are released at the free surface. The particles are

Collision of a Raindrop with Cloud Droplets

separated by constant horizontal distances that are small compared with the wavelength. When a particle reaches the bottom, its x coordinate is recorded. After experimenting with a large number of sand particles released at various time instants, statistics can be obtained that show the distribution of particle density at the bottom. You may choose any combination of a, d, h, and k for the numerical experiment.

1.7 Flight Path of a Glider—A Graphical Presentation

In the previous sections numerical results were printed out for each program. Some of the data taken from the output were recorded by hand as points on a graph. After connecting these points by using smooth lines, we transformed the numerical data into a set of curves.

It is often more convenient to have the result of a program plotted directly by the computer. The simplest way to achieve this is to represent the data points by some specified characters printed in a two-dimensional coordinate system. Because characters can be printed only on particular lines and columns, they merely represent the approximate positions of the data points. This method will not be used to prepare graphs in which accuracy is required.

To illustrate this, we will present the numerical result for the flight paths of a glider in the form of an on-line graph. Figure 1.7.1 shows a glider of mass m flying at a velocity w, which makes an angle θ with the horizontal x-axis. The aerodynamic forces acting on the glider in the directions normal and parallel to the flight path are called the lift L and drag D, respectively. If u and v are the velocity components, the equations of motion of the center of mass, ignoring the added mass and the angular motion about the mass center, are

$$m \frac{du}{dt} = -L \sin \theta - D \cos \theta \qquad (1.7.1)$$

$$m \frac{dv}{dt} = -mg + L \cos \theta - D \sin \theta \qquad (1.7.2)$$

in which $\sin \theta = v/w$ and $\cos \theta = u/w$. The lift and drag of the aircraft as a whole can be expressed in terms of a lift coefficient c_l and a drag coefficient c_d, so that

$$L = c_l \tfrac{1}{2} \rho w^2 S$$
$$D = c_d \tfrac{1}{2} \rho w^2 S \qquad (1.7.3)$$

where ρ is the density of air and S is the projected wing area.

Suppose at an initial instant $t = 0$ the velocity is w_0 and the inclination angle

Dynamics of a Body Moving through a Fluid Medium

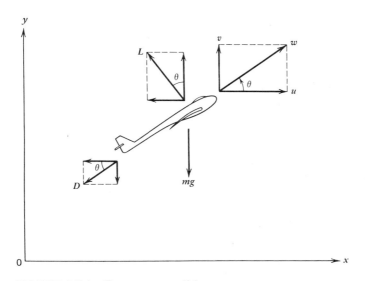

FIGURE 1.7.1 Forces on a glider.

is θ_0. Using w_0 as the reference velocity, w_0/g as the reference time, and w_0^2/g as the reference length, we can construct dimensionless velocity components U, V, dimensionless time T, and dimensionless coordinates X, Y. Upon substitution from (1.7.3) and by using trigonometric relations, (1.7.1) and (1.7.2), expressed in dimensionless form, become

$$\frac{d^2X}{dT^2} = -A(U^2 + V^2)^{1/2}(BU + V) \tag{1.7.4}$$

$$\frac{d^2Y}{dT^2} = -1 + A(U^2 + V^2)^{1/2}(U - BV) \tag{1.7.5}$$

where $U = dX/dT$ and $V = dY/dT$. The two dimensionless parameters A and B represent the ratios $c_l \frac{1}{2}\rho w_0^2 S/mg$ and c_d/c_l, respectively. Since the lift and drag coefficients are functions of angle of attack controlled by the pilot, both A and B are generally variables. In the following computation A and B are assumed to be constant. An example for such a case is a model glider that has wings fixed at a constant angle of attack.

A glider having $A = 1.5$ and $B = 0.06$ is considered in Program 1.7. Starting from the origin of the coordinate system with the same initial velocity, flight paths are computed for $\theta_0 = -90°$ and $180°$ by using the fourth-order Runge-Kutta method. At $\theta_0 = -90°$ the glider dives vertically, and at $\theta_0 = 180°$ it flies upside down in the negative x direction. In order to be used in the plotting subroutines, the coordinates of the glider have to be stored as the elements of

Flight Path of a Glider—A Graphical Presentation **63**

some one-dimensional arrays. At the nth time step the coordinates on the first flight path are called $x_1(n)$ and $y_1(n)$, and those on the second flight path are called $x_2(n)$ and $y_2(n)$.

The plotting subroutines to be used in Program 1.7 and also in some later programs are called PLOTN, available at the University of Colorado Computing Center. At other institutions similar subroutines with some variations are probably available. PLOTN consists of four subroutines, which are described separately as follows.

PLOT1 (NSCALE, NHL, NSBH, NVL, NSBV, HCHAR, VCHAR) sets up the grid spacing and determines the form of the numbers to be printed as axis labels. These numbers are controlled by NSCALE, which is a one-dimensional array containing four integer values, I, J, K, and L, in order. The printed values of the ordinate are 10^I times the actual values, and have J digits following the decimal point. K and L play the same roles for the printed values

List of Principal Variables in Program 1.7

Program Symbol	Definition
A	Ratio of initial lift to weight of the glider
B	Ratio of drag to lift of the glider
DT	Dimensionless time increment
F1, F2	Right-hand sides of (1.7.4) and (1.7.5), respectively
LCHAR	Characters for the vertical label of the graph
M	A counter for changing angle θ_0
N	A counter for time steps
NSCALE	An argument of subroutine PLOT1
PI	π
PLOT1, PLOT2	
PLOT3, PLOT4	Subroutines used for preparing on-line graphs
T	Dimensionless time
THETA0	Initial angle between flight path and x-axis, θ_0, radians
U, U0	Dimensionless horizontal velocity of glider and its initial value, respectively
V, V0	Dimensionless vertical velocity of glider and its initial value, respectively
W0	Dimensionless initial velocity of glider
X, Y	Coordinates of glider
X1, Y1	Coordinates of glider for $\theta_0 = -90°$
X2, Y2	Coordinates of glider for $\theta_0 = 180°$

```
C                    ***** PROGRAM 1.7 *****

C        GRAPHICAL PRESENTATION OF THE PATHS OF A GLIDER

         DIMENSION LCHAR(3),NSCALE(4),X1(200),Y1(200),X2(200),Y2(200)
         EXTERNAL F1,F2
         COMMON A,B

C        ..... ASSIGN VALUES TO THE FOUR ARGUMENTS OF THE ARRAY NSCALE.....
         DATA NSCALE/ 0, 1, 0, 1 /

C        ..... SET UP COORDINATE AXES AND PREPARE THE GRID .....
         CALL PLOT1( NSCALE, 4, 10, 7, 17, 1H-, 1HI )
         CALL PLOT2( 5., -1., 1., -2., .TRUE. )

C        ..... ASSIGN VALUES TO A,B, AND TO TIME INCREMENT DT .....
         A = 1.5
         B = 0.06
         DT = 0.05

C        ..... COMPUTE TWO FLIGHT PATHS STARTING FROM THE ORIGIN WITH
C              THE SAME INITIAL VELOCITY BUT IN DIFFERENT DIRECTIONS .....
         W0 = 1.
         PI = 3.14159
         DO 4 M=1,2
         THETA0 = -PI/2.
         IF ( M.EQ.2 ) THETA0=PI
         U0 = W0 * COS( THETA0 )
         V0 = W0 * SIN( THETA0 )

C        ..... INTEGRATE EQUATIONS (1.7.4) AND (1.7.5) USING THE FOURTH-
C              ORDER RUNGE-KUTTA METHOD. X1,Y1 ARE RESPECTIVELY THE
C              HORIZONTAL AND VERTICAL COORDINATES OF THE GLIDER FOR
C              M=1, AND X2,Y2 ARE THOSE FOR M=2 .....
         T = 0.
         X = 0.
         Y = 0.
         U = U0
         V = V0
         DO 2 N=1,200
         IF ( M.EQ.2 )  GO TO 1
         X1( N ) = X
         Y1( N ) = Y
         GO TO 2
       1 X2( N ) = X
         Y2( N ) = Y
       2 CALL KUTTA( X, Y, U, V, T, DT, F1, F2 )

C        ..... PLACE PLOTTING CHARACTER ( 0 FOR M=1, * FOR M=2 ) INTO THE
C              COORDINATE SYSTEM AT SELECTED LOCATIONS ALONG FLIGHT PATH...
         IF ( M.EQ.2 ) GO TO 3
         CALL PLOT3( 1H0, X1, Y1, 1, 200, 3 )
         GO TO 4
       3 CALL PLOT3( 1H*, X2, Y2, 1, 200, 3 )
       4 CONTINUE

C        ..... PRINT GRAPH AND INSERT LABELS FOR ORDINATE AND ABSCISSA.....
         WRITE(6,10)
         LCHAR(1) = 1HY
         CALL PLOT4( LCHAR, 1 )
         WRITE(6,20)
```

```
   10 FORMAT( 1H1 )
   20 FORMAT(/ 69X, 1HX // 36X,
      A        53HFIGURE 1.7.2   FLIGHT PATHS OF A GLIDER STARTING FROM/
      B        51X,38HTHE ORIGIN HAVING A=1.5, B=.06, W0=1.0  //
      C        53X,34H0 IS USED FOR THETA0 = -90 DEGREES /
      D        53X,34H* IS USED FOR THETA0 = 180 DEGREES  )

      END




      FUNCTION F1( X, Y, U, V, T )
C     ..... IT REPRESENTS THE RIGHT-HAND SIDE OF EQUATION (1.7.4) .....

      COMMON A,B
      F1 = -A * SQRT(U*U+V*V) * (B*U+V)
      RETURN
      END




      FUNCTION F2( X, Y, U, V, T )
C     ..... IT REPRESENTS THE RIGHT-HAND SIDE OF EQUATION (1.7.5) .....

      COMMON A,B
      F2 = -1. + A * SQRT(U*U+V*V) * (U-B*V)
      RETURN
      END




      SUBROUTINE KUTTA( X, Y, U, V, T, H, F1, F2 )
C     ..... TAKEN FROM PROGRAM 1.4 .....
```

of the abscissa. NHL is the integer number of horizontal grid lines on the graph, and NSBH is the integer number of spaces between two neighboring horizontal grid lines. NVL and NSBV are the similar numbers for the vertical grid lines. HCHAR and VCHAR are, respectively, the characters used to form horizontal and vertical grid lines.

PLOT2 (XMAX, XMIN, YMAX, YMIN, NSCL) prepares the grid. XMAX, XMIN and YMAX, YMIN in floating-point form specify the maximum and minimum values of the abscissa and ordinate, respectively. The logical value FALSE should be assigned to the argument NSCL, and PLOT1 should not be called if the standard grid is desired, which has eleven vertical and six horizontal grid lines and ten spaces between both vertical and horizontal lines. Otherwise the logical value TRUE is assigned to NSCL.

PLOT3 (PCHAR, X, Y, NFIRST, NLAST, NPOINT) places the specified plotting character assigned to PCHAR in the appropriate positions corresponding to the points (X(N), Y(N)). N starts from the integer value NFIRST

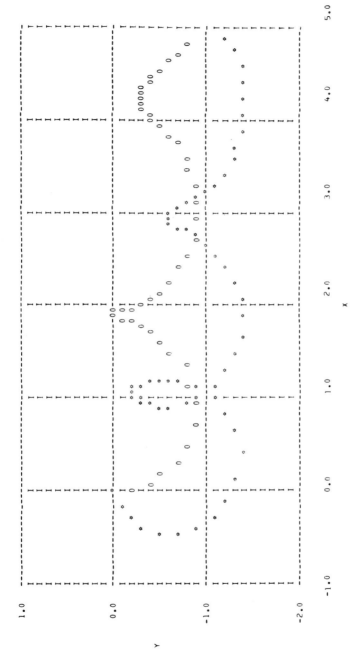

FIGURE 1.7.2 Flight paths of a glider starting from the origin having A = 1.5, B = .06, WO = 1.0.
O is used for THETAO = −90 degrees
* is used for THETAO = 180 degrees

and increases continuously by an amount NPOINT until it reaches or exceeds the value NLAST. Points falling outside the limits of the grid are ignored and will not be shown.

PLOT4 (LCHAR, NCHAR) writes the image of the completed graph on the printer. LCHAR is a single-dimensioned array having three elements, each containing up to ten characters. NCHAR is the number of the valid characters in LCHAR to be printed at the left of the graph as a vertical label. They will be printed one character to a line and centered in the middle of the graph.

On the printer the size of one space in the vertical direction is larger than that in the horizontal direction. Care must be taken in calling PLOT1 to set up a grid having the same scale in both directions. The scales of the abscissa and ordinate used in Program 1.7 are approximately equal.

The two flight paths are printed in Fig. 1.7.2 by the computer. If numerical values of the coordinates of the points along the paths are needed, some additional WRITE and FORMAT statements can be inserted at appropriate locations in Program 1.7.

If there were no drag and the glider were weightless, the lift would always be perpendicular to the flight direction resulting in a circular path. The weight changes the path into a coil-shaped curve drifting along the x-axis with an undamped amplitude. These are analogous to the paths of a charged particle moving in a uniform magnetic field with and without a body force normal to the field lines. In the presence of a drag force the kinetic energy is dissipated continuously, and the average altitude of the glider is a decreasing function of time.

In the case $\theta_0 = -90°$, the glider first dives and then climbs upward. When it reaches a position slightly higher than its original height, most of the kinetic energy has been used up in overcoming the frictional and gravitational forces. The low speed cannot generate enough lift to support the weight, so that the glider stalls and then turns itself into a diving position to pick up speed again. Such a trajectory is often observed when a schoolchild tries out a paper glider. On the other hand, in the case $\theta_0 = 180°$, the initial lift force is in the same direction as that of the weight and thus helps to accelerate the glider. When the glider reaches the first horizontal position, it has gained such a high speed that it has sufficient kinetic energy to go through a loop.

When a thrust force is added to (1.7.1) and (1.7.2), they become the equations of motion for an airplane. With some modifications Program 1.7 can be used to simulate taking off, climbing, and other two-dimensional aircraft maneuvers.

chapter two
INVISCID FLUID FLOWS

All the problems in this chapter are concerned with flows in the absence of viscosity. The justification for ignoring viscosity and a brief review of incompressible irrotational flows can be found in Section 2.1. The numerical solution of boundary-value problems involving linear second-order ordinary differential equations is taken up in Section 2.2, and its application to a radial flow is discussed in Section 2.3.

To solve the problem of flow past a body, two different approaches may be taken. Bodies of various shapes can be generated either by superimposing elementary flows or by transforming from known flow patterns using the conformal mapping technique. These inverse methods are described in Sections 2.4 and 2.7, respectively. On the other hand, direct methods are used to solve for the flow around a body of prescribed shape. The von Kármán's method, in which singularities are placed along the centerline of an axisymmetric body, and the panel method, in which singularities are distributed on the surface of a two-dimensional body, are respectively introduced in Sections 2.6 and 2.8.

The basic operations in vector algebra can conveniently be programmed in the form of individual subprograms and can be attached to any program containing such operations. This is done in Section 2.5 to find the motion of a pair of trailing vortices above a runway.

The remaining six sections deal specifically with the second-order partial differential equations with two independent variables. These equations are first classified in Section 2.9. Numerical methods are devised in Section 2.10 for solving elliptic equations, and their computational stabilities are examined. An example is shown to compute the flow induced by a concentrated vorticity in a rectangular domain. In Section 2.11, techniques for handling irregular and derivative boundary conditions are illustrated. The numerical solution of linear hyperbolic equations and its stability criterion are considered in Section 2.12, and an application to solve the problem of propagation and reflection of

69

a sound wave is given in Section 2.13. The propagation of a finite-amplitude wave is studied in Section 2.14 using a numerical scheme constructed for solving nonlinear hyperbolic equations.

2.1 Incompressible Potential Flows

Real fluids are all viscous. Viscosity is caused by the redistribution of excessive momenta among neighboring fluid molecules through the action of intermolecular collisions. Thus a viscous force is exerted on the surface of a fluid element where a local velocity gradient is present. It may be either a shearing force tangent to the surface, such as the one found in a boundary layer, or a normal force that exists, for example, within a shock wave.

The importance of the viscous force in comparison with the inertial force is represented by the Reynolds number, which is the ratio of a characteristic inertial force to a characteristic viscous force in a flow field. Because of the low viscosity of air and water, the Reynolds numbers of most flows of practical interest are usually very high; in other words, in these flows the viscous forces are very small compared with the inertia.

For a high-Reynolds number flow past a streamlined body from which the flow does not separate, Prandtl (1904) postulated that the influence of viscosity is confined to a very thin boundary layer in the immediate neighborhood of the solid wall, and that in the region outside of the boundary layer the flow behaves as if there were no viscosity. Prandtl's postulation has been proved to be a powerful tool in solving many practical flow problems. For instance, the inviscid flow theory predicts extremely well the lift and pressure distribution on an airfoil for angles of attack below the value at which the flow starts to separate from the body, although the drag has to be found by solving the boundary-layer equations.

It is known that vorticities are generated by the shearing viscous forces, so that the boundary-layer flow is a rotational one. On the other hand, in the absence of viscous and other rotational forces, the originally irrotational flow far upstream will remain so in the region outside the boundary layer. Letting $\mathbf{V}$ denote the velocity field in this region, the irrotationality condition states that the vorticity vanishes; that is,

$$\nabla \times \mathbf{V} = 0 \tag{2.1.1}$$

The preceding equation is automatically satisfied if a *velocity potential* ϕ is introduced such that

$$\mathbf{V} = \nabla \phi \tag{2.1.2}$$

For this reason irrotational flows are also called *potential flows*. As a result of introducing the velocity potential, the velocity vector generally having three components is replaced by a single scalar quantity ϕ.

Furthermore, if the fluid is assumed incompressible, the velocity field must also satisfy the *continuity equation* for such a fluid.

$$\nabla \cdot \mathbf{V} = 0 \qquad (2.1.3)$$

Upon substitution from (2.1.2), (2.1.3) becomes

$$\nabla^2 \phi = 0 \qquad (2.1.4)$$

where $\nabla^2 \equiv \nabla \cdot \nabla$ is the *Laplacian operator*, and (2.1.4) is called the *Laplace equation*.

Using Prandtl's postulation, the problem of an incompressible flow past a given streamlined body at high Reynolds numbers is to be solved as follows. For the external region a velocity potential is to be found by solving the Laplace equation (2.1.4), which satisfies both the boundary condition prescribed far upstream and the requirement that the fluid velocity be tangent to the body surface. The latter condition is justified by the fact that the boundary layer around the body is thin. The velocity distribution of the external flow along the body surface is then used as the outer boundary condition for the boundary-layer flow. This inner flow satisfies the no-slip condition at the rigid wall and is a solution to the governing boundary-layer equations to be shown in the next chapter.

In the present chapter the boundary-layer flow is disregarded completely. Once the velocity field in the external region is determined from the kinematic considerations just described, the pressure field p can be computed from a dynamic relation, called the *Euler equation*.

$$\rho \frac{D\mathbf{V}}{Dt} = -\nabla p \qquad (2.1.5)$$

It is the equation of motion for inviscid fluids with body forces omitted. In this equation ρ is the density and the operator

$$\frac{D}{Dt} \equiv \frac{\partial}{\partial t} + \mathbf{V} \cdot \nabla \qquad (2.1.6)$$

is the *total* or *substantial derivative* that gives the rate of change by following a fluid particle.

It is easier to compute the pressure from the integrated form of the Euler equation (i.e., the *Bernoulli equation*) than from the Euler equation itself. Under the conditions of incompressible fluid and irrotational velocity field, the Bernoulli equation has the form

$$\rho \frac{\partial \phi}{\partial t} + p + \tfrac{1}{2}\rho V^2 = H \qquad (2.1.7)$$

Incompressible Potential Flows

where the constant of integration, H, is the *Bernoulli constant*. For a steady flow (2.1.7) simply states that the sum of the hydrostatic pressure, p, and the dynamic pressure, $\frac{1}{2}\rho V^2$, is a constant at any point in the flow. In this case H is the *stagnation pressure*, or the pressure measured at a point where the fluid velocity vanishes.

For two-dimensional planar flows, (2.1.2) and (2.1.4) can be expressed in Cartesian coordinates as

$$u = \frac{\partial\phi}{\partial x}, \qquad v = \frac{\partial\phi}{\partial y} \tag{2.1.8}$$

and

$$\frac{\partial^2\phi}{\partial x^2} + \frac{\partial^2\phi}{\partial y^2} = 0 \tag{2.1.9}$$

where u and v are, respectively, the velocity components along the x- and y-axes. In polar coordinates, however, they have the expressions

$$u_r = \frac{\partial\phi}{\partial r}, \qquad u_\theta = \frac{1}{r}\frac{\partial\phi}{\partial\theta} \tag{2.1.10}$$

and

$$\frac{\partial^2\phi}{\partial r^2} + \frac{1}{r}\frac{\partial\phi}{\partial r} + \frac{1}{r^2}\frac{\partial^2\phi}{\partial\theta^2} = 0 \tag{2.1.11}$$

where u_r is the velocity component in the radial direction and u_θ is the velocity component in the azimuthal direction.

Instead of velocity potential, *stream function* is sometimes used for analyzing planar or axisymmetric incompressible flows. The continuity equation (2.1.3) suggests that the velocity vector can also be derived from a vector potential through the relationship

$$\mathbf{V} = \nabla \times (\mathbf{i}_z\psi) \tag{2.1.12}$$

where $\mathbf{i}_z$ is a unit vector along the z-axis, which is perpendicular to the plane of fluid motion, and ψ is called the stream function, which represents the volume of fluid passing per unit time between a given point and a reference streamline (see, for example, Kuethe and Chow, 1976, Section 2.6). *Streamlines*, to which the instantaneous velocity vectors are tangent, are described by the equations $\psi = $ constant. In terms of stream function, the irrotationality relation (2.1.1) becomes

$$\nabla \times \nabla \times (\mathbf{i}_z\psi) = 0 \tag{2.1.13}$$

(2.1.12) and (2.1.13), when written in Cartesian coordinates, have the expressions

$$u = \frac{\partial\psi}{\partial y}, \qquad v = -\frac{\partial\psi}{\partial x} \tag{2.1.14}$$

Inviscid Fluid Flows

$$\frac{\partial^2 \psi}{\partial x^2} + \frac{\partial^2 \psi}{\partial y^2} = 0 \tag{2.1.15}$$

or, in polar coordinates, the expressions

$$u_r = \frac{1}{r}\frac{\partial \psi}{\partial \theta}, \qquad u_\theta = -\frac{\partial \psi}{\partial r} \tag{2.1.16}$$

$$\frac{\partial^2 \psi}{\partial r^2} + \frac{1}{r}\frac{\partial \psi}{\partial r} + \frac{1}{r^2}\frac{\partial^2 \psi}{\partial \theta^2} = 0 \tag{2.1.17}$$

It turns out that, similar to the velocity potential, the stream function for planar incompressible flows is also governed by the Laplace equation. However, the nonsteady term in the Bernoulli equation cannot be written in terms of ψ in a simple form similar to that shown in (2.1.7) in terms of ϕ.

The continuity equation (2.1.3) does not apply in regions where fluid mass is created or destroyed. The formulation is now generalized to include the possible sources and sinks in the flow field. Consider an infinitesimal control volume $\Delta x\, \Delta y\, \Delta z$ fixed in the Cartesian coordinate system as sketched in Fig. 2.1.1, in which the z dimension is not shown for the purpose of simplicity. The distribution of sources is represented by $q(x, y, z, t)$, which is the volume of fluid created per unit time from a unit volume located at the point (x, y, z). A simple arithmetic procedure computing the volume fluxes across the sur-

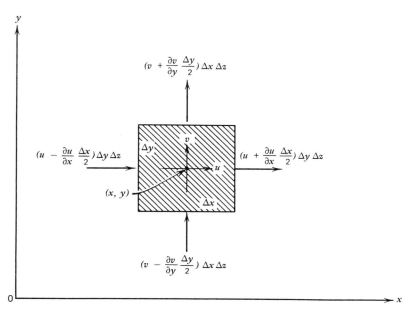

FIGURE 2.1.1 Flow through a control volume.

Incompressible Potential Flows

faces, as indicated in Fig. 2.1.1 (the fluxes in the z direction can easily be added), shows that the net volume of fluid flowing out of the control volume per unit time is

$$\left(\frac{\partial u}{\partial x} + \frac{\partial v}{\partial y} + \frac{\partial w}{\partial z}\right) \Delta x \, \Delta y \, \Delta z$$

For an incompressible fluid, this amount of fluid must be equal to $q \, \Delta x \, \Delta y \, \Delta z$, or the amount produced by all the sources contained within the volume. Thus the resulting continuity equation is

$$\frac{\partial u}{\partial x} + \frac{\partial v}{\partial y} + \frac{\partial w}{\partial z} = q(x, y, z, t) \tag{2.1.18}$$

or, in vector notation,

$$\nabla \cdot \mathbf{V} = q \tag{2.1.19}$$

Sinks are represented by negative values of q. In the absence of sources or sinks, $q = 0$, and the continuity equation (2.1.3) is recovered.

When the velocity vector is expressed in terms of velocity potential, (2.1.19) becomes

$$\nabla^2 \phi = q \tag{2.1.20}$$

which is in the form of a *Poisson equation*.

Despite the fact that both Laplace and Poisson equations are linear, analytical solution for flow past a body of arbitrary shape is by no means simple. Various numerical and seminumerical solution techniques will be introduced in the following sections.

An extensive discussion on potential flows is out of the scope of this book. Such a task can be found in any introductory book on fluid dynamics.

2.2 Numerical Solution of Second-Order Ordinary Differential Equations— Boundary-Value Problems

In connection with the numerical solution of ordinary differential equations, various initial-value problems were treated in Chapter One. In another class of fluid dynamic problems a solution is to be found that satisfies not only the differential equation throughout some domain of its independent variable, but also some conditions on boundaries of that domain. These problems are called *boundary-value problems.*

We consider here the numerical solution of a *linear* second-order ordinary differential equation of the form

$$\frac{d^2 f}{dx^2} + A(x)\frac{df}{dx} + B(x)f = D(x) \tag{2.2.1}$$

for the domain $x_{min} \le x \le x_{max}$, subject to the boundary conditions that the values of f are given at the end points of that range, at $x = x_{min}$ and x_{max}. The method developed here will be modified later for solving problems involving derivative boundary conditions.

As the first step in our numerical method, the differential equation is to be approximated by an equation of the finite-difference form. Similar to what was done in solving the initial-value problems, the axis of the independent variable is again subdivided into small intervals of constant width h. If the number of intervals in the given range of x is $n + 1$, and if the end points of the intervals are labeled according to the index notation shown in Fig. 2.2.1, starting with x_0 at the left end of the range, then the following relations hold:

$$x_i = x_0 + ih, \qquad i = 0, 1, 2, \cdots, n + 1 \qquad (2.2.2)$$

in which $x_0 = x_{min}$ and $x_{n+1} = x_{max}$. Values of the function f evaluated at these end points are also named in index notation according to the definition $f_i \equiv f(x_i)$. It is desired to approximate the derivatives of f at an arbitrary point x_i by expressions containing values of f evaluated in the neighborhood of x_i.

Let x_j be such a neighboring point with $j = i + m$, where m is a positive or negative integer. For small values of h the function evaluated at x_j may be

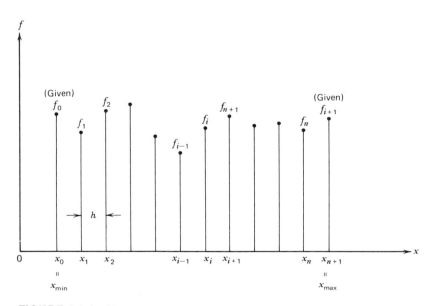

FIGURE 2.2.1 Numerical solution of an ordinary boundary-value problem.

Numerical Solution of Second-Order Ordinary Differential Equations expanded in a Taylor's series about x_i:

$$f_j \equiv f(x_i + mh)$$

$$= f_i + mhf'_i + \frac{(mh)^2}{2!} f''_i + \frac{(mh)^3}{3!} f'''_i + \cdots \qquad (2.2.3)$$

where a prime denotes a differentiation with respect to x. By assigning the values -1 and $+1$ to m in (2.2.3), we obtain two expressions for $f(x)$ at immediate neighboring points of x_i.

$$f_{i-1} = f_i - hf'_i + \tfrac{1}{2}h^2 f''_i - \tfrac{1}{6}h^3 f'''_i + \cdots \qquad (2.2.4)$$

$$f_{i+1} = f_i + hf'_i + \tfrac{1}{2}h^2 f''_i + \tfrac{1}{6}h^3 f'''_i + \cdots \qquad (2.2.5)$$

There are many different ways of computing a particular derivative. For example, f'_i may be solved for from (2.2.5) to give

$$f'_i = \frac{1}{h}(f_{i+1} - f_i) - \tfrac{1}{2}hf''_i - \tfrac{1}{6}h^2 f'''_i - \cdots$$

With a truncation error $O(hf''_i)$, it is approximated by

$$f'_i \doteq \frac{1}{h}(f_{i+1} - f_i) \qquad (2.2.6)$$

This a *forward-difference formula* for the first-order derivative, so named because it involves the values of the function at x_i and at a point ahead of it. On the other hand, if we use (2.2.4), a *backward-difference formula* of the form

$$f'_i \doteq \frac{1}{h}(f_i - f_{i-1}) \qquad (2.2.7)$$

is obtained whose truncation error is of the same order of magnitude as the one associated with the previous formula, except for a difference in sign. Thus the approximations (2.2.6) and (2.2.7) will give similar accuracies in a numerical computation. A more accurate approximation, having an error of $O(h^2 f'''_i)$, is obtained by subtracting (2.2.4) from (2.2.5); this has the form

$$f'_i \doteq \frac{1}{2h}(f_{i+1} - f_{i-1}) \qquad (2.2.8)$$

Because of its appearance on the right-hand side, it is called a *central-difference formula* for f'. Formulas giving any desired accuracies can be obtained by assigning additional values to m in (2.2.3) and using the same technique just demonstrated.

To approximate the second-order derivative of f, we show only a central-difference formula

$$f''_i \doteq \frac{1}{h^2} (f_{i+1} - 2f_i + f_{i-1}) \tag{2.2.9}$$

having an error $O(h^2 f_i^{iv})$. It is obtained by adding (2.2.4) and (2.2.5) to eliminate f'_i. Formulas of various truncation errors for computing derivatives up to the fourth order are tabulated by McCormick and Salvadori (1964, Section 3.2).

We now return to the differential equation (2.2.1). To allow errors of the same order of magnitude in the two derivatives shown in that equation, we use the central-difference formulas (2.2.8) and (2.2.9). Thus, at a point x_i, the differential equation (2.2.1) is approximated by the difference equation

$$\frac{1}{h^2} (f_{i+1} - 2f_i + f_{i-1}) + \frac{1}{2h} A_i(f_{i+1} - f_{i-1}) + B_i f_i = D_i$$

where A_i, B_i and D_i denote, respectively, $A(x_i)$, $B(x_i)$, and $D(x_i)$. Multiplying the equation by h^2 and grouping similar terms, we obtain

$$\left(1 - \frac{h}{2} A_i\right) f_{i-1} + (-2 + h^2 B_i) f_i + \left(1 + \frac{h}{2} A_i\right) f_{i+1} = h^2 D_i$$

It can be written in a more convenient form

$$C_{i1} f_{i-1} + C_{i2} f_i + C_{i3} f_{i+1} = C_{i4} \tag{2.2.10}$$

by calling

$$C_{i1} = 1 - \frac{h}{2} A_i$$

$$C_{i2} = -2 + h^2 B_i$$

$$C_{i3} = 1 + \frac{h}{2} A_i$$

$$C_{i4} = h^2 D_i \tag{2.2.11}$$

The preceding coefficients are known constants at any *interior* point in the specified range of x. (2.2.10), applied at $i = 1, 2, \cdots, n$, gives the set of n linear algebraic equations to be solved simultaneously for the n unknowns f_i.

Special attention is needed when $i = 1$ and n are assigned in (2.2.10). These two equations are

$$C_{11} f_0 + C_{12} f_1 + C_{13} f_2 = C_{14}$$

and

$$C_{n1} f_{n-1} + C_{n2} f_n + C_{n3} f_{n+1} = C_{n4}$$

Since f_0 and f_{n+1} are known from boundary conditions, the two involved terms on the left side are constant and therefore are moved to the right. To simplify

Numerical Solution of Second-Order Ordinary Differential Equations   **77** the appearance of our program, we assign the value of $(C_{14} - C_{11}f_0)$ to C_{14} and the value of $(C_{n4} - C_{n3}f_{n+1})$ to C_{n4}. This action may be expressed symbolically as

$$C_{14} \leftarrow (C_{14} - C_{11}f_0)$$
$$C_{n4} \leftarrow (C_{n4} - C_{n3}f_{n+1}) \tag{2.2.12}$$

After this rearrangement, C_{11} and C_{n3} disappear from the left side and the set of n simultaneous equations becomes, when expressed in matrix notation,

$$
\begin{bmatrix}
C_{12} & C_{13} & 0 & 0 & \cdots & 0 & 0 & 0 \\
C_{21} & C_{22} & C_{23} & 0 & \cdots & 0 & 0 & 0 \\
0 & C_{31} & C_{32} & C_{33} & \cdots & 0 & 0 & 0 \\
\cdot & \cdot & \cdot & \cdot & \cdots & \cdot & \cdot & \cdot \\
\cdot & \cdot & \cdot & \cdot & \cdots & \cdot & \cdot & \cdot \\
0 & 0 & 0 & 0 & \cdots & C_{n-1,1} & C_{n-1,2} & C_{n-1,3} \\
0 & 0 & 0 & 0 & \cdots & 0 & C_{n1} & C_{n2}
\end{bmatrix}
\begin{pmatrix}
f_1 \\ f_2 \\ f_3 \\ \cdot \\ \cdot \\ f_{n-1} \\ f_n
\end{pmatrix}
=
\begin{pmatrix}
C_{14} \\ C_{24} \\ C_{34} \\ \cdot \\ \cdot \\ C_{n-1,4} \\ C_{n4}
\end{pmatrix}
\tag{2.2.13}
$$

The coefficient matrix on the left-hand side of (2.2.13) is called a *tridiagonal matrix*, whose nonvanishing elements form a band three elements wide along the diagonal. This particular set of equations can be solved by using the *Gaussian elimination method*. According to this method, we multiply the second equation in (2.2.13) by C_{12} and the first by C_{21}, and then take the difference of the two to eliminate f_1. The resulting equation is

$$(C_{22}C_{12} - C_{21}C_{13})f_2 + C_{23}C_{12}f_3 = (C_{24}C_{12} - C_{21}C_{14})$$

In this equation, if we call the coefficient of f_2 by the name C_{22}, that of f_3 by the name C_{23}, and the group on the right-hand side by the name C_{24}, and if the final equation is used to replace the second equation in (2.2.13), the form of (2.2.13) will remain unchanged except that C_{21} is replaced by zero. Working with the new second equation and the third equation in (2.2.13), C_{31} can be eliminated by using the same technique. The same process is repeated until $C_{n-1,1}$ is eliminated. In summary, the eliminating processes are achieved by successively renaming the coefficients according to the following assignments.

$$
\left.
\begin{aligned}
C_{i2} &\leftarrow (C_{i2}C_{i-1,2} - C_{i1}C_{i-1,3}) \\
C_{i3} &\leftarrow C_{i3}C_{i-1,2} \\
C_{i4} &\leftarrow (C_{i4}C_{i-1,2} - C_{i1}C_{i-1,4})
\end{aligned}
\right\}
\quad i = 2, 3, \cdots, n - 1 \tag{2.2.14}
$$

At this stage, all the equations in (2.2.13) are in the simple form of having only two terms on the left-hand side. The value of f_n can immediately be found by solving simultaneously the last two equations in (2.2.13). Thus

$$f_n = \frac{C_{n4}C_{n-1,2} - C_{n1}C_{n-1,4}}{C_{n2}C_{n-1,2} - C_{n1}C_{n-1,3}} \tag{2.2.15}$$

It can easily be verified that the remaining unknowns can be calculated, in a backward order, from the recurrence formula

$$f_j = \frac{C_{j4} - C_{j3}f_{j+1}}{C_{j2}}, \qquad j = n-1, n-2, \ldots, 2, 1 \tag{2.2.16}$$

The numerical method just described is now summarized as follows. The differential equation (2.2.1) is first replaced by a set of difference equations (2.2.10), whose coefficients are calculated according to (2.2.11). Boundary conditions are taken care of by executing the actions shown in (2.2.12). Finally, the unknowns f_i, with $i = 1, 2, \ldots, n$, are computed by following the procedure represented by (2.2.14) to (2.2.16). The last part of the computation will be programmed under a subroutine named TRID, which will appear in the next program.

2.3 Radial Flow Caused by Distributed Sources and Sinks

The numerical method just described for solving boundary-value problems is applied in this section to solve a flow problem with a known solution, so that the accuracy of the method can be tested. The method will be used later in Section 3.4 on problems whose solutions cannot be obtained analytically.

Consider a radial flow in the domain $r_0 \leq r \leq 4r_0$, within which there is an axisymmetric distribution of sources whose strength increases linearly with r. For such a problem the governing equation (2.1.20) becomes

$$\left(\frac{d^2}{dr^2} + \frac{1}{r}\frac{d}{dr}\right)\phi = q_0 \frac{r}{r_0} \tag{2.3.1}$$

where q_0 is the source strength at $r = r_0$. The radial velocity is computed from (2.1.10).

$$u_r = \frac{d\phi}{dr} \tag{2.3.2}$$

It is always more convenient to work with dimensionless equations. If u_0 represents a characteristic velocity, the following dimensionless variables can be introduced.

$$R = \frac{r}{r_0}, \qquad \Phi = \frac{\phi}{r_0 u_0}, \qquad U = \frac{u_r}{u_0} \tag{2.3.3}$$

It follows that in the domain $1 \leqslant R \leqslant 4$, the flow is described by the dimensionless equations

$$\left(\frac{d^2}{dR^2} + \frac{1}{R}\frac{d}{dR}\right)\Phi = \alpha R \tag{2.3.4}$$

and

$$U = \frac{d\Phi}{dR} \tag{2.3.5}$$

in which $\alpha = q_0 r_0 / u_0$ is a dimensionless source strength.

Equation (2.3.4) has a general solution of the form

$$\Phi = \tfrac{1}{9}\alpha R^3 + \beta \ln R + \gamma$$

The first term on the right-hand side is the particular solution representing the flow caused by the source distribution. The homogeneous solution consists of two terms: the first represents a line source located at $R = 0$ having strength β, and the second is just a constant. Since the velocity field is not affected by the constant term in Φ, γ may conveniently be set to zero. Thus, for $\alpha = 9$ and $\beta = -24$, the expression

$$\Phi = R^3 - 24 \ln R \tag{2.3.6}$$

is the exact solution of the equation

$$\frac{d^2\Phi}{dR^2} + \frac{1}{R}\frac{d\Phi}{dR} = 9R \tag{2.3.7}$$

satisfying the boundary conditions

$$\begin{aligned} \Phi &= 1 & &\text{at } R = 1 \\ \Phi &= 64 - 24 \ln 4 & &\text{at } R = 4 \end{aligned} \tag{2.3.8}$$

The corresponding dimensionless radial velocity is, from (2.3.5),

$$U = 3R^2 - \frac{24}{R} \tag{2.3.9}$$

To solve numerically the boundary-value problem consisting of the differential equation (2.3.7) and boundary conditions (2.3.8), the equation is first compared with the standard form (2.2.1) to obtain

$$A(R) = \frac{1}{R}$$

$$B(R) = 0$$

$$D(R) = 9R \tag{2.3.10}$$

The independent variable in these relations has already been changed from

Inviscid Fluid Flows

the original x to R. The details of the numerical computation are depicted in Program 2.1, in which the range of R between 1 and 4 is divided equally into 30 intervals. Once the values of Φ evaluated at the interior points are obtained following the procedure described in the previous section, the radial velocity, which is the first-order derivative of Φ according to (2.3.5), is computed using the central-difference formula

$$U_i = \frac{\Phi_{i+1} - \Phi_{i-1}}{2h} \qquad (2.3.11)$$

The numerical results for Φ and U are then printed out to compare with the exact solutions represented by (2.3.6) and (2.3.9).

The output of Program 2.1 shows that the errors of the numerical solution are less than 1% in Φ and 0.2% in U, which may be considered satisfactory in most engineering problems. Results of higher accuracies can be obtained by increasing the number of intervals at the cost of longer computing time. You may verify this statement by comparing the data obtained by varying the number of intervals in Program 2.1.

List of Principal Variables in Program 2.1

Program Symbol	Definition
(Main Program)	
C	Coefficient matrices defined in (2.2.13)
EXPHI	Exact solution for Φ
EXU	Exact solution for U
H	Size of intervals
INTR'/L	Number of intervals in the concerned range of R
N	Number of interior grid points, n, in the range
NM1	$n-1$
PHI	Numerical solution for Φ
PHINP1	Value of Φ at R_{n+1} (i.e., at R_{max})
PHI0	Value of Φ at R_0 (i.e., at R_{min})
R	Radial position
RMAX, RMIN	Upper and lower ends of the range
U	Numerical solution for radial velocity
(Subroutine TRID)	
C	Coefficient matrices defined in (2.2.13)
F	Solution vector
N	Number of simultaneous equations to be solved

Radial Flow Caused by Distributed Sources and Sinks

```
C          ***** PROGRAM 2.1 *****

C          SOLVING A BOUNDARY-VALUE PROBLEM CONCERNING THE AXISYMMETRIC
C          FLOW CAUSED BY A SOURCE DISTRIBUTION AND A LINE SINK AT ORIGIN

           DIMENSION R(29),C(29,4),PHI(29),EXPHI(29),U(29),EXU(29)
           DATA RMIN,RMAX,INTRVL / 1.0, 4.0, 30 /

C          ..... USE ARITHMETIC FUNCTIONS TO REPRESENT (2.3.10) .....
           A(X) = 1./X
           B(X) = 0.
           D(X) = 9.*X

C          ..... N AND H DEFINED IN FIG. 2.2.1 ARE COMPUTED .....
           N = INTRVL - 1
           H = (RMAX-RMIN) / FLOAT(INTRVL)

C          ..... COMPUTE R(I), C(I,J), AND EXACT VALUES OF PHI AND U .....
           RADIUS = RMIN + H
           DO 1 I=1,N
           R(I) = RADIUS
           C(I,1) = 1. - H/2.*A(R(I))
           C(I,2) = -2. + H*H*B(R(I))
           C(I,3) = 1. + H/2.*A(R(I))
           C(I,4) = H*H*D(R(I))
           EXPHI(I) = R(I)**3 - 24.*ALOG(R(I))
           EXU(I) = 3.*R(I)**2 - 24./R(I)
         1 RADIUS = RADIUS + H

C          ..... SPECIFY BOUNDARY CONDITIONS AND MODIFY SOME OF
C                THE COEFFICIENTS ACCORDING TO (2.2.12) .....
           PHIO = 1.0
           PHINP1 = 64. - 24.*ALOG(4.)
           C(1,4) = C(1,4) - C(1,1)*PHIO
           C(1,1) = 0.
           C(N,4) = C(N,4) - C(N,3)*PHINP1
           C(N,3) = 0.

C          ..... COMPUTE PHI AND U AT INTERIOR POINTS .....
           CALL TRID( C, PHI, N )
           U(1) = (PHI(2)-PHIO) / (2.*H)
           NM1 = N-1
           DO 2 J=2,NM1
         2 U(J) = (PHI(J+1)-PHI(J-1)) / (2.*H)
           U(N) = (PHINP1-PHI(NM1)) / (2.*H)

C          ..... PRINT RESULT IN COMPARISON WITH EXACT SOLUTION .....
           WRITE(6,10)
           WRITE(6,11) ((I,R(I),PHI(I),EXPHI(I),U(I),EXU(I)), I=1,N)

        10 FORMAT(1H1 ///// 6X, 1HI, 3X, 4HR(I), 6X, 6HPHI(I), 3X,
          A          8HEXPHI(I), 6X, 4HU(I), 5X, 6HEXU(I) / )
        11 FORMAT( 5X, I2, F7.2, 2X, 2F10.3, 2X, 2F10.3 )

           END
```

```fortran
      SUBROUTINE TRID( C, F, N )
C     ..... SOLVING A SET OF N LINEAR SIMULTANEOUS EQUATIONS DESCRIBED
C           BY (2.2.13) HAVING A TRIDIAGONAL COEFFICIENT MATRIX.
C           COEFFICIENT MATRICES ARE REPRESENTED BY C(I,J) AND THE
C           SOLUTION IS STORED IN THE ONE-DIMENSIONAL ARRAY F(I) .....

      DIMENSION C(N,4), F(N)

C     ..... ELIMINATE THE FIRST TERM OF EACH OF THE EQUATIONS (EXCEPT
C           THE FIRST AND THE LAST ONE) ACCORDING TO (2.2.14) .....
      NM1 = N-1
      DO 1 I=2,NM1
      C(I,2) = C(I,2)*C(I-1,2) - C(I,1)*C(I-1,3)
      C(I,3) = C(I,3)*C(I-1,2)
    1 C(I,4) = C(I,4)*C(I-1,2) - C(I,1)*C(I-1,4)

C     ..... COMPUTE SOLUTION USING (2.2.15) AND (2.2.16) .....
      F(N) = ( C(N,4)*C(N-1,2)-C(N,1)*C(N-1,4) ) /
     A       ( C(N,2)*C(N-1,2)-C(N,1)*C(N-1,3) )
      DO 2 K=1,NM1
      J = N-K
    2 F(J) = ( C(J,4)-C(J,3)*F(J+1) ) / C(J,2)
      RETURN
      END
```

I	R(I)	PHI(I)	EXPHI(I)	U(I)	EXU(I)
1	1.10	-.954	-.956	-18.220	-18.188
2	1.20	-2.644	-2.648	-15.702	-15.680
3	1.30	-4.095	-4.100	-13.407	-13.392
4	1.40	-5.325	-5.331	-11.274	-11.263
5	1.50	-6.350	-6.356	-9.258	-9.250
6	1.60	-7.177	-7.184	-7.325	-7.320
7	1.70	-7.815	-7.822	-5.451	-5.448
8	1.80	-8.267	-8.275	-3.615	-3.613
9	1.90	-8.538	-8.545	-1.802	-1.802
10	2.00	-8.628	-8.636	-.000	-.000
11	2.10	-8.538	-8.545	1.802	1.801
12	2.20	-8.267	-8.275	3.612	3.611
13	2.30	-7.815	-7.823	5.437	5.435
14	2.40	-7.180	-7.187	7.282	7.280
15	2.50	-6.359	-6.366	9.152	9.150
16	2.60	-5.350	-5.356	11.052	11.049
17	2.70	-4.149	-4.155	12.984	12.981
18	2.80	-2.753	-2.759	14.951	14.949
19	2.90	-1.158	-1.164	16.957	16.954
20	3.00	.639	.633	19.003	19.000
21	3.10	2.642	2.637	21.091	21.088
22	3.20	4.857	4.852	23.223	23.220
23	3.30	7.287	7.283	25.400	25.397
24	3.40	9.937	9.933	27.624	27.621
25	3.50	12.812	12.809	29.896	29.893
26	3.60	15.916	15.914	32.216	32.213
27	3.70	19.255	19.253	34.586	34.584
28	3.80	22.833	22.832	37.007	37.004
29	3.90	26.656	26.656	39.479	39.476

2.4 Inverse Method I: Superposition of Elementary Flows

In the previous numerical example, we look directly for a solution to the governing equation that satisfies a set of assumed boundary conditions. Such boundary-value problems are common in studying potential flows. For example, in the problem of a uniform flow past a body of given shape, a solution to the Laplace equation is to be found that satisfies both the uniform flow condition far away and the condition that the flow be tangent to the surface of the body. However, to solve such a problem utilizing a *direct method* is not always simple, despite the fact that the governing differential equation is linear. A numerical technique for solving two-dimensional problems directly will be introduced in Section 2.10, but an *inverse method* is described here that solves the flow problems using a different approach.

The Laplace equation, (2.1.15) or (2.1.17), governing the stream function for two-dimensional potential flows is in such a simple form that some elementary solutions to this equation can easily be found. Each of these solutions represents a physically possible elementary potential flow. Because the sum of a linear combination of any number of solutions to a linear differential equation is also a solution to the same equation, infinitely many solutions can thus be generated. Some of the flows represented by the linear combination of solutions may simulate those caused by moving bodies of various shapes through a fluid. Thus, in the inverse method, instead of finding the solution for the flow past a body of a prescribed shape, flows around different bodies are constructed through various combinations of elementary flows and are filed. Based on these typical flow patterns, a body close to the desired shape can generally be constructed through an appropriate combination of elementary flows. The inverse method is also often used in solving problems in electromagnetism and heat conduction whose governing equations are in the form of either the Laplace or the Poisson equation.

There are four elementary two-dimensional flows that are commonly encountered in fluid mechanics: the *uniform flow*, the *line source* (or *sink*), the *line vortex*, and the *doublet*. Their stream functions are listed below, and the corresponding flow patterns are sketched in Fig. 2.4.1.

a. Uniform flow $\qquad\qquad \psi = U(y \cos \alpha - x \sin \alpha) \qquad\qquad$ (2.4.1)

b. Line source $\qquad\qquad \psi = \dfrac{\Lambda}{2\pi} \tan^{-1} \dfrac{y - y_0}{x - x_0}$

$$= \dfrac{\Lambda}{2\pi} \theta' \qquad\qquad (2.4.2)$$

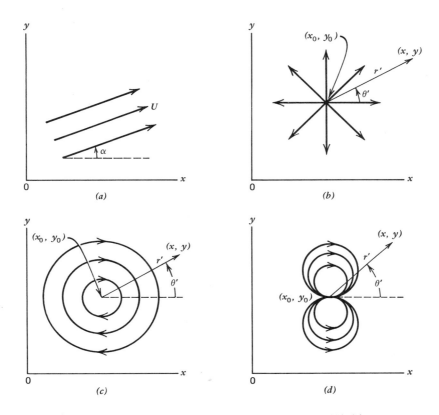

FIGURE 2.4.1 Elementary flows. (a) Uniform flow. (b) Line source.
(c) Line vortex. (d) Doublet.

c. Line vortex

$$\psi = \frac{\Gamma}{2\pi} \ln[(x - x_0)^2 + (y - y_0)^2]^{1/2}$$

$$= \frac{\Gamma}{2\pi} \ln r' \tag{2.4.3}$$

d. Doublet

$$\psi = -\frac{\kappa}{2\pi} \frac{y - y_0}{(x - x_0)^2 + (y - y_0)^2}$$

$$= -\frac{\kappa}{2\pi} \frac{\sin \theta'}{r'} \tag{2.4.4}$$

Expression (2.4.1) is the stream function for a uniform flow of speed U that makes an angle α with the x-axis. Special cases of horizontal and vertical uniform flows are deduced by letting $\alpha = 0°$ and $90°$, respectively.

Superposition of Elementary Flows

Stream function (2.4.2) describes a source situated at (x_0, y_0) having a strength Λ, which represents the volume of fluid per unit time streaming from a unit length of the line source. Thus a negative value of Λ corresponds to a line sink. Fluid velocity along a streamline radiating away from the source center has the magnitude $\Lambda/2\pi r'$, where r' is the radial distance from that center. It follows that the flow speed becomes infinity as r' approaches zero.

The stream function shown in (2.4.3) gives the circular flow pattern of a line vortex centered at (x_0, y_0). The tangential velocity has the magnitude $\Gamma/2\pi r'$, which is a constant along a given streamline. The product of the circumference of a circular streamline and the speed along it is equal to the *circulation* Γ, which is used to denote the strength of a vortex flow. In aerodynamics the clockwise circulation is considered positive. Again, the speed at the vortex center is unbounded.

When a source of strength Λ at $(x_0 - \Delta x, y_0)$ is added to a sink of strength $-\Lambda$ at $(x_0 + \Delta x, y_0)$, a new flow field is obtained. Furthermore, by letting Δx approach zero while keeping the product $2\Delta x\Lambda$ a constant κ, the stream function (2.4.4) for a doublet at (x_0, y_0) is obtained. The streamlines are circles passing through the point (x_0, y_0) with centers on the straight line $x = x_0$. The same flow pattern can be produced by superimposing a vortex at $(x_0, y_0 + \Delta y)$ to a vortex of opposite circulation at $(x_0, y_0 - \Delta y)$, and then letting Δy approach zero. κ is called the strength of the doublet. The velocity at the center of a doublet is infinitely large.

Except the uniform flow, the other three elementary flows just discussed have the same property, that the velocity becomes infinity when the center is approached. Because of this property, they are sometimes called *singularities*. They do not cause any mathematical difficulties if each of the singular points is within a domain that is considered to be occupied by a body. For example, when a source is placed in a uniform stream, a half-body is generated in the flow. If the source is replaced by a doublet, a circular cylinder is generated instead. The singularity in each case is enclosed within the boundary of a rigid body, and the flow outside the body is therefore free of singularities. Detail descriptions of elementary flows and their syntheses can be found in most textbooks on fluid mechanics (e.g., Chapter 4 of Kuethe and Chow, 1976).

In general, bodies of infinite extension are generated by sources or sinks. On the other hand, doublets or a group of sources and sinks of vanishing total strength are used to form bodies of closed boundary. The inverse method of superposition of elementary flows is simple in principle, but to figure out the body shape and the flow pattern is usually tedious and time-consuming if only manual calculators are used. To make this method easier to work with, a computer program is constructed that plots some representative streamlines of a flow that consists of any number of elementary flows.

Consider a two-dimensional stream function of the general form

$$\psi = f(x, y) \tag{2.4.5}$$

The flow pattern within a rectangular space bounded between x_{min} and x_{max} in the x direction and between y_{min} and y_{max} in the y direction is to be plotted. The space is subdivided, as shown in Fig. 2.4.2, by vertical lines at a constant distance Δx apart, and by horizontal lines at a distance Δy apart. The sizes of Δx and Δy are not necessarily the same. Grid points that are formed at the intersections of these two sets of perpendicular lines have coordinates (x_i, y_j), where $i = 1, 2, \cdots, m$ and $j = 1, 2, \cdots, n$ according to the notation of Fig. 2.4.2. The values of the stream function evaluated at the grid points are called $\psi_{i,j}$ which are computed for all values of i and j from the relation

$$\psi_{i,j} = f(x_i, y_j) \tag{2.4.6}$$

In the output of the program, a graph will be shown that displays the points where the *vertical* grid lines intersect certain particular streamlines. If the number of vertical lines is reasonably large, the printed points will trace out the approximate shape of the flow pattern. Since the graph will be constructed by calling subroutines PLOTN, described in Section 1.7, the positions of some of the printed points are only approximate. For this reason, the coordinates of all the points to appear on the graph will also be tabulated in the output.

To search for all the points at which a vertical grid line intersects with the

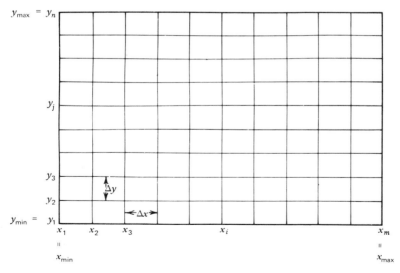

FIGURE 2.4.2 Rectangular grid system.

Superposition of Elementary Flows

streamline $\psi = \psi_a$, let us consider an arbitrary distribution of the stream function along the grid line at x_i, as sketched in Fig. 2.4.3. If the height of each spike represents the local value of the stream function, the streamline $\psi = \psi_a$ will go through an interval so that at one end of the interval the spike is higher than ψ_a, and at the other end the spike is shorter than ψ_a. In other words, if P represents the difference between the stream function at the left end of an interval and ψ_a, and Q represents the difference at the right end, the streamline $\psi = \psi_a$ goes through the interval if the product of P and Q is negative. The y-coordinate of the intersection point is obtained approximately by subtracting $\Delta y |Q|/(|P|+|Q|)$ from the y-coordinate of the point on the right end of that interval, if the variation of ψ within the small interval is approximated by a straight line (see Fig. 2.4.3). All intersection points on the grid line can be located after every interval on that line has been examined.

The searching process along a specified streamline is executed by calling the subroutine SEARCH, which performs the previously mentioned action one vertical line after another, starting from the first one at x_1. The counter k is used to indicate the order of the points thus found. xx_k and yy_k are the coordinates of the kth point, and the total number of points found on a streamline within the given domain is k_{max}.

For flows having streamlines nearly parallel to the y-axis, it is more convenient to show the intersection points of those streamlines and the horizontal grids. This can easily be achieved by changing a few statements in the present subroutine.

The example shown in Program 2.2 deals with a flow constructed by adding to a uniform horizontal stream five doublets of different strengths placed on the x-axis at $x/l = -1.0$, -0.5, 0, 0.5, and 1.0, respectively, where l is a

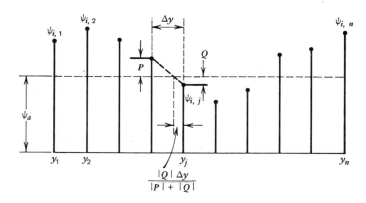

FIGURE 2.4.3 Distribution of stream function along a vertical grid line at x_i.

reference length. The stream function of the resultant flow is, from (2.4.1) and (2.4.4),

$$\psi = Uy - \sum_{i=1}^{5} \frac{\kappa_i}{2\pi} \frac{y}{(x - d_i l)^2 + y^2} \tag{2.4.7}$$

where κ_i are the strengths of the five doublets, and the five values of d_i are, in order, -1.0, -0.5, 0, 0.5, and 1.0. This expression can be made dimensionless

List of Principal Variables in Program 2.2

Program Symbol	Definition
(Main Program)	
CONST	An array storing the values of stream function for the streamlines to be plotted
DX, DY	Interval sizes Δx and Δy, respectively
KMAX	Total number of points to be plotted along a streamline
LCHAR	Characters for the vertical label of the graph
M, N	Numbers of vertical and horizontal lines in the grid system, respectively
PCHAR	Plotting characters
PLTC	An array storing plotting characters
PSI	Values of stream function Ψ at grid points
PSIA	Value of stream function for the streamline to be plotted, Ψ_a
X, Y	Coordinates of grid points
XMAX, XMIN	Maximum and minimum values of abscissa
XX, YY	Coordinates of points found in subroutine SEARCH
YMAX, YMIN	Maximum and minimum values of ordinate
(Subroutine SEARCH)	
P	Difference between stream function at the left end of an interval and Ψ_a
Q	Difference between stream function at the right end of an interval and Ψ_a
(Function F)	
C	An array storing dimensionless strengths of doublets
D	An array storing dimensionless positions of doublets on x-axis

Superposition of Elementary Flows

```
C                   ***** PROGRAM 2.2 *****

C           PLOTTING THE FLOW RESULTING FROM SUPERPOSITION
C           OF A UNIFORM STREAM AND FIVE DOUBLETS

        DIMENSION X(50),Y(50),PSI(50,50),CONST(5),XX(100),YY(100),
       A          PLTC(5),NSCALE(4),LCHAR(3)
        COMMON M,N,DY

C       ..... ASSIGN INPUT DATA .....
        DATA XMAX,XMIN,YMAX,YMIN / 3., -3., 2., -2. /
        DATA CONST / 0., -0.5, -1.0, 0.5, 1.0 /
        DATA PLTC / 1H0, 1H1, 1H2, 1H3, 1H4 /
        DATA NSCALE / 0, 1, 0, 1 /
        DX = 0.2
        DY = 0.2
        M = 31
        N = 21

C       ..... COMPUTE COORDINATES FOR GRID POINTS AND EVALUATE
C             THE VALUES OF STREAM FUNCTION THERE .....
        X(1) = XMIN
        DO 1 I=2,M
      1 X(I) = X(I-1) + DX
        Y(1) = YMIN
        DO 2 J=2,N
      2 Y(J) = Y(J-1) + DY
        DO 3 I=1,M
        DO 3 J=1,N
      3 PSI(I,J) = F(X(I),Y(J))

C       ..... SET UP THE COORDINATE AXES TO BE PLOTTED, CONTAINING THREE
C             HORIZONTAL AND THREE VERTICAL LINES .....
        CALL PLOT1( NSCALE, 3, 20, 3, 50, 1H-, 1HI )
        CALL PLOT2( XMAX, XMIN, YMAX, YMIN, .TRUE. )

C       ..... SEARCH FOR POINTS ON 5 STREAMLINES ALONG WHICH THE STREAM
C             FUNCTION HAS THE VALUES STORED IN CONST.  THEIR COORDINATES
C             ARE PRINTED AND ARE REPRESENTED ON THE GRAPH BY 5 DIFFERENT
C             PLOTTING CHARACTERS STORED IN PLTC .....
        WRITE(6,10)
        DO 4 L=1,5
        PSIA = CONST(L)
        CALL SEARCH( X, Y, PSI, PSIA, XX, YY, KMAX )
        WRITE(6,11)  ( PSIA, ((XX(K),YY(K)),K=1,KMAX) )
        PCHAR = PLTC(L)
        CALL PLOT3( PCHAR, XX, YY, 1, KMAX, 1 )
      4 CONTINUE

C       ..... PRINT GRAPH ON A NEW PAGE WITHOUT A LABEL ON THE LEFT .....
        WRITE(6,10)
        LCHAR(1) = 10H
        CALL PLOT4( LCHAR, 0 )
        WRITE(6,12)

     10 FORMAT( 1H1, //// )
     11 FORMAT( //10X39HTHE POINTS (X,Y) ON THE STREAMLINE PSI=,
       A          F5.2, / (6(4X, F5.2, 2H ,, F5.2 )) )
     12 FORMAT( // 34X, 30HFIGURE 2.4.4   FLOW GENERATED ,
       A          38HBY A UNIFORM STREAM AND FIVE DOUBLETS. )

        END
```

```
      SUBROUTINE SEARCH( X, Y, PSI, PSIA, XX, YY, KMAX )
C     ..... SEARCHING THROUGH VERTICAL GRID LINES FOR POINTS (XX,YY) ON
C           A STREAMLINE ASSUMING THE VALUE PSIA.  PSI CONTAINS VALUES
C           OF STREAM FUNCTION AT GRID POINTS (X,Y).  KMAX IS THE TOTAL
C           NUMBER OF POINTS THUS FOUND .....

      DIMENSION X(50),Y(50),PSI(50,50),XX(100),YY(100)
      COMMON M,N,DY

C     ..... EXAMINE EVERY INTERVAL ON EACH OF THE M VERTICAL GRID LINES.
C           A POINT (XX,YY) IS LOCATED AT A GRID POINT IF THE LOCAL PSI
C           DIFFERS FROM THE DESIRED VALUE BY 0.00001.  OTHERWISE WE
C           LOOK FOR THE INTERVALS ACROSS WHICH P*Q≤0, AND THEN COMPUTE
C           THE COORDINATES (XX,YY) WITHIN THOSE INTERVALS .....
      K = 0
      I = 0
    1 J = 1
      I = I + 1
      IF( I.GT.M ) GO TO 7
    2 P = PSI(I,J) - PSIA
      IF( ABS(P).LE.0.00001 )  GO TO 6
    3 J = J + 1
      IF( J.GT.N )  GO TO 1
      Q = PSI(I,J) - PSIA
      IF( P*Q )  5,5,4
    4 P = Q
      GO TO 3
    5 K = K + 1
      XX(K) = X(I)
      YY(K) = Y(J) - DY*ABS(Q)/(ABS(P)+ABS(Q))
      P = Q
      GO TO 3
    6 K = K + 1
      XX(K) = X(I)
      YY(K) = Y(J)
      J = J + 1
      IF( J.GT.N )  GO TO 1
      GO TO 2
    7 KMAX = K
      RETURN
      END




      FUNCTION F( X, Y )
C     ..... IT REPRESENTS RIGHT-HAND SIDE OF EQUATION (2.4.8) WITH THE
C           DOUBLETS SHIFTED TO THE RIGHT THROUGH A SHORT DISTANCE .....

      DIMENSION C(5),D(5)
      DATA C / 0.15, 0.3, 0.2, 0.1, 0.05 /
      DATA D / -1.0, -0.5, 0., 0.5, 1.0 /
      F = Y
      DO 1 I=1,5
    1 F = F - C(I)*Y/((X-D(I)-1.0E-6)**2+Y**2)
      RETURN
      END
```

THE POINTS (X,Y) ON THE STREAMLINE PSI= 0.00

X	Y	X	Y	X	Y	X	Y	X	Y	X	Y
-3.00	.00	-2.80	.00	-2.60	.00	-2.40	.00	-2.20	.00	-2.00	.00
-1.80	.00	-1.60	.00	-1.40	-.31	-1.40	-.00	-1.40	.31	-1.20	-.54
-1.20	-.00	-1.20	.54	-1.00	-.65	-1.00	-.00	-1.00	.65	-.80	-.72
-.80	-.00	-.80	.72	-.60	-.76	-.60	-.00	-.60	.76	-.40	-.78
-.40	-.00	-.40	.78	-.20	-.77	-.20	-.00	-.20	.77	.00	-.74
.00	-.00	.00	.74	.20	-.69	.20	-.00	.20	.69	.40	-.62
.40	-.00	.40	.62	.60	-.54	.60	-.00	.60	.54	.80	-.44
.80	-.00	.80	.44	1.00	-.36	1.00	-.00	1.00	.36	1.20	-.22
1.20	-.00	1.20	.22	1.40	.00	1.60	.00	1.80	.00	2.00	.00
2.20	.00	2.40	.00	2.60	.00	2.80	.00	3.00	.00		

THE POINTS (X,Y) ON THE STREAMLINE PSI= -.50

X	Y	X	Y	X	Y	X	Y	X	Y	X	Y
-3.00	-.56	-2.80	-.58	-2.60	-.59	-2.40	-.62	-2.20	-.65	-2.00	-.69
-1.80	-.75	-1.60	-.82	-1.40	-.90	-1.20	-.97	-1.00	-1.02	-1.00	.12
-1.00	.34	-.80	-1.07	-.80	.14	-.80	.40	-.60	-1.10	-.60	.08
-.60	.49	-.40	-1.11	-.40	.08	-.40	.51	-.20	-1.10	-.20	.12
-.20	.47	.00	-1.09	.00	.09	.00	.43	.20	-1.06	.20	.17
.20	.30	.40	-1.02	.40	.20	.40	.21	.60	-.97	.80	-.92
1.30	-.86	1.20	-.80	1.40	-.74	1.60	-.68	1.80	-.64	2.00	-.61
2.20	-.59	2.40	-.57	2.60	-.56	2.80	-.55	3.00	-.54		

THE POINTS (X,Y) ON THE STREAMLINE PSI=-1.00

X	Y	X	Y	X	Y	X	Y	X	Y	X	Y
-3.00	-1.11	-2.80	-1.13	-2.60	-1.15	-2.40	-1.17	-2.20	-1.20	-2.00	-1.24
-1.80	-1.28	-1.60	-1.32	-1.40	-1.36	-1.20	-1.40	-1.00	-1.43	-.80	-1.46
-.60	-1.47	-.60	.16	-.60	.29	-.40	-1.48	-.40	.15	-.40	.30
-.20	-1.48	.00	-1.47	.00	.18	.00	.24	.20	-1.46	.40	-1.43
.60	-1.40	.80	-1.37	1.00	-1.33	1.20	-1.30	1.40	-1.26	1.60	-1.22
1.80	-1.19	2.00	-1.16	2.20	-1.14	2.40	-1.12	2.60	-1.10	2.80	-1.09
3.00	-1.08										

THE POINTS (X,Y) ON THE STREAMLINE PSI= .50

X	Y	X	Y	X	Y	X	Y	X	Y	X	Y
-3.00	.56	-2.80	.58	-2.60	.59	-2.40	.62	-2.20	.65	-2.00	.69
-1.80	.75	-1.60	.82	-1.40	.90	-1.20	.97	-1.00	-.34	-1.00	-.12
-1.00	1.02	-.80	-.40	-.80	-.14	-.80	1.07	-.60	-.49	-.60	-.08
-.60	1.10	-.40	-.51	-.40	-.08	-.40	1.11	-.20	-.47	-.20	-.12
-.20	1.10	.00	-.43	.00	-.09	.00	1.09	.20	-.30	.20	-.17
.20	1.06	.40	-.21	.40	-.20	.40	1.02	.60	.97	.80	.92
1.00	.86	1.20	.80	1.40	.74	1.60	.68	1.80	.64	2.00	.61
2.20	.59	2.40	.57	2.60	.56	2.80	.55	3.00	.54		

THE POINTS (X,Y) ON THE STREAMLINE PSI= 1.00

X	Y	X	Y	X	Y	X	Y	X	Y	X	Y
-3.00	1.11	-2.80	1.13	-2.60	1.15	-2.40	1.17	-2.20	1.20	-2.00	1.24
-1.80	1.28	-1.60	1.32	-1.40	1.36	-1.20	1.40	-1.00	1.43	-.80	1.46
-.60	-.29	-.60	.16	-.60	1.47	-.40	-.30	-.40	-.15	-.40	1.48
-.20	1.48	.00	-.24	.00	-.18	.00	1.47	.20	1.46	.40	1.43
.60	1.40	.80	1.37	1.00	1.33	1.20	1.30	1.40	1.26	1.60	1.22
1.80	1.19	2.00	1.16	2.20	1.14	2.40	1.12	2.60	1.10	2.80	1.09
3.00	1.08										

in the simpler form

$$\Psi = Y - \sum_{i=1}^{5} \frac{c_i Y}{(X - d_i)^2 + Y^2} \tag{2.4.8}$$

in which $\Psi = \psi/Ul$, $X = x/l$, $Y = y/l$, and $c_i = \kappa_i/2\pi l^2 U$. The shape of the body and the flow around it are plotted on the dimensionless spatial coordinates for the five dimensionless strengths 0.15, 0.3, 0.2, 0.1, and 0.05, assigned in order to c_i. The right-hand side of (2.4.8) represents the function f in (2.4.5). Remember that a doublet is a singularity having not only an infinite velocity, but also an infinitely large value of stream function at its center. In the present case, the centers of the assumed doublets coincide with five grid

Inviscid Fluid Flows

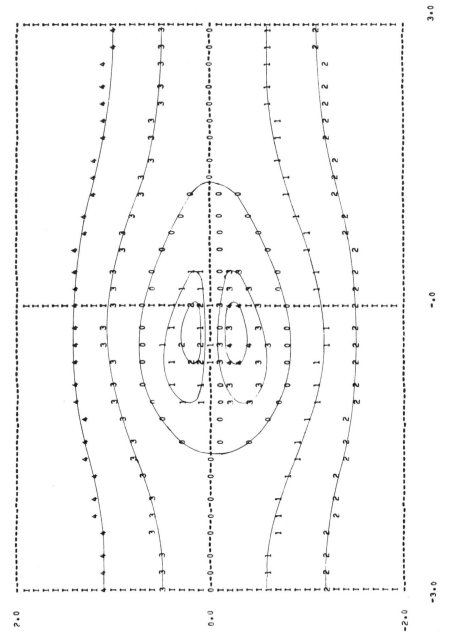

FIGURE 2.4.4 Flow generated by a uniform stream and five doublets.

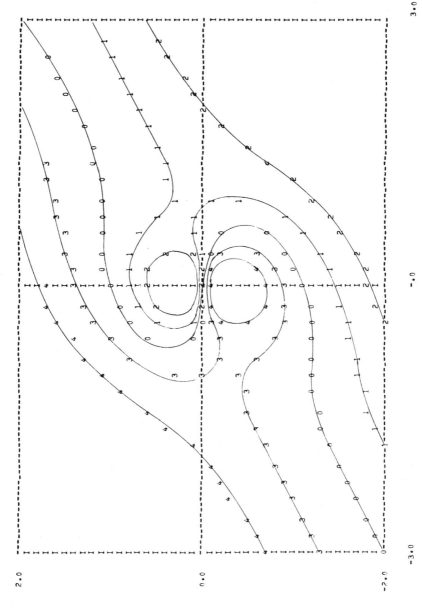

FIGURE 2.4.5 Superposition of a doublet at the origin and a uniform stream making a 30° angle with the x-axis.

94

points at which the stream function will be evaluated. The computational difficulty can be avoided by shifting the doublets to the right from their original positions through a tiny dimensionless distance of 10^{-6}, as shown in the function subprogram of Program 2.2. By doing so the effects on flow pattern are so small that they cannot be observed.

On the computer-printed graph, lines are drawn by hand to connect symbols of the same kind so that a more definite flow pattern is shown. An egg-shaped body having all of the singularities enclosed within its body boundary is generated in Fig. 2.4.4 by Program 2.2. The body is wider at the section where the stronger doublets are located. This result provides some guide for the construction of bodies symmetric about the x-axis, for example, symmetric airfoils at zero angle of attack.

Program 2.2 may be extended further to compute the pressure distribution around a body or along any streamline in a flow field. Once the positions of the points along that curve are located, the velocity components there can be calculated using either (2.1.14) or (2.1.16). Pressure is finally obtained after substituting the velocity components into the steady Bernoulli equation.

Superposition of elementary flows does not always end up with flow past a rigid body. In the previously mentioned example of a flow past a circular cylinder obtained by adding a doublet to a uniform stream, if the horizontal uniform stream is replaced by one making a 30° angle with the x-axis, the resultant flow has the pattern shown in Fig. 2.4.5. The figure is an output of Program 2.2 with the function F modified accordingly. In this flow a well-defined body containing the doublet cannot be found.

Problem 2.1 To become familiar with the inverse method of superposition of elementary flows, plot, by modifying Program 2.2, the flow consisting of

1. A uniform stream and a source.

2. A uniform stream and a sink.

3. A source on the negative x-axis and a doublet at the origin of the coordinate system.

4. A horizontal uniform stream and two doublets of equal strengths on the x-axis separated by a distance d.

Verify that a smooth oval-shaped body is generated in the flow for a small value of d, the body has the form of a dumbbell for a larger value of d and, finally, the body degenerates into two separate bodies when d is further increased.

Superposition of Elementary Flows   **95**

Hint: Special care must be taken when sources or sinks appear in a flow whose stream function contains a function in the general form of $\tan^{-1}(y/x)$. The Fortran library function ATAN(Y/X) returns an angle between $-\pi/2$ and $+\pi/2$ and, therefore, is not appropriate to be used in the present program. However, if the function ATAN2(Y, X) is used, the angle returned varies from 0 to π in the first and second quadrants and from $-\pi$ to -0 in the third and fourth quadrants. It is more convenient to normalize such a function by dividing it by π. The stream function of the combined flow should be examined to determine which streamline or streamlines may form a body.

Problem 2.2 Plot the streamlines of a flow formed by the combination of a vortex of circulation Γ at $(0, d)$ and another of circulation $-\Gamma$ at $(0, -d)$. The flow pattern will show symmetry about the x-axis, with its upper half describing the streamlines of a vortex distorted by a flat surface at $y = 0$.

This result demonstrates the following. The flow around a vortex in the presence of a flat plane can be obtained by adding to the original vortex its mirror image with respect to the plane surface. As another example, the flow field caused by a source at $(-d, 0)$ in the presence of a vertical wall formed by the y-axis is constructed by combining the stream function of the original source and that of its mirror image, which is a source of the same strength situated at $(d, 0)$. This technique for constructing flows near a flat surface is called the *method of images.*

Problem 2.3 Plot the upper half of the flow consisting of a horizontal uniform stream, a doublet on the y-axis, and its image with respect to the plane $y = 0$. It simulates the flow around a closed body in the presence of a plate parallel to the flow far ahead.

Compute the pressure distribution along the upper surface of the flat plate, which shows that the plate is acted on by a net force in a direction toward the body. It follows that the same net force is exerted on the body but in an opposite direction.

2.5 Kinematics of Line Vortices and Vector Manipulations

The line vortex, one of the elementary solutions to the Laplace equation considered in Section 2.4, does exist in the real world. It appears in the form of a hurricane or a tornado in the atmosphere, or in the form of a swirling filament above the drain of a bathtub. Except for a core region within which the velocity is retarded by viscous forces, the flow fields of these real vortices are described closely by the stream function (2.4.3), which gives a circumferential velocity inversely proportional to the radial distance from the center.

Aeronautical engineers are more interested in the pair of vortices trailing behind an airplane. These trailing vortices, generated by the pressure difference between the lower and upper surfaces of a lifting wing, are separated by a distance approximately equal to the wingspan of the aircraft. They have circulations of the same magnitude but of opposite signs. Streamlines enclosing the two central cores revolve in such a way that the flow is downward in the region between the cores. Since viscosity affects the flow mainly in the small core regions, the trailing vortices are found to be very persistent, and their strength is decreased only slightly at a distance as large as 25 km behind a large aircraft. During take-off or landing, these long-lasting vortices left behind by a large aircraft, moving under mutual interactions above the runway for a period of time, may cause serious hazards for a following smaller aircraft. In fact, many crashes attributable to such a phenomenon have been reported (Chigier, 1974).

The motion of a pair of trailing vortices on a runway will be studied in this section. The result may provide important information, such as the time required for the vortices to move out of the edges of the runway, or the waiting time needed by the following aircraft for using the same runway to be sure that the vortex-induced velocity on the runway is within a specified safety margin.

Let us first consider the velocity field induced by a line vortex of circulation Γ centered at $\mathbf{r}_0$ (refer to Fig. 2.5.1). Aerodynamicists usually assign a positive value to the circulation of a clockwise-revolving vortex, because a positive lift would be generated on such a vortex if it is placed in a flow moving to the right. The centerline of this vortex is assumed to be parallel to the z-axis. Let $\mathbf{i}_z$ denote the unit vector along the positive z direction. At a point of position vector $\mathbf{r}$, the induced velocity is in the x-y plane and is given by

$$\mathbf{V} = -\frac{\Gamma}{2\pi}\frac{\mathbf{i}_z \times (\mathbf{r} - \mathbf{r}_0)}{|\mathbf{r} - \mathbf{r}_0|^2} \tag{2.5.1}$$

The expression is obtained by applying the *Biot-Savart law*, which was originally derived for electromagnetic fields. Here the vector strength $-\mathbf{i}_z\Gamma$ of the vortex line is analogous to the electric current along a line conductor, and the velocity $\mathbf{V}$ is analogous to the magnetic field intensity induced by that current at the same position. The direction of the vector strength of a line vortex is determined from the direction of circulation by using the right-hand rule; therefore, the vector strength of the vortex shown in Fig. 2.5.1 is opposite to the direction of $\mathbf{i}_z$.

Kinematics of Line Vortices and Vector Manipulations $\qquad$ **97**

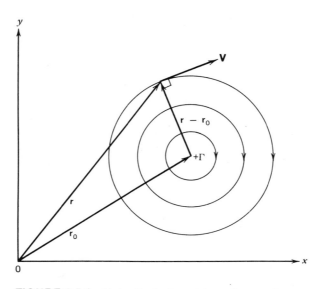

FIGURE 2.5.1 Velocity induced by a line vortex.

According to (2.5.1), the velocity induced at a point is perpendicular to the radial line connecting this point and the line vortex, and its magnitude is inversely proportional to the distance of this point from the vortex center.

Now consider a pair of trailing vortices left behind by an aircraft on a runway. Suppose that at time $t = 0$ they are at a height h above the ground and are separated by a distance b equal to the wingspan of the aircraft. According to the method of images described in Problem 2.2, the flow field is simulated by the upper half of the flow constructed by adding to the vortex pair their mirror images with respect to the ground. The configuration at the initial instant is shown in Fig. 2.5.2. The instantaneous velocity of the vortex at $\mathbf{r}_A$, induced at that location by the other three, is obtained by applying (2.5.1) three times; hence

$$\mathbf{V}_A = \frac{\Gamma}{2\pi} \mathbf{i}_z \times \left(\frac{\mathbf{r}_A - \mathbf{r}_B}{|\mathbf{r}_A - \mathbf{r}_B|^2} - \frac{\mathbf{r}_A - \mathbf{r}_C}{|\mathbf{r}_A - \mathbf{r}_C|^2} + \frac{\mathbf{r}_A - \mathbf{r}_D}{|\mathbf{r}_A - \mathbf{r}_D|^2} \right) \qquad (2.5.2)$$

Through a small time interval Δt, this vortex will move from its original position at time t to a new position

$$\mathbf{r}_A(t + \Delta t) = \mathbf{r}_A(t) + \mathbf{V}_A(t)\,\Delta t \qquad (2.5.3)$$

The new position vector of the vortex at $\mathbf{r}_B$ is simply the mirror image of $\mathbf{r}_A(t + \Delta t)$ with respect to the y-axis, and the image of the pair will also move accordingly. The motion of the first vortex is then recomputed, and the procedure is repeated for any desired length of time. Because the curved

98   Inviscid Fluid Flows

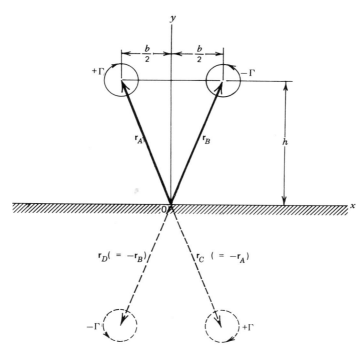

FIGURE 2.5.2 Trailing vortices above runway at $t = 0$.

trajectories of the vortices are approximated in this way by many straight segments, the size of Δt must be made small enough to insure allowable computational errors.

The strength of trailing vortices is proportional to the total weight W of the aircraft. Based on a simple horseshoe vortex line model of the finite wing, the trailing vortices are considered as the continuation of the bound vortex inside the wing, but are bent in the downstream direction at the wing tips. Thus the circulation around each trailing vortex has the same magnitude as that around the wing. If ρ is the density of air, U the speed of the aircraft, and L' the lift per unit span of the wing, then the circulation Γ around the wing can be computed using the infinite-span approximation from *Kutta-Joukowski theorem* (Kuethe and Chow, 1976, Chapter 4).

$$L' = \rho U \Gamma \qquad (2.5.4)$$

But $L' = W/b$ from its definition. The preceding equation then reduces to

$$\Gamma = \frac{W}{b \rho U} \qquad (2.5.5)$$

Kinematics of Line Vortices and Vector Manipulations

Program 2.3 is concerned specifically with the wake of a Boeing 747 aircraft. The span of this jumbo jet is approximately 60 m, and the initial height of the vortices is assumed to be 100 m above the runway. Based on a total weight of 4×10^6 N and a take-off speed of 67 m/s (approximately 240 km/hr), the magnitude of the circulation is 815 m²/s at sea level, where $\rho = 1.22$ kg/m³. The remaining computations involve only the two vector equations (2.5.2) and (2.5.3), which require the following five basic operations: the vectorial addition and subtraction, computing the magnitude of a vector, the product of a scalar and a vector and, finally, the cross product of two vectors. These operations are performed by calling the subprograms VADD, VSBT, VMAG, PDSV, and VCROSS attached to the main program.

A vector is represented in Program 2.3 as a one-dimensional array having three elements, which are used to store, in order, the x, y, and z components of that vector. For instance, the unit vector $\mathbf{i}_z$ is given the name UNIVZ in the program, and assigned to it are the three values 0, 0, and 1. When vector $\mathbf{A}$ is added to vector $\mathbf{B}$, the sum of their corresponding elements, that is, $A(1) + B(1)$, $A(2) + B(2)$, and $A(3) + B(3)$, are first calculated and the values are stored in the three elements of the resultant vector $\mathbf{C}$. These are exactly what the subroutine VADD is doing. The statements contained in the other four subprograms are all self-explanatory and do not need further clarification.

List of Principal Variables in Program 2.3

Program Symbol	Definition
(Main Program)	
B	Wingspan of the airplane, b, m
DRA	$\Delta \mathbf{r}_A$, equal to $\mathbf{V}_A \Delta t$, m
DT	Time increment Δt, s
GAMMA	Magnitude of circulation Γ, m²/s
H	Initial height h of vortices above runway, m
M	Time step counter
RA, RB, RC, RD	Position vectors $\mathbf{r}_A$, $\mathbf{r}_B$, $\mathbf{r}_C$, $\mathbf{r}_D$ of the vortex pair and their images, m
RAB, RAC, RAD	Representing vectors $\mathbf{r}_A - \mathbf{r}_B$, $\mathbf{r}_A - \mathbf{r}_C$, and $\mathbf{r}_A - \mathbf{r}_D$, respectively, m
T	Time t, s
UNIVZ	Unit vector along z-axis, $\mathbf{i}_z$, m
VA	$\mathbf{V}_A$, velocity of the vortex at $\mathbf{r}_A$, m/s
V1, ..., V6	Dummy vectors used to facilitate vectorial operations

```
C                    ***** PROGRAM 2.3 *****

C          COMPUTING THE MOTION OF TWO TRAILING VORTICES OF STRENGTH
C          GAMMA, WHICH INITIALLY ARE AT A HEIGHT H ABOVE THE RUNWAY AND
C          ARE SEPARATED BY A DISTANCE B, THE WINGSPAN OF THE AIRCRAFT

           DIMENSION UNIVZ(3),RA(3),RB(3),RC(3),RD(3),VA(3),RAB(3),RAC(3),
          A          RAD(3),V1(3),V2(3),V3(3),V4(3),V5(3),V6(3),DRA(3)

C          ..... INITIALIZE THE PROBLEM AND PRINT HEADINGS .....
           DATA (UNIVZ(I),I=1,3) / 0.0, 0.0, 1.0 /
           DATA B,H,GAMMA / 60., 100., 815. /
           DATA T,DT,RA(3),RB(3),RC(3),RD(3) / 0.0, 0.05, 4*0.0 /
           RA(1) = -B/2.
           RA(2) = H
           WRITE(6,4)

C          ..... LET M BE THE NUMBER OF TIME STEPS.  AT EACH TIME STEP RB
C                IS OBTAINED FROM RA, RC AND RD ARE IMAGES OF RB AND RA .....
           DO 3 M=1,2001
           RB(1) = -RA(1)
           RB(2) =  RA(2)
           RC(1) =  RB(1)
           RC(2) = -RB(2)
           RD(1) =  RA(1)
           RD(2) = -RA(2)

C          ..... PRINT THE INITIAL CONDITION AND THE RESULT
C                EVERY 40 TIME STEPS THEREAFTER .....
           IF( M.EQ.1 ) GO TO 1
           IF( (M-1)/40*40 .NE. (M-1) ) GO TO 2
         1 WRITE(6,5) T,RA(1),RA(2),RB(1),RB(2)

C          ..... COMPUTE VA BASED ON FORMULA (2.5.2) .....
         2 CALL VSBT( RA, RB, RAB )
           CALL PDSV( 1./VMAG(RAB)**2, RAB, V1 )
           CALL VSBT( RA, RC, RAC )
           CALL PDSV( 1./VMAG(RAC)**2, RAC, V2 )
           CALL VSBT( RA, RD, RAD )
           CALL PDSV( 1./VMAG(RAD)**2, RAD, V3 )
           CALL VSBT( V1, V2, V4 )
           CALL VADD( V4, V3, V5 )
           CALL VCROSS( UNIVZ, V5, V6 )
           CALL PDSV( GAMMA/(2.*3.14159), V6, VA )

C          ..... COMPUTE NEW POSITION RA USING FORMULA (2.5.3),
C                AND GIVE AN INCREMENT TO T .....
           CALL PDSV( DT, VA, DRA )
           CALL VADD( RA, DRA, RA )
           T = T + DT
         3 CONTINUE

         4 FORMAT( 1H1 //// 8X, 1HT, 12X, 5HRA(1), 6X, 5HRA(2), 10X, 5HRB(1),
          A        6X, 5HRB(2) / 6X, 5H(SEC), 11X, 3H(M), 8X, 3H(M),
          B        12X, 3H(M), 8X, 3H(M) / )
         5 FORMAT( F11.2, 4X, 2F11.2, 4X, 2F11.2 )

           END
```

```
      SUBROUTINE VADD( A, B, C )
C     ..... VECTORIAL ADDITION OF A AND B TO FORM VECTOR C .....
      DIMENSION A(3),B(3),C(3)
      DO 1 I=1,3
    1 C(I) = A(I) + B(I)
      RETURN
      END




      SUBROUTINE VSBT( A, B, C )
C     ..... VECTORIAL SUBTRACTION OF B FROM A TO FORM VECTOR C .....
      DIMENSION A(3),B(3),C(3)
      DO 1 I=1,3
    1 C(I) = A(I) - B(I)
      RETURN
      END




      FUNCTION VMAG( V )
C     ..... COMPUTING THE MAGNITUDE OF VECTOR V .....
      DIMENSION V(3)
      VMAG = SQRT( V(1)**2 + V(2)**2 + V(3)**2 )
      RETURN
      END




      SUBROUTINE PDSV( A, B, C )
C     ..... PRODUCT OF SCALAR A AND VECTOR B TO FORM VECTOR C .....
      DIMENSION B(3),C(3)
      DO 1 I=1,3
    1 C(I) = A * B(I)
      RETURN
      END




      SUBROUTINE VCROSS( A, B, C )
C     ..... CROSS PRODUCT OF A AND B TO FORM VECTOR C .....
      DIMENSION A(3),B(3),C(3)
      C(1) = A(2)*B(3) - A(3)*B(2)
      C(2) = A(3)*B(1) - A(1)*B(3)
      C(3) = A(1)*B(2) - A(2)*B(1)
      RETURN
      END
```

102

T (SEC)	RA(1) (M)	RA(2) (M)	RB(1) (M)	RB(2) (M)
0.00	-30.00	100.00	30.00	100.00
2.00	-30.11	96.05	30.11	96.05
4.00	-30.24	92.16	30.24	92.16
6.00	-30.39	88.31	30.39	88.31
8.00	-30.55	84.52	30.55	84.52
10.00	-30.74	80.80	30.74	80.80
12.00	-30.96	77.15	30.96	77.15
14.00	-31.21	73.59	31.21	73.59
16.00	-31.50	70.11	31.50	70.11
18.00	-31.83	66.73	31.83	66.73
20.00	-32.22	63.47	32.22	63.47
22.00	-32.67	60.33	32.67	60.33
24.00	-33.20	57.33	33.20	57.33
26.00	-33.81	54.48	33.81	54.48
28.00	-34.52	51.79	34.52	51.79
30.00	-35.35	49.28	35.35	49.28
32.00	-36.32	46.95	36.32	46.95
34.00	-37.42	44.81	37.42	44.81
36.00	-38.70	42.86	38.70	42.86
38.00	-40.14	41.11	40.14	41.11
40.00	-41.77	39.55	41.77	39.55
42.00	-43.60	38.16	43.60	38.16
44.00	-45.62	36.96	45.62	36.96
46.00	-47.83	35.90	47.83	35.90
48.00	-50.23	34.99	50.23	34.99
50.00	-52.81	34.21	52.81	34.21
52.00	-55.57	33.53	55.57	33.53
54.00	-58.48	32.95	58.48	32.95
56.00	-61.53	32.46	61.53	32.46
58.00	-64.72	32.03	64.72	32.03
60.00	-68.03	31.66	68.03	31.66
62.00	-71.45	31.35	71.45	31.35
64.00	-74.97	31.07	74.97	31.07
66.00	-78.57	30.84	78.57	30.84
68.00	-82.25	30.63	82.25	30.63
70.00	-86.00	30.45	86.00	30.45
72.00	-89.82	30.29	89.82	30.29
74.00	-93.69	30.15	93.69	30.15
76.00	-97.61	30.03	97.61	30.03
78.00	-101.58	29.92	101.58	29.92
80.00	-105.59	29.83	105.59	29.83
82.00	-109.63	29.74	109.63	29.74
84.00	-113.71	29.66	113.71	29.66
86.00	-117.82	29.60	117.82	29.60
88.00	-121.95	29.53	121.95	29.53
90.00	-126.11	29.48	126.11	29.48
92.00	-130.30	29.43	130.30	29.43
94.00	-134.50	29.38	134.50	29.38
96.00	-138.72	29.34	138.72	29.34
98.00	-142.96	29.30	142.96	29.30
100.00	-147.22	29.26	147.22	29.26

When the dot product of two vectors is needed in some other applications, a subprogram similar to those shown here can easily be constructed.

The trajectory of the vortex pair computed by Program 2.3 is also plotted as solid curves in Fig. 2.5.3. It shows that the two vortices first move almost vertically downward, with small influence from their images because of the large height, and then start to spread out as the ground is approached. When the distance between the two becomes large, their mutual interactions diminish so that each of them moves under the strong influence of its own image. During this period they move away from each other at a speed

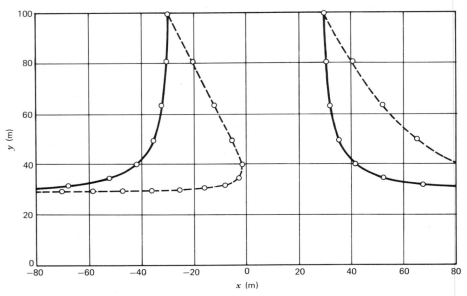

FIGURE 2.5.3 Paths of trailing vortices above runway in a quiet atmosphere (solid curves), and those in the presence of a 1 m/s cross wind (dashed curves). Circles indicate positions at 10-s intervals.

approximately equal to $\Gamma/4\pi\, d$, where d is the height of the vortices above the ground ($\doteq b/2$). The result also shows that the trailing vortices of this airplane will stay in the vicinity of the runway for a few minutes. An analytical description of the trajectory was given by Lamb (1932, p. 223) for the case of the vortex pair starting from an infinite height.

The flow pattern at any instant may be plotted, with the help of Program 2.2, by specifying the strengths and positions of the four vortices. In the meantime, if so desired, the velocity field and the pressure distribution on the ground can also be computed.

Inviscid Fluid Flows

Problem 2.4 Compute the motion of two line vortices separated by a unit distance for each of the following cases.

1. The vortex strengths are in the same direction and of the same magnitude. Show that they will move along a same circular path having a diameter of unit length.

2. The same as case 1 except the direction of one of the vortices is reversed.

3. The vortex strengths are in the same direction, but one is twice as strong as the other. Show that they will move along two concentric circular paths, with the stronger vortex on the inner circle.

4. The same as case 3 except the direction of one of the vortices is reversed.

Problem 2.5 The vortex trajectory above the ground changes in the presence of a cross wind, whose effect is to carry the two vortices in its direction at the wind speed. Write a program to include a horizontal wind of constant speed U in the positive x direction, and run it for several values of U while keeping the initial condition the same as that in Program 2.3.

The result for $U = 1$ m/s is plotted as dashed curves in Fig. 2.5.3 for comparison. It shows that this wind causes one of the vortices to stay above the runway for a much longer period. For very strong winds, both vortices are blown to one side of the runway and will not come back.

2.6 von Kármán's Method for Approximating Flow Past Bodies of Revolution

Although the method of superposition of elementary flows described in Section 2.4 is a powerful tool for generating bodies of various shapes in a flow field, when using this method to construct the flow around a given body, it may require an effort to adjust the positions and strengths of the elementary flows so that the shape of the resulting body is close to what is desired. To avoid this difficulty, a numerical technique was developed by von Kármán (1927) that provides a systematic means for calculating the strengths of a group of sources placed at fixed locations. Even though the method was originally devised for computing the flow around the axisymmetric hull of a dirigible, it can also be applied to planar two-dimensional flow problems. von

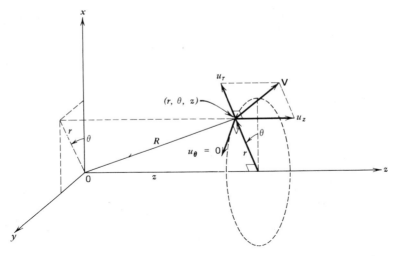

FIGURE 2.6.1 Cylindrical coordinate system for axisymmetric flows.

Kármán's method will be derived in this section and its usage will be illustrated by solving a flow problem with a known analytical solution.

It is natural to adopt the cylindrical coordinate system shown in Fig. 2.6.1 for analyzing axisymmetric flows. For a steady incompressible flow a stream function ψ can still be defined whose value at a given point is proportional to the volume of fluid flowing per unit time through the circular tube whose surface passes through that point. Velocity components are related to the stream function through the following equations.

$$u_r = -\frac{1}{r}\frac{\partial \psi}{\partial z}, \qquad u_z = \frac{1}{r}\frac{\partial \psi}{\partial r} \tag{2.6.1}$$

In terms of stream function, the irrotationality condition has the form

$$\frac{\partial^2 \psi}{\partial r^2} - \frac{1}{r}\frac{\partial \psi}{\partial r} + \frac{\partial^2 \psi}{\partial z^2} = 0 \tag{2.6.2}$$

which, unlike the planar flows described in Section 2.1, is not a Laplace equation.

Elementary solutions to the preceding linear equation can again be sought and examined. For example,

$$\psi = \tfrac{1}{2}Ur^2 \tag{2.6.3}$$

represents a uniform stream of speed U flowing along the z-axis, and the solution

$$\psi = -\frac{mz}{\sqrt{r^2 + z^2}} \qquad (2.6.4)$$

is the stream function of a *point source* at the origin of the coordinate system. By using (2.6.1), it can easily be shown that, corresponding to the stream function (2.6.4), the total velocity $\mathbf{V}$ is pointed radially outward from the origin and its magnitude is equal to m/R^2, where $R = \sqrt{r^2 + z^2}$ is the radial distance from the origin. Thus the constant $m(= VR^2)$ is proportional to the volume of fluid originating from the source per unit time, and therefore is called the strength of the point source. A negative value of m represents the strength of a point sink. Similar to the two-dimensional case, a doublet can be constructed for the axisymmetric flow by letting a point source-sink pair approach each other along the z-axis while keeping the product of the strength and the distance in between a constant μ. The doublet so formed at the origin has the stream function

$$\psi = -\frac{\mu r^2}{(r^2 + z^2)^{3/2}} \qquad (2.6.5)$$

Various bodies of revolution can be generated in a flow by linear combinations of elementary flows with proper strengths and at proper locations.

To approximate the flow resulting from a uniform axial stream of speed U past a given body of revolution, point sources or sinks are distributed uniformly within each of the n properly chosen segments inside the body along the axis of symmetry. The segments, labeled $1, 2, \cdots, n$, as shown in Fig. 2.6.2, generally have different lengths and contain sources or sinks of different densities. The number of segments that are employed depends on the desired degree of accuracy. The end points of segment j are designated $P'_j(0, z'_j)$ and $P'_{j+1}(0, z'_{j+1})$, respectively. The length of this segment is $s_j(= z'_{j+1} - z'_j)$, along which sources of a constant strength q_j per unit length are distributed. The stream function of the flow induced at any point (r, z) by the sources within a small interval $d\zeta$ located at $(0, \zeta)$ on this segment is, from (2.6.4),

$$d\psi = -\frac{q_j(z - \zeta)\, d\zeta}{\sqrt{r^2 + (z - \zeta)^2}}$$

Integration of the right-hand side from z'_j to z'_{j+1} gives the induced stream function at (r, z) caused by the whole source distribution on segment j, which is of the form

$$q_j[\sqrt{r^2 + (z - z'_{j+1})^2} - \sqrt{r^2 + (z - z'_j)^2}]$$

von Kármán's Method

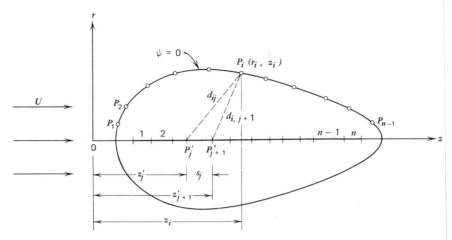

FIGURE 2.6.2 Uniform flow past a body of revolution.

The stream function at a point (r, z) in the flow field consists of two parts, one caused by the uniform stream and the other by the source distributions. Thus

$$\psi(r, z) = \tfrac{1}{2}Ur^2 + \sum_{j=1}^{n} q_j[\sqrt{r^2 + (z - z'_{j+1})^2} - \sqrt{r^2 + (z - z'_j)^2}] \qquad (2.6.6)$$

Now $n - 1$ points $P_1, P_2, \cdots, P_{n-1}$ are selected on the surface of the prescribed body. At a surface point $P_i(r_i, z_i)$ the stream function is, from the general expression (2.6.6),

$$\psi(r_i, z_i) = \tfrac{1}{2}Ur_i^2 + \sum_{j=1}^{n} q_j[\sqrt{r_i^2 + (z_i - z'_{j+1})^2} - \sqrt{r_i^2 + (z_i - z'_j)^2}] \qquad (2.6.7)$$

The left-hand side of the above equation is zero because the point P_i is on the body surface. And, from Fig. 2.6.2, the two terms contained within the brackets in (2.6.7) represent the two distances $d_{i,j+1}$ and d_{ij}, measured respectively from the surface point P_i to the end points P'_{j+1} and P'_j of segment j. With the notation that

$$\Delta d_{ij} = d_{ij} - d_{i,j+1} = \sqrt{r_i^2 + (z_i - z'_j)^2} - \sqrt{r_i^2 + (z_i - z'_{j+1})^2} \qquad (2.6.8)$$

and after replacing q_j/U by Q_j, (2.6.7) becomes

$$\sum_{j=1}^{n} \Delta d_{ij} Q_j = \tfrac{1}{2}r_i^2 \qquad (2.6.9)$$

in which all quantities but the Q_js are known for specified locations of both the source segments and surface points.

Since Q is proportional to the volume of fluid produced per unit time from

108   Inviscid Fluid Flows

sources within a unit length of a segment, it requires that the total strength of the sources or sinks must be zero in order to form a closed body. That is,

$$\sum_{j=1}^{n} s_j Q_j = 0 \tag{2.6.10}$$

Combining the $n-1$ equations obtained by evaluating (2.6.9) at all surface points and the equation (2.6.10), we have a complete set of n simultaneous algebraic equations for the unknown source strengths.

$$\begin{bmatrix} \Delta\, d_{11} & \Delta\, d_{12} & \cdots & \Delta\, d_{1n} \\ \Delta\, d_{21} & \Delta\, d_{22} & \cdots & \Delta\, d_{2n} \\ \cdot & \cdot & \cdots & \cdot \\ \Delta\, d_{n-1,1} & \Delta\, d_{n-1,2} & \cdots & \Delta\, d_{n-1,n} \\ s_1 & s_2 & \cdots & s_n \end{bmatrix} \begin{pmatrix} Q_1 \\ Q_2 \\ \cdot \\ Q_{n-1} \\ Q_n \end{pmatrix} = \begin{pmatrix} r_1^2/2 \\ r_2^2/2 \\ \cdot \\ r_{n-1}^2/2 \\ 0 \end{pmatrix} \tag{2.6.11}$$

Once Q_js are known, the stream function expressed by (2.6.6) becomes determined. According to (2.6.1), the corresponding velocity components are

$$u_r = -\sum_{j=1}^{n} \frac{UQ_j}{r} \left[\frac{z - z'_{j+1}}{\sqrt{r^2 + (z - z'_{j+1})^2}} - \frac{z - z'_j}{\sqrt{r^2 + (z - z'_j)^2}} \right] \tag{2.6.12}$$

$$u_z = U \left\{ 1 + \sum_{j=1}^{n} Q_j \left[\frac{1}{\sqrt{r^2 + (z - z'_{j+1})^2}} - \frac{1}{\sqrt{r^2 + (z - z'_j)^2}} \right] \right\} \tag{2.6.13}$$

They show that the velocities induced by the source distributions are diminishing far away from the body.

The pressure field can be computed from the Bernoulli equation (2.1.7) with the nonsteady term dropped. If P is used to denote the pressure far away from the body, the Bernoulli constant can be evaluated, and (2.1.7) becomes

$$p + \tfrac{1}{2}\rho(u_r^2 + u_z^2) = P + \tfrac{1}{2}\rho U^2$$

Instead of the pressure, it is sometimes more convenient to use the dimensionless *pressure coefficient* c_p, defined as the excess pressure over the free-stream value divided by the free-stream dynamic pressure. Thus

$$c_p = \frac{p - P}{\tfrac{1}{2}\rho U^2} = 1 - \left(\frac{u_r}{U}\right)^2 - \left(\frac{u_z}{U}\right)^2 \tag{2.6.14}$$

and its value at any point can be computed after substitution of the velocity components at that point obtained from (2.6.12) and (2.6.13).

To complete von Kármán's method, we need an algorithm for solving the simultaneous equations (2.6.11). Usually there is more than one library subroutine available in any reasonable computer system for the solution of a set of linear algebraic equations. However, we like to construct our own subprograms, which are used not only in the present section but also in some later ones. For this purpose, a generalized system of equations having the following form is considered.

$$\sum_{j=1}^{n} C_{ij}x_j = A_i, \qquad i = 1, 2, \cdots, n$$

It may also be expressed in matrix form as

$$[C_{ij}](x_j) = (A_i) \qquad (2.6.15)$$

in which $[C_{ij}]$ is an $n \times n$ matrix of the coefficients, and both (x_j) and (A_i) are column matrices of size n. In the problems we are dealing with here, the determinant of the coefficient matrix, designated as $|C_{ij}|$, does not vanish. With this property the system can be solved by employing *Cramer's rule*, which can be stated as follows.

The expression for x_k, the kth element in the solution column matrix (x_j), is the ratio of two determinants, the denominator being $|C_{ij}|$, and the numerator being the determinant of the matrix obtained by replacing the kth column in $[C_{ij}]$ by the column matrix (A_i).

There are many different methods for evaluating the determinant of a square matrix. The procedure we are going to follow is called the *pivotal condensation technique*, which demands the elimination of all the elements below the diagonal elements in the determinant. When this is achieved, the value of the determinant is simply the product of the diagonal elements.

We start with the original form of an arbitrary nth-order determinant $|a_{ij}|$ and write its value as

$$\text{Determ} = \begin{vmatrix} a_{11} & a_{12} & \cdots & a_{1n} \\ a_{21} & a_{22} & \cdots & a_{2n} \\ \cdot & \cdot & \cdots & \cdot \\ a_{n1} & a_{n2} & \cdots & a_{nn} \end{vmatrix}$$

As the first step, all the elements will be eliminated from the first column except the first one, which is at the pivotal position at this moment. This is done by multiplying the rows $i = 2, 3, \cdots, n$ by a_{11} and subtracting from them the first row multiplied respectively by a_{i1}. The value of the determinant is unchanged if each of the $n - 1$ rows below the first is divided by a_{11}. Therefore

$$\text{Determ} = \begin{vmatrix} a_{11} & a_{12} & & \cdots & a_{1n} \\ 0 & a_{22} - a_{12}a_{21}/a_{11} & & \cdots & a_{2n} - a_{1n}a_{21}/a_{11} \\ \cdot & \cdot & & \cdots & \cdot \\ 0 & a_{n2} - a_{12}a_{n1}/a_{11} & & \cdots & a_{nn} - a_{1n}a_{n1}/a_{11} \end{vmatrix}$$

If the elements below the first row are renamed according to the assignment statements

$$a_{ij} \leftarrow (a_{ij} - a_{1j}a_{i1}/a_{11}), \qquad \begin{array}{l} i = 2, 3, \cdots, n \\ j = 2, 3, \cdots, n \end{array}$$

the above equation can be written as

$$\text{Determ} = a_{11} \begin{vmatrix} a_{22} & a_{23} & \cdots & a_{2n} \\ a_{32} & a_{33} & \cdots & a_{3n} \\ \cdot & \cdot & \cdots & \cdot \\ a_{n2} & a_{n3} & \cdots & a_{nn} \end{vmatrix}$$

and the order of the determinant on the right-hand side is reduced by one. The same technique is again used to eliminate the elements below the pivotal element a_{22} in the first column of the above determinant. After renaming the elements, the value of the determinant is then expressed as the product of a_{11}, a_{22}, and a determinant of order $n - 2$.

After the procedure has been repeated m times, the value of the determinant takes the general form

$$\text{Determ} = a_{11} \cdot a_{22} \cdots a_{mm}|a_{ij}| \tag{2.6.16}$$

The elements in the determinant of order $n - m$ on the right-hand side have been computed according to the following relationships.

$$a_{ij} \leftarrow (a_{ij} - a_{mj}a_{im}/a_{mm}), \qquad \begin{array}{l} i = m + 1, \cdots, n \\ j = m + 1, \cdots, n \end{array} \tag{2.6.17}$$

All the elements shown on the right of the arrow are those obtained in the previous step. When the value of m becomes $n - 1$, the determinant on the right side of (2.6.16) has only one element left. At this final step of condensation, the required value is found to be

$$\text{Determ} = a_{11} \cdot a_{22} \cdots a_{nn} \tag{2.6.18}$$

Cramer's rule is programmed under a subroutine called CRAMER (C, A, X, N), whose arguments are named according to (2.6.15). C is the coefficient matrix, A is the constant vector on the right-hand side of that equation, X is the solution vector, and N is the number of unknowns or the size of the system. To compute the determinant of a matrix within this subroutine, a function DETERM (ARRAY, N) is called on, in which ARRAY and N are, respectively, the name and order of the matrix. The value of DETERM is calculated from (2.6.18) after the computations (2.6.17) have been executed for $m = 1, 2, \ldots, n - 1$. Both subprograms are attached to Program 2.4, which is constructed for solving a particular problem using von Kármán's method.

List of Principal Variables in Program 2.4

Program Symbol	Definition
(Main Program)	
A	Radius of the sphere, a
CONST	Column matrix on the right side of (2.6.11)
CP, CPEX	Pressure coefficients and their exact values, respectively, at the surface points
D	d_{ij} defined in (2.6.8)
DD	Coefficient matrix on the left side of (2.6.11)
PHI	Angular positions of the surface points defined in (2.6.22)
PI	π
Q	Q_j, source strengths q_j divided by U
R, Z	r_i and z_i, respectively, coordinates of the surface points
S	Lengths of source segments, s_j
U	Free stream speed
UR, UZ	Velocity components u_r and u_z at the surface points
V, VEX	Magnitudes of fluid velocity and their exact values, respectively, at the surface points
ZZ	z'_j, coordinates of the end points of source segments on the z-axis

```
C                    ***** PROGRAM 2.4 *****
C           APPROXIMATING THE FLOW OF A UNIFORM STREAM PAST
C           A SPHERE BY THE USE OF VON KARMAN,S METHOD

      DIMENSION CONST(10),CP(9),CPEX(9),D(9,11),DD(10,10),PHI(9),Q(10),
     A          R(9),S(10),V(9),VEX(9),Z(9),ZZ(11)
      PI = 3.1415927
      A  = 1.0
      U  = 1.0

C     ..... COMPUTE THE POSITIONS ON THE Z-AXIS OF THE END POINTS OF
C           SOURCE SEGMENTS, WHICH ARE OF THE SAME LENGTH OF 0.16A .....
      ZZ(1) = -0.8 * A
      DO 1 I=2,11
      IM1 = I-1
      ZZ(I) = ZZ(IM1) + 0.16*A
    1 S(IM1) = 0.16 * A

C     ..... COMPUTE THE COORDINATES OF THE SURFACE POINTS
C           ACCORDING TO (2.6.22 AND 23) .....
      DO 2 I=1,9
      PHI(I) = ( 10. - FLOAT(I) ) * PI / 10.
      R(I) = A * SIN(PHI(I))
    2 Z(I) = A * COS(PHI(I))
```

```
C          ..... COMPUTE D(I,J) AND DD(I,J) ACCORDING TO (2.6.8) .....
           DO 3 I=1,9
           DO 3 J=1,11
         3 D(I,J) = SQRT( R(I)**2+(Z(I)-ZZ(J))**2 )
           DO 4 I=1,9
           DO 4 J=1,10
         4 DD(I,J) = D(I,J) - D(I,J+1)

C          ..... IN (2.6.11), CALL THE COEFFICIENT MATRIX DD(I,J) AND CALL
C                THE COLUMN MATRIX ON THE RIGHT CONST(I).  PRINT OUT THESE
C                TWO MATRICES AFTER ALL ELEMENTS HAVE BEEN COMPUTED .....
           WRITE(6,10)
           DO 5 J=1,10
         5 DD(10,J) = S(J)
           DO 6 I=1,9
         6 CONST(I) = R(I)**2 / 2.0
           CONST(10) = 0.0
           DO 7 I=1,10
         7 WRITE(6,11)   (DD(I,J),J=1,10), CONST(I)

C          ..... SOLVE FOR SOURCE STRENGTHS USING CRAMER,S RULE,
C                THEN PRINT OUT THEIR VALUES .....
           CALL CRAMER( DD, CONST, Q, 10 )
           WRITE(6,12)   Q

C          ..... COMPUTE VELOCITY AND PRESSURE COEFFICIENT AT EACH SURFACE
C                POINT USING (2.6.12, 13, AND 14) AND THE EXACT VALUES USING
C                (2.6.20 AND 21).  PRINT OUT THE RESULTS FOR COMPARISON .....
           WRITE(6,13)
           DO 9 I=1,9
           UR = 0.0
           UZ = 1.0
           DO 8 J=1,10
           UR = UR + Q(J)/R(I)*((Z(I)-ZZ(J+1))/D(I,J+1)-(Z(I)-ZZ(J))/D(I,J))
         8 UZ = UZ + Q(J)*(1./D(I,J+1)-1./D(I,J))
           V(I) = U * SQRT(UR**2+UZ**2)
           VEX(I) = 1.5*U*R(I)/A
           CP(I) = 1. - (V(I)/U)**2
           CPEX(I) = 1. - (VEX(I)/U)**2
         9 WRITE(6,14)   R(I),Z(I),V(I),VEX(I),CP(I),CPEX(I)

        10 FORMAT( 1H1//9X45HTHE SQUARE MATRIX ON THE LEFT AND THE COLUMN ,
           A          19HMATRIX ON THE RIGHT / 10X17HOF (2.6.11) ARE : // )
        11 FORMAT( 10F7.3, 3X, F7.3 )
        12 FORMAT( ////// 19X, 27HSOURCE STRENGTHS Q(I) ARE,
           A          20HFOR I = 1,2,...,10 : // 10F8.3 )
        13 FORMAT( ////// 9X46HCOMPARISON OF NUMERICAL RESULT WITH THE EXACT ,
           A          19HAT 9 SURFACE POINTS // 12X4HR(I), 6X4HZ(I), 8X4HV(I),
           B          4X6HVEX(I), 7X5HCP(I), 4X7HCPEX(I) / 12X4H----, 6X4H----,
           C          8X4H----, 4X6H------, 7X5H-----, 4X7H------- )
        14 FORMAT( 7X, 2F10.4, 3X, 2F9.4, 3X, 2F10.4 )

           END



           FUNCTION DETERM( ARRAY, N )
C          ..... DETERM IS THE VALUE OF THE DETERMINANT OF AN NXN MATRIX
C                CALLED ARRAY.  IT IS REPLACED FIRST BY THE MATRIX A SO THAT
C                WHEN RETURNED TO THE MAIN PROGRAM THE MATRIX ARRAY RETAINS
C                ITS ORIGINAL FORM .....
           DIMENSION ARRAY(10,10),A(10,10)
           DO 1 I=1,N
           DO 1 J=1,N
         1 A(I,J) = ARRAY(I,J)
```

113

```
C        ..... EXECUTE (2.6.17) FOR PIVOTAL CONDENSATION .....
         M = 1
       2 K = M+1
         DO 3 I=K,N
         RATIO = A(I,M)/A(M,M)
         DO 3 J=K,N
       3 A(I,J) = A(I,J) - A(M,J)*RATIO
         IF( M.EQ.N-1 )  GO TO 4
         M = M+1
         GO TO 2

C        ..... COMPUTE THE VALUE OF THE DETERMINANT BY USING (2.6.18) .....
       4 DETERM = 1.0
         DO 5 L=1,N
       5 DETERM = DETERM * A(L,L)
         RETURN
         END

         SUBROUTINE CRAMER( C, A, X, N )
C        ..... SOLVING A SET OF N SIMULTANEOUS LINEAR ALGEBRAIC EQUATIONS
C              (2.6.15) BY USING CRAMER,S RULE.  C IS THE COEFFICIENT
C              MATRIX, A IS THE CONSTANT VECTOR ON THE RIGHT, AND X IS
C              THE SOLUTION VECTOR .....
         DIMENSION C(10,10),CC(10,10),A(10),X(10)
         DENOM = DETERM( C, N )
         DO 3 K=1,N
         DO 1 I=1,N
         DO 1 J=1,N
       1 CC(I,J) = C(I,J)
         DO 2 I=1,N
       2 CC(I,K) = A(I)
       3 X(K) = DETERM( CC, N ) / DENOM
         RETURN
         END
```

THE SQUARE MATRIX ON THE LEFT AND THE COLUMN MATRIX ON THE RIGHT
OF (2.6.11) ARE :

-.094	-.125	-.139	-.147	-.151	-.153	-.155	-.156	-.157	-.157	.048
-.024	-.062	-.091	-.111	-.124	-.133	-.140	-.144	-.147	-.149	.173
.026	-.005	-.036	-.063	-.085	-.102	-.114	-.124	-.131	-.136	.327
.063	.041	.015	-.012	-.037	-.060	-.080	-.095	-.108	-.117	.452
.093	.078	.059	.037	.013	-.013	-.037	-.059	-.078	-.093	.500
.117	.108	.095	.080	.060	.037	.012	-.015	-.041	-.063	.452
.136	.131	.124	.114	.102	.085	.063	.036	.005	-.026	.327
.149	.147	.144	.140	.133	.124	.111	.091	.062	.024	.173
.157	.157	.156	.155	.153	.151	.147	.139	.125	.094	.048
.160	.160	.160	.160	.160	.160	.160	.160	.160	.160	0.000

SOURCE STRENGTHS Q(I) ARE, FOR I = 1,2,....10 :

.012   -.142   .873   -4.385   29.200   -29.200   4.385   -.873   .142   -.012

COMPARISON OF NUMERICAL RESULT WITH THE EXACT AT 9 SURFACE POINTS

R(I)	Z(I)	V(I)	VEX(I)	CP(I)	CPEX(I)
----	----	----	------	-----	-------
.3090	-.9511	.4634	.4635	.7853	.7851
.5878	-.8090	.8817	.8817	.2226	.2226
.8090	-.5878	1.2135	1.2135	-.4726	-.4726
.9511	-.3090	1.4266	1.4266	-1.0352	-1.0351
1.0000	-.0000	1.5000	1.5000	-1.2500	-1.2500
.9511	.3090	1.4266	1.4266	-1.0352	-1.0351
.8090	.5878	1.2135	1.2135	-.4726	-.4726
.5878	.8090	.8817	.8817	.2226	.2226
.3090	.9511	.4634	.4635	.7853	.7851

114

In using the pivotal condensation method to find the value of a determinant, the element at the pivotal position must not vanish at any step of execution, as indicated by (2.6.17). In order to obtain a more accurate result, (2.6.17) indicates that the pivotal element a_{mm} should be the largest among all elements in the same column and the same row. Our function DETERM could be improved by interchanging the pivotal column or row with another until the pivotal element assumes the maximum value.

Various methods for solving the system (2.6.15) can be found in books on numerical analyses. For example, a subroutine is shown in Section 5.4 of Carnahan, Luther, and Wilkes (1969) that is programmed for solving simultaneous algebraic equations using the method of matrix inversion with the maximum pivot strategy. It is a more efficient subroutine; however, its algorithm is much more complicated than that shown in the subroutines presented here.

To test the accuracy of von Kármán's method, it will be applied to a problem having a known solution. Such a problem can be formed by adding a uniform flow to a point doublet, described respectively by (2.6.3) and (2.6.5). By assigning the value $Ua^3/2$ to the strength μ of the doublet, the stream function of the combined flow becomes

$$\psi = \tfrac{1}{2}Ur^2\left(1 - \frac{a^3}{R^3}\right) \tag{2.6.19}$$

where $R(=\sqrt{r^2 + z^2})$ is the radial distance from the origin. It shows that the streamline (actually the stream surface) $\psi = 0$ consists of the z-axis, along which $r = 0$, and the surface of a sphere of radius a. It can be verified, with the help of (2.6.1), that the magnitude of the velocity at a point (r, z) on the sphere is

$$|\mathbf{V}| = \frac{3}{2}U\frac{r}{a} \tag{2.6.20}$$

The pressure coefficient at that point is, according to (2.6.14),

$$c_p = 1 - \frac{9}{4}\left(\frac{r}{a}\right)^2 \tag{2.6.21}$$

For the numerical scheme of Program 2.4, sources are distributed within ten segments of equal lengths on the z-axis between $z = \pm 0.8a$, and nine equally spaced points are chosen on the upper surface of the sphere, as shown in Fig. 2.6.3. If φ_i denotes the angle between the z-axis and the radial position of the surface point $P_i(r_i, z_i)$, the nine values of φ_i are

$$\varphi_i = (10 - i)\frac{\pi}{10}, \qquad i = 1, 2, \cdots, 9 \tag{2.6.22}$$

von Kármán's Method

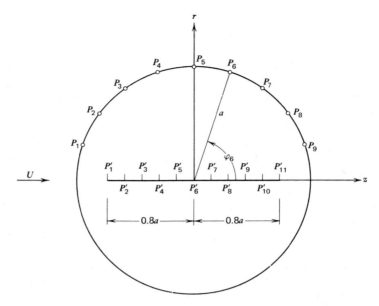

FIGURE 2.6.3 Source segments and surface points used in program 2.4.

from which the coordinates are computed through the relationships

$$r_i = a \sin \varphi_i \qquad \text{and} \qquad z_i = a \cos \varphi_i \qquad (2.6.23)$$

After the source strengths are determined, the velocities and pressure coefficients at the surface points will be calculated and compared with the exact values represented by (2.6.20) and (2.6.21).

It is more convenient to nondimensionalize length and velocity by using the characteristic quantities a and U, respectively. This is achieved by letting $a = 1$ and $U = 1$ in Program 2.4.

The output of Program 2.4 shows that the velocity and pressure coefficient obtained by using von Kármán's method are almost identical to those computed from the exact solution, except at a few points where discrepancies appear at the fourth decimal place. In von Kármán's original work the body of a ZR III dirigible was generated in a uniform flow by using a group of sources near the bow and a separate group of sinks near the stern. The tacit assumption was that the contributions of the bow sources to the pressure distribution over the aft section and that of the stern sinks over the forward section would be negligible. The calculated pressure coefficient was in good

116   Inviscid Fluid Flows

agreement with the measured pressure coefficient without counting the discrepancy in the neighborhood of the gondola.

Program 2.4 can be modified to compute the flow around a body of revolution of any shape. The user needs only to respecify the positions of both source segments and surface points and to change the affected statements if the number of source segments is different from 10, the number presently used in that program.

Problem 2.6 It works as well if the source segments of Fig. 2.6.2 are replaced by discrete point sources of unknown strengths. Write a program for this purpose and use it to solve the sphere problem stated in Program 2.4.

Problem 2.7 Extend von Kármán's method to solve the problem of uniform flow past a two-dimensional closed body of arbitrary shape, which is symmetric about its centerline parallel to the oncoming flow.

In this case line sources and sinks described by (2.4.2) will be distributed within small intervals along the centerline. In the evaluation of the integral over a source segment to find the stream function at a point, it is easier to replace the integrand $\tan^{-1}[(y - y_0)/(x - x_0)]$ by $\cot^{-1}[(x - x_0)/(y - y_0)]$.

After a procedure has been developed for the numerical computation, it is to be applied to calculate the fluid speed, V, and pressure coefficient, c_p, on the surface of a two-dimensional wing at zero angle of attack, whose cross section is a symmetric NACA 0009 airfoil. The shapes and characteristics of this and many other airfoils designated by the National Advisory Committee for Aeronautics are described in the book by Abbott and von Doenhoff (1949). The following table shows the dimensionless coordinates of some selected points on the airfoil and the dimensionless speeds there, in which c is the chord length of the wing and U is the free-stream speed in the x direction.

x/c	0.	0.050	0.100	0.200	0.300	0.400	0.500	0.600	0.700	0.800	0.900	1.000
y/c	0.	0.027	0.035	0.043	0.045	0.044	0.040	0.034	0.027	0.020	0.011	0.
V/U	0.	1.140	1.144	1.137	1.119	1.100	1.082	1.061	1.043	1.018	0.982	0.

The tabulated values of V/U, obtained by employing a different method, are included for comparison with the numerical result.

2.7   Inverse Method II: Conformal Mapping

In considering steady two-dimensional potential flows in Section 2.1, it was shown that the velocity components could be derived from either the velocity potential or the stream function, both of which satisfy the Laplace equation. The relationships involved there are repeated as follows.

$$u = \frac{\partial \phi}{\partial x} = \frac{\partial \psi}{\partial y} \tag{2.7.1}$$

$$v = \frac{\partial \phi}{\partial y} = -\frac{\partial \psi}{\partial x} \tag{2.7.2}$$

$$\frac{\partial^2 \phi}{\partial x^2} + \frac{\partial^2 \phi}{\partial y^2} = 0 \tag{2.7.3}$$

$$\frac{\partial^2 \psi}{\partial x^2} + \frac{\partial^2 \psi}{\partial y^2} = 0 \tag{2.7.4}$$

The $\phi = $ constant and $\psi = $ constant curves are called, respectively, the *equipotential lines* and the *streamlines* of the flow. At a point where an equipotential line intersects a streamline, the vectors that are respectively normal to the two curves are perpendicular to each other, since

$$\nabla \phi \cdot \nabla \psi = \frac{\partial \phi}{\partial x} \frac{\partial \psi}{\partial x} + \frac{\partial \phi}{\partial y} \frac{\partial \psi}{\partial y}$$
$$= (u)(-v) + (u)(v) = 0$$

Thus the equipotential lines and the streamlines of a two-dimensional flow form two families of mutually orthogonal curves in the x-y plane, except at stagnation points where the directions of the normal vectors are not defined.

We now introduce a complex variable $z = x + iy$, where $i = \sqrt{-1}$ is called the imaginary unit, and construct with velocity potential and stream function a complex function

$$w(z) = \phi(x, y) + i\psi(x, y) \tag{2.7.5}$$

If w is a single-valued function of z, as assumed in the above form, and if its derivative of the first order, dw/dz, exists throughout the region occupied by the fluid, then the derivative at any point in that region can be computed regardless of the way in which the increment in z tends to zero. Let $\Delta z = \Delta x + i\,\Delta y$ be such an increment. To find the derivative we may choose to let Δy equal zero and then let Δx approach zero (see Churchill, 1948, Section

18). By choosing such a path, $\Delta z = \Delta x$ and

$$\frac{dw}{dz} = \lim_{\Delta x \to 0} \frac{\phi(x + \Delta x, y) - \phi(x, y) + i\psi(x + \Delta x, y) - i\psi(x, y)}{\Delta x}$$

$$= \frac{\partial \phi}{\partial x} + i \frac{\partial \psi}{\partial x} \qquad (2.7.6)$$

Or, alternatively, we may choose to let Δx equal zero and then let Δy approach zero. Thus $\Delta z = i \Delta y$ and

$$\frac{dw}{dz} = \lim_{\Delta y \to 0} \frac{\phi(x, y + \Delta y) - \phi(x, y) + i\psi(x, y + \Delta y) - i\psi(x, y)}{i \Delta y}$$

$$= -i \frac{\partial \phi}{\partial y} + \frac{\partial \psi}{\partial y} \qquad (2.7.7)$$

Equating the right-hand members of (2.7.6) and (2.7.7) results in

$$\frac{\partial \phi}{\partial x} = \frac{\partial \psi}{\partial y} \quad \text{and} \quad \frac{\partial \phi}{\partial y} = -\frac{\partial \psi}{\partial x} \qquad (2.7.8)$$

These equations are called the *Cauchy-Riemann conditions*, which are the necessary conditions for the existence of the derivative of the function $w(z)$. But these conditions are satisfied automatically by (2.7.1) and (2.7.2) derived for a potential flow. It means that $\phi + i\psi$, the function on the right-hand side of (2.7.5), is an *analytic function* of z, that is, a function that is differentiable for the values of z in a region of the z plane. If a function is analytic in the neighborhood of a point but not at the point itself, this point is called a singular point, or a singularity, of that function. Furthermore, (2.7.3) and (2.7.4) are obtained after eliminating either ψ or ϕ from the Cauchy-Riemann conditions (2.7.8), indicating that both real and imaginary parts of an analytic function satisfy the Laplace equation.

Conversely, if we choose an arbitrary analytic function $w(z)$, the real and imaginary parts of the function then automatically qualify as the velocity potential and the stream function, respectively, of a potential flow in the x-y plane. The function w shown in (2.7.5) is called the *complex potential*, whose derivative is related to the velocity components through the following relationship, derived from either (2.7.6) or (2.7.7).

$$\frac{dw}{dz} = u - iv \qquad (2.7.9)$$

Hence the magnitude of the velocity vector is obtained by taking the absolute value of the above expression.

As an example, let us consider a simple analytic function

$$w(z) = Uz\, e^{-i\alpha} \qquad (2.7.10)$$

Inverse Method II: Conformal Mapping

in which U and α are constants. After being separated into real and imaginary parts, (2.7.10) becomes

$$w = U(x \cos \alpha + y \sin \alpha) + iU(y \cos \alpha - x \sin \alpha)$$

The imaginary part is precisely the stream function of the uniform flow described by (2.4.1) and Fig. 2.4.1a. It follows that the real part represents the velocity potential and the right side of (2.7.10) represents the complex potential of that uniform flow. To find the fluid velocity components, we simply use (2.7.9) to obtain $u = U \cos \alpha$ and $v = U \sin \alpha$. It can also be shown in a similar fashion that the functions $(\Lambda/2\pi) \log (z - z_0)$, $(i\Gamma/2\pi) \log (z - z_0)$, and $\kappa/2\pi(z - z_0)$ are, respectively, the complex potentials of the source, vortex, and doublet described by (2.4.2) to (2.4.4), where $z_0 = x_0 + iy_0$. Each of these three functions has a singular point at z_0, where the first derivative of the function (or the velocity) is unbounded.

We have thus arrived at another inverse method, through which we can generate as many potential flows as we can write analytic functions. The principle of superposition of elementary flows can still be applied in generating new flows. For example, the sum of Uz, the complex potential of a uniform horizontal flow, and $\kappa/2\pi z$, the complex potential of a doublet at the origin, represents the complex potential of a uniform flow past a circular cylinder of radius $\sqrt{\kappa/2\pi U}$ centered at the origin.

There exists another powerful method, called the *method of conformal mapping*, that is used to generate new flow patterns by mapping a known flow in the x-y plane into a new x'-y' plane through various transformations between the two sets of coordinates. Suppose that the complex potential $w(z)$ in (2.7.5) is known for a given flow. Let the variable $z'(= x' + iy')$ be related to z through the relationship

$$z = f(z') \tag{2.7.11}$$

where f is an analytic function of z'. Upon the use of (2.7.11), w becomes an analytic function of z' because the derivative

$$\frac{dw}{dz'} = \frac{dw}{dz} \frac{df}{dz'} \tag{2.7.12}$$

exists, due to the existence of the two derivatives on the right-hand side of (2.7.12). After being transformed into the z' plane, the complex potential may be grouped into real and imaginary parts according to the new coordinates, so that

$$w[f(z')] = \phi'(x', y') + i\psi'(x', y') \tag{2.7.13}$$

It follows from the properties of analytic functions that ϕ' and ψ' must satisfy the Cauchy-Riemann conditions and, therefore, the Laplace equations.

$$\frac{\partial^2\phi'}{\partial x'^2} + \frac{\partial^2\phi'}{\partial y'^2} = 0$$

$$\frac{\partial^2\psi'}{\partial x'^2} + \frac{\partial^2\psi'}{\partial y'^2} = 0$$

Thus the $\phi' = $ constant and $\psi' = $ constant lines remain mutually orthogonal in the x'-y' plane after the transformation, and (2.7.13) represents the complex potential of a flow in the x'-y' plane.

Furthermore, because (2.7.5) and (2.7.13) represent the same complex function, we must have

$$\phi(x, y) = \phi'(x', y')$$

$$\psi(x, y) = \psi'(x', y')$$

except that some additive constants may appear in these equations. Hence, under the transformation (2.7.11), a streamline $\psi = c$ and a perpendicular equipotential line $\phi = k$ in the x-y plane map, respectively, into a streamline $\psi' = c$ and an equipotential line $\phi' = k$ in the x'-y' plane. The two curves, although distorted in shape, remain perpendicular to each other after the transformation. For this reason the mapping is said to be conformal. Theorems related to conformal mapping and their proofs are referred to Chapter 8 of the book by Churchill (1948).

To illustrate the use of this method, we consider the complex potential

$$\begin{aligned} w(z) &= Uz \\ &= Ux + iUy \end{aligned} \tag{2.7.14}$$

of a uniform stream with speed U flowing along the positive x direction. By choosing a specific form

$$z = z'^2 \tag{2.7.15}$$

for the transformation (2.7.11), the complex potential becomes

$$\begin{aligned} w(z') &= Uz'^2 \\ &= U(x'^2 - y'^2) + i2Ux'y' \end{aligned} \tag{2.7.16}$$

Thus, the streamlines $y = c$ and the equipotential lines $x = k$ in the x-y plane are mapped into streamlines $2x'y' = c$ and equipotential lines $x'^2 - y'^2 = k$ in the x'-y' plane, as shown, respectively, by the solid and dashed lines in Fig. 2.7.1. The pattern describes the flow past a 90° corner.

In general the function on the right side of (2.7.11) may be expanded in a series of positive and negative powers of z' and, in this respect, the trans-

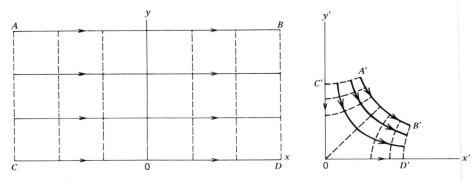

FIGURE 2.7.1 Mapping of a uniform flow in x-y plane into x'-y' plane by the transformation $z = z'^2$.

formation (2.7.15) is merely a special case of the series expansion. If two terms of the following particular form are retained in the series, the relation

$$z = z' + \frac{a^2}{z'} \tag{2.7.17}$$

is called the *Joukowski transformation*. Under this transformation the complex potential of a uniform flow, (2.7.14), becomes

$$w(z') = U\left(z' + \frac{a^2}{z'}\right) \tag{2.7.18}$$

which is the complex potential for a uniform flow past a circular cylinder of radius a. The same result was obtained earlier by adding the complex potential of a uniform flow to that of a doublet.

The Joukowski transformation is probably one of the most widely used transformations in solving potential flow problems. For instance, the transformation

$$z = z' + \frac{b^2}{z'} \tag{2.7.19}$$

where $b^2 < a^2$, maps a circle of radius a with center at the origin of the x', y' coordinates into an ellipse in the x-y plane (Fig. 2.7.2). The same transformation maps two circles intersecting the positive x'-axis at $x' = b$, with the center of one circle on the y'-axis and that of the other off the coordinate axes, into a circular arc and a so-called *Joukowski airfoil* in the x-y plane (Fig. 2.7.3). The detailed description of these mappings can be found in Chapter 4 of Yih (1969).

Suppose that the flow about the elliptic cylinder of Fig. 2.7.2 is to be sought.

Inviscid Fluid Flows

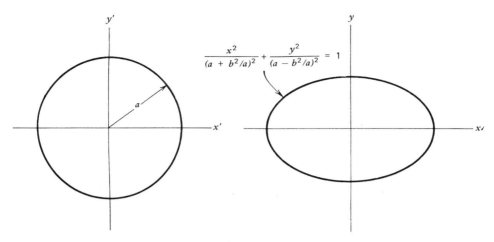

$$\frac{x^2}{(a + b^2/a)^2} + \frac{y^2}{(a - b^2/a)^2} = 1$$

FIGURE 2.7.2 Mapping of a circle in x'-y' plane into an ellipse in x-y plane by the Joukowski transformation $z = z' + b^2/z'$.

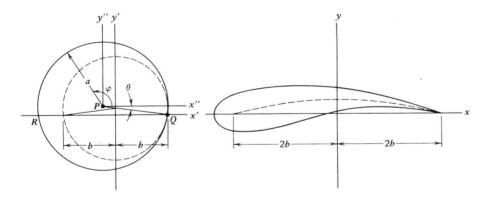

FIGURE 2.7.3 Mapping of two circles in x'-y' plane into a circular arc and an airfoil in x-y plane by the Joukowski transformation $z = z' + b^2/z'$.

A relationship that expresses z' as a function of z may first be obtained by taking the inverse of (2.7.19). Upon substitution of this relationship into the expression (2.7.18) for the complex potential of the flow about a circular cylinder, we obtain the complex potential of the flow about the elliptic cylinder. Streamlines can then be plotted by assigning various constant values to the stream function, or the imaginary part of the transformed complex potential. This procedure, however, involves tedious analytical work, especially when separating a complicated function into real and imaginary parts. For

Inverse Method II: Conformal Mapping   **123**

this reason we will choose a different approach by solving the problem numerically.

In the computer program to be constructed, we will find the coordinates of points along some streamlines of the flow before the mapping. This given flow is usually in a relatively simple form, and its characteristics are well known. Some common examples are the uniform flow and the flow around a circle. These points will be mapped according to the specified transformation and will be plotted on the transformed plane. Such a plot will display the pattern of the transformed flow if the number of points is large. In a similar fashion the equipotential lines, velocity, or pressure distribution of the transformed flow may also be plotted.

Our illustrative example is concerned with a uniform flow U in the positive x direction past the Joukowski airfoil shown in Fig. 2.7.3. The airfoil is the image in the x-y plane of the circle of radius a centered at the point P in the x'-y' plane and, in particular, its sharp trailing edge at $x = 2b$ is the image of the point Q at $x' = b$. The flow about the airfoil is obtained from the flow of uniform speed U past the circle using the transformation (2.7.19).

The magnitude of the velocity $\mathbf{V}$ in the x-y plane is related to that of $\mathbf{V}'$ in the x'-y' plane through the equation

$$\left|\frac{dw}{dz}\right| = \left|\frac{dw}{dz'}\right| \bigg/ \left|\frac{dz}{dz'}\right|$$

By the use of (2.7.9) and (2.7.19), it becomes

$$V = \frac{V'}{\left|1 - \left(\dfrac{b}{z'}\right)^2\right|} \tag{2.7.20}$$

This equation shows that the flow speed at the trailing edge of the airfoil becomes infinite if the speed at Q (where $z' = b$) is of a finite magnitude. Physically such a situation is not allowed. Based on observations of bodies with a sharp trailing edge moving through a fluid, the *Kutta condition* states that the flow speed at the trailing edge must be zero if the trailing-edge angle is finite, or the speed there must be finite if the trailing-edge angle is zero. In the present case of a Joukowski airfoil, the Kutta condition is fulfilled only when the speed at Q vanishes. In other words, the airfoil will create about itself a clockwise circulation Γ whose strength is just sufficient to make the point Q a stagnation point on the circular cylinder in the x'-y' plane. Let $z'_P(=x'_P + iy'_P)$ be the position of the point P in the x'-y' plane and let θ be the angle between the line PQ and the direction of the free stream; as indicated in Fig. 2.7.3, the magnitude of the circulation is (Kuethe and Chow, 1976, Section 4.7)

$$\Gamma = 4\pi a U \sin \theta$$

Since $\sin \theta = y'_P/a$, this equation becomes

$$\Gamma = 4\pi y'_P U \tag{2.7.21}$$

The flow about the circle is then constructed by adding to the uniform stream a doublet and a line vortex. In the x'', y'' coordinate system whose origin coincides with P, the complex potential of the resultant flow is

$$w = U\left[z'' + \frac{a^2}{z''} + i2y'_P \log\left(\frac{z''}{a}\right)\right] \tag{2.7.22}$$

where $z'' = x'' + iy''$, and a constant $-i2y'_P \log a$ has been added so that the value of the stream function on the circle remains unchanged after the introduction of the vortex. The complex potential can easily be expressed in terms of z' by using the transformation

$$z'' = z' - z'_P \tag{2.7.23}$$

The shape of the airfoil is controlled by varying a, b and the coordinates of the center P; a and b determine the thickness envelope and chord length, respectively, while the height of P determines the maximum camber of the airfoil. By inspecting Fig. 2.7.3, these variables are related through the equation

$$a^2 = y'^2_P + (b - x'_P)^2$$

For a given value of a, we may assume a set of values for two of the quantities on the right-hand side of the previous equation, say, for b and y'_P, and compute the value of the third, or,

$$x'_P = b - \sqrt{a^2 - y'^2_P} \tag{2.7.24}$$

x'_P must be negative in order to produce an airfoil having its sharp edge on the downstream side, similar to the one shown in Fig. 2.7.3. If y'_P is chosen to be zero, the circle will be mapped into an uncambered airfoil symmetric about the x-axis.

The values $U = 1$ m/s, $a = 1$ m, $b = 0.8$ m, and $y'_P = 0.199$ m are used in Program 2.5. A rectangular grid system is first defined in the x'-y' plane bounded by x'_{min}, x'_{max}, y'_{min}, and y'_{max}. Stream function is computed at all the grid points using the imaginary part of (2.7.22). The subroutine SEARCH, which was constructed in Program 2.2, is then called to locate the points along some representative streamlines. Another subroutine named MAPPNG is constructed for the purpose of mapping these points from the x'-y' plane into the x-y plane through Joukowski transformation (2.7.19). This subroutine also serves the purpose of throwing away the points falling inside the circle in the x'-y' plane and those outside of the space bounded by x_{min}, x_{max}, y_{min}, and y_{max} in the x-y plane, which are the boundaries of the region to be

List of Principal Variables in Program 2.5

Program Symbol	Definition
(Main Program)	
A	a, radius of the circle in the x'-y' plane that maps into Joukowski airfoil
B	b, the constant in the transformation (2.7.19)
CI	i or $\sqrt{-1}$, the imaginary unit
CONST	A one-dimensional array containing the values assigned to the streamlines to be plotted
CP	c_p, pressure coefficient
DPHI	Increment in angle φ
DX1, DY1	Grid sizes in x' and y' directions, respectively
M, N	Numbers of grid lines in x' and y' directions, respectively
PHI	φ, the angle shown in Fig. 2.7.3
PI	π
PSI	ψ, stream function
TL	A one-dimensional array containing the characters used for the title of the graph
U	Speed of the uniform flow
V	Magnitude of fluid velocity
X, Y	x and y, coordinates in x-y plane
X1, Y1	x' and y', coordinates in x'-y' plane
XL, YL	One-dimensional arrays containing, respectively, the characters used for labeling the x- and y-axes of the graph
XMIN, XMAX, YMIN, YMAX	Limiting values on the x- and y-axes, respectively, used to specify the region to be plotted
XMIN1, XMAX1, YMIN1, YMAX1	Limiting values on the x'- and y'-axes, respectively, used to specify the region of the flow before transformation
XP1, YP1	x'_P and y'_P, respectively, the coordinates of point P in Fig. 2.7.3
Z	$z(= x + iy)$
Z1	$z'(= x' + iy')$
Z2	$z''(= z' - z'_P)$
ZP1	$z'_P(= x'_P + iy'_P)$

```
C                    ***** PROGRAM 2.5 *****

C           TRANSFORMATION FROM THE FLOW PAST A CIRCULAR CYLINDER TO
C           THAT AROUND A JOUKOWSKI AIRFOIL THROUGH CONFORMAL MAPPING


            COMPLEX CI,Z,Z1,Z2,ZP1
            COMMON M,N,DY1 /GROUP/A,B,ZP1
            DIMENSION X(200),Y(200),X1(121),Y1(121),PSI(121,121),CONST(10),
           A          XL(8),YL(8),TL(8),PHI(200),CP(200)
            DATA XMAX,XMIN,YMAX,YMIN/ 3., -3., 3., -3. /
            DATA XMAX1,XMIN1,YMAX1,YMIN1/ 3., -3., 3., -3. /


C           ..... COMPUTE XP1 AND ZP1 FROM GIVEN VALUES OF A, B, U, AND YP1...
            A   = 1.0
            B   = 0.8
            U   = 1.0
            CI  = CMPLX( 0., 1. )
            YP1 = 0.199
            XP1 = B - SQRT(A**2-YP1**2)
            ZP1 = CMPLX( XP1, YP1 )


C           ..... DEFINE GRID POINTS IN Z1-PLANE AND
C                 EVALUATE STREAM FUNCTION THERE .....
            DX1 = 0.05
            DY1 = 0.05
            M = (XMAX1-XMIN1)/DX1 + 1
            N = (YMAX1-YMIN1)/DY1 + 1
            X1(1) = XMIN1
            DO 1 I=2,M
          1 X1(I) = X1(I-1) + DX1
            Y1(1) = YMIN1
            DO 2 J=2,N
          2 Y1(J) = Y1(J-1) + DY1
            DO 3 I=1,M
            DO 3 J=1,N
            Z1 = CMPLX( X1(I), Y1(J) )
            Z2 = Z1 - ZP1
          3 PSI(I,J) = U*AIMAG( Z2 + A**2/Z2 + CI*2.*YP1*CLOG(Z2/A) )


C           ..... SET UP A GRAPH FOR PLOTTING FLOW PATTERN .....
            XL(5) = 10H     X - AXI
            XL(6) = 10HS
            YL(5) = 10HY - AXIS
            TL(1) = 10HFIGURE 2.7
            TL(2) = 10H.4  FLOW A
            TL(3) = 10HROUND A JO
            TL(4) = 10HUKOWSKI AI
            TL(5) = 10HRFOIL.
            CONST(1) = -2.0
            CALL EZLABL( 5HF10.0, 0, 5HF10.0, 6, 0, 6, 0 )


C           ..... SEARCH FOR POINTS ON THE FIRST STREAMLINE IN Z1-PLANE .....
            CALL SEARCH( X1, Y1, PSI, CONST(1), X, Y, KMAX )
```

```
C       ..... MAP SOME SELECTED POINTS ONTO THE Z-PLANE, AND PLOT THE
C             FIRST CURVE .....
        CALL MAPPNG( X, Y, KMAX, XMIN, XMAX, YMIN, YMAX, NPOINT )
        CALL EZPLOT(X,Y,NPOINT,1H.,XMIN,XMAX,YMIN,YMAX,XL,YL,TL,1,11)

C       ..... PLOT NINE ADDITIONAL CURVES .....
        DO 4 K=2,10
        CONST(K) = CONST(K-1) + 0.5
        CALL SEARCH( X1, Y1, PSI, CONST(K), X, Y, KMAX )
        CALL MAPPNG( X, Y, KMAX, XMIN, XMAX, YMIN, YMAX, NPOINT )
        CALL NXCURV( X, Y, NPOINT, 1H. )
      4 CONTINUE

C       ..... SPECIFY THE ANGLE PHI FOR 200 POINTS ON THE CIRCLE WHICH
C             TRANSFORMS INTO JOUKOWSKI AIRFOIL.  THEN FIND THE POSITION,
C             SPEED, AND PRESSURE COEFFICIENT AT THESE POINTS AFTER THE
C             TRANSFORMATION .....
        PI = 4. * ATAN(1.)
        DPHI = PI/100.
        DO 7 L=1,200
        IF( L.EQ.1 ) GO TO 5
        PHI(L) = PHI(L-1) + DPHI
        GO TO 6
      5 PHI(L) = 0.0
      6 Z2 = A * CEXP(CI*PHI(L))
        Z1 = Z2 + ZP1
        Z  = Z1 + B**2/Z1
        X(L) = REAL(Z)
        Y(L) = AIMAG(Z)
        V = U*CABS( (1.-(A/Z2)**2+CI*2.*YP1/Z2)/(1.-(B/Z1)**2) )
        CP(L) = 1. - (V/U)**2
      7 CONTINUE

C       ..... MORE POINTS ARE PLOTTED ON THE CONTOUR OF THE AIRFOIL .....
        CALL NXCURV( X, Y, 200, 1H. )

C       ..... PLOT PRESSURE COEFFICIENT ON ANOTHER GRAPH .....
        YL(4) = 10H   CP : PRE
        YL(5) = 10HSSURE COEF
        YL(6) = 10HFICIENT
        TL(2) = 10H.5   PRESSU
        TL(3) = 10HRE DISTRIB
        TL(4) = 10HUTION AROU
        TL(5) = 10HND THE AIR
        TL(6) = 10HFOIL.
        CALL EZLABL( 5HF10.0, 0, 5HF10.0, 6, 0, 5, 0 )
        CALL EZPLOT(X,CP,200,1H.,XMIN,XMAX,-3.,2.,XL,YL,TL,1,1)

        END




        SUBROUTINE MAPPNG( X1, Y1, K1, XMIN, XMAX, YMIN, YMAX, NPOINT )
C       ..... THIS SUBROUTINE EXAMINES EACH OF THE K1 POINTS (X1,Y1) IN
C             Z1-PLANE AND THROWS AWAY THE POINTS INSIDE THE CIRCLE THAT
C             MAPS INTO THE AIRFOIL.  AFTER MAPPING THE REMAINING POINTS
C             ONTO Z-PLANE THROUGH JOUKOWSKI TRANSFORMATION, ONLY THOSE
C             WHICH ARE WITHIN THE FRAME BOUNDED BY XMIN, XMAX, YMIN, AND
C             YMAX ARE RETAINED.  THESE POINTS, OF TOTAL NUMBER NPOINT,
C             ARE RETURNED TO THE MAIN PROGRAM AS POINTS OF COORDINATES
C             (X1,Y1) IN Z-PLANE .....
```

```
      DIMENSION X1(K1),Y1(K1)
      COMMON /GROUP/A,B,ZP1
      COMPLEX Z,Z1,Z2,ZP1
      NPOINT = 0
      DO 1 I=1,K1
      Z1 = CMPLX( X1(I), Y1(I) )
      Z2 = Z1 - ZP1
      R = CABS(Z2)
      IF( R.LT.A )  GO TO 1
      Z = Z1 + B**2/Z1
      X = REAL(Z)
      Y = AIMAG(Z)
      IF( X.LT.XMIN .OR. X.GT.XMAX .OR.
     A    Y.LT.YMIN .OR. Y.GT.YMAX )  GO TO 1
      NPOINT = NPOINT + 1
      X1(NPOINT) = X
      Y1(NPOINT) = Y
    1 CONTINUE
      RETURN
      END
```

```
      SUBROUTINE SEARCH( X, Y, PSI, PSIA, XX, YY, KMAX )
C     ..... TAKEN FROM PROGRAM 2.2 WITH THE DIMENSION STATEMENT
C           REPLACED BY THE ONE SHOWN BELOW .....
      DIMENSION X(121),Y(121),PSI(121,121),XX(200),YY(200)
```

plotted. The image of the flow pattern will be displayed on a microfilm, which shows more accurate positions of the points than what would be printed by calling the subroutines PLOTN.

There are several methods for generating graphs on microfilm at the University of Colorado Computing Center. For simplicity we follow the one that involves the least effort, although the frames so produced are always square in shape.

To place the image of a set of points on a frame bounded by x_{min} and x_{max} on the horizontal axis and by y_{min} and y_{max} on the vertical axis, and to specify the titles for the coordinate system and that for the entire plot, the following subroutine will be called.

EZPLOT(X, Y, N, PS, XMIN, XMAX, YMIN, YMAX, XL, YL, TL, LT, NC)

Among the arguments of this subroutine, X and Y are arrays containing, respectively, the horizontal and vertical coordinates of the N points to be plotted by the plotting symbol PS. XL, YL, and TL are arrays containing characters, which are used for labeling the x-axis, the y-axis, and the plot title, respectively. LT is used to specify the plot type. For example, LT = 1

Inverse Method II: Conformal Mapping   **129**

indicates that the scales on both the x- and y-axes are linear, while LT = 2 indicates that the scale on the x-axis is linear but that on the y-axis is logarithmic. Finally, NC is the total number of curves to be plotted on this coordinate system. The first curve is plotted after EZPLOT has been called, but each of the remaining NC − 1 curves will be plotted only when the following subroutine is called:

NXCURV(X, Y, N, PS)

where the arguments have the same meanings as those in EZPLOT.

Numerical annotation of coordinate axes is accomplished by calling

EZLABL(XFMT, ORIENT, YFMT, MJX, MNX, MJY, MNY)

in which XFMT and YFMT are E or F formats with a field width of 10, used respectively for describing the scale numbers on the x- and y-axes. ORIENT is used to determine the orientation of the x-axis. It is set to equal 0 for horizontal and 1 for vertical direction. MJX and MNX are the numbers of major and minor divisions on the x-axis, whereas MJY and MNY are those on the y-axis.

In addition to the flow pattern, we are also interested in the forces acting on the airfoil. The net pressure force is a lift in the direction of the positive y-axis, and its magnitude per unit span is $\rho U \Gamma$ according to (2.5.4), or $4\pi\rho y'_P U^2$ after using (2.7.21). The surface pressure, however, varies with location and will be computed and plotted on a separate graph according to the following procedure

Two hundred evenly spaced points are selected on the circular cylinder of Fig. 2.7.3. The coordinates of these surface points in the x''-y'' plane are calculated by varying the angle φ, starting from the value zero, in the expression

$$z'' = a\, e^{i\varphi}$$

By using the transformations $z' = z'' + z'_P$ and (2.7.19), these points are mapped successively into the x-y plane. In this way the exact positions of the points on the airfoil are obtained, whereas the subroutine SEARCH would return only the approximate positions. Another advantage of using this method is that the points that are on the streamline $\psi = 0$ but not on the airfoil are automatically excluded.

To find the speed V at the surface of the airfoil, we use the formula

$$V = \left|\frac{dw}{dz}\right| = \left|\frac{dw}{dz''}\right|\left|\frac{dz''}{dz'}\right| \bigg/ \left|\frac{dz}{dz'}\right|$$

Upon substitution from (2.7.19), (2.7.22), and (2.7.23), it reduces to

$$V = U \left|\frac{1 - (a/z'')^2 + i2y'_P/z''}{1 - (b/z')^2}\right| \tag{2.7.25}$$

Once the speed at a point becomes known, the pressure coefficient (or the dimensionless pressure difference) at that point can be computed according to

$$c_p = \frac{p - P}{\frac{1}{2}\rho U^2} = 1 - \left(\frac{V}{U}\right)^2 \qquad (2.7.26)$$

which is obtained by using (2.6.14) and the Bernoulli equation. The values of c_p are plotted as a function of the x-coordinate of the 200 surface points.

Figure 2.7.4 shows the shape of the airfoil, which is described by the streamline $\psi = 0$. The coordinates of the surface points can easily be printed out if so desired. The plot reveals that the space between the body and the streamline $\psi = 0.5$ is narrower than that between the body and the streamline $\psi = -0.5$, indicating that the speed on the upper surface is higher than that on

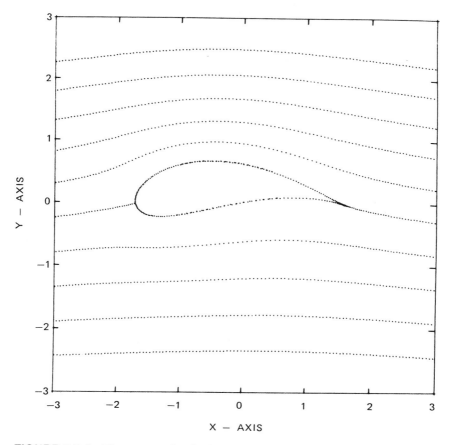

FIGURE 2.7.4 Flow around a Joukowski airfoil.

Inverse Method II: Conformal Mapping

131

the lower. The corresponding pressure distributions on the upper and lower surfaces of the airfoil are represented, respectively, by the lower and upper branches of the curve plotted in Fig. 2.7.5. At the forward stagnation point where $V = 0$, the pressure coefficient takes on the maximum value of unity according to (2.7.26). At the sharp trailing edge the speed is finite but not zero, although it is transformed from the stagnation point Q in the x'-y' plane. The area enclosed by the pressure distribution curve is proportional to the total lift of the airfoil.

This program may be used not only to generate airfoils of different shapes by varying the values assigned to b and y'_P, but also, with some modifications, to plot flow patterns obtained by mapping from various known flows through arbitrary transformations.

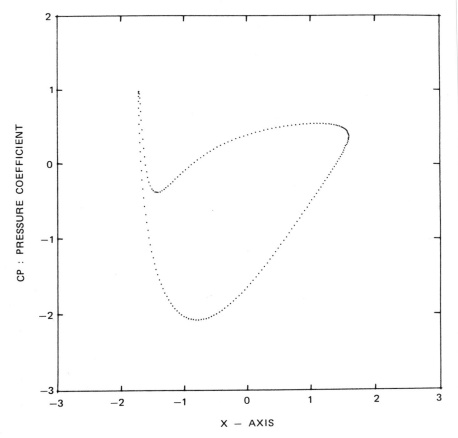

FIGURE 2.7.5 Pressure distribution around the airfoil.

Inviscid Fluid Flows

Problem 2.8   The transformation

$$z = z' + \frac{1}{z'} \qquad (2.7.27)$$

maps the circle $|z'| = 1$ into the line segment on the x-axis between the points $x = -2$ and $x = 2$, and the region outside the circle into the entire x-y plane. It is known that

$$w = U\left(z'e^{-i\alpha} + \frac{e^{i\alpha}}{z'}\right) \qquad (2.7.28)$$

represents the complex potential for a uniform flow of speed U past a cylinder of unit radius centered at $z' = 0$. The flow makes an angle α with the x'-axis at an infinite distance from the cylinder. Under the transformation (2.7.27), the complex potential (2.7.28) then becomes that for a uniform flow past a flat plate at an angle of attack α. Plot the streamlines of this transformed flow and the pressure distribution on the upper and lower surfaces of the plate. Note that the velocity and therefore the pressure is unbounded at the edges of the plate.

Problem 2.9   If the Kutta condition is to be satisfied on the flat plate described in the previous problem, a circulation is required to move the rear stagnation point on the circle in the x'-y' plane to the point $(1, 0)$, which maps into the trailing edge of the plate. Plot the streamlines and pressure distribution and compare them with those obtained for the case without a circulation.

Problem 2.10   It can be verified that the transformation

$$z' = \frac{h}{\pi}[\sqrt{z^2 - 1} + \log(\sqrt{z^2 - 1} - z)]$$

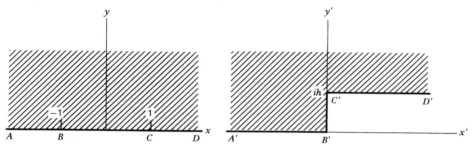

FIGURE 2.7.6   The transformation $z' = (h/\pi)[\sqrt{z^2 - 1} + \log(\sqrt{z^2 - 1} - z)]$.

Inverse Method II: Conformal Mapping   **133**

maps the x-axis into a step in the x'-y' plane and the region $y > 0$ into the space above the step. Plot the streamlines of a flow of uniform speed U past the step and the pressure distribution along the surface $A'B'C'D'$ shown in Fig. 2.7.6.

2.8 Panel Method

Several methods have been introduced so far for solving the problem of uniform flow past bodies. A body of unspecified shape may be generated, according to Section 2.4, by adding to the uniform flow a linear combination of singularities including sources, sinks, doublets, and vortices. To generate an axisymmetric body (or a two-dimensional symmetric body) of given shape by distributing singularities along its axis of symmetry, von Kármán's method of Section 2.6 outlines a procedure for determining the strengths of the singularities placed at given locations. Furthermore, the method of conformal mapping described in Section 2.7 provides another means of constructing bodies of a variety of shapes by varying the transformation between two coordinate systems.

In practical applications, however, it is often necessary to calculate flow about bodies of complicated configurations such as an airplane with wing, tail, and engine appendages. These problems, which cannot be solved satisfactorily by any of the previously mentioned methods, stimulated some aeronautical engineers to develop the *panel method* for handling bodies of arbitrary shape. In this method the surface of the concerned body is first covered by a finite number of small areas called panels, each of which is distributed by singularities of a certain kind that have an undetermined uniform density. The distributed singularities are used to deflect the oncoming stream so that it will flow around the body. In general, source or doublet panels are used on the nonlifting surfaces, and vortex panels are used on the lifting surfaces of a body. The requirement that the oncoming flow be tangent to every panel at a particular location gives a set of equations, which is used to compute the singularity densities on the panels. Thus the flow, consisting of a uniform flow and the flow induced by the singularities on a finite number of panels, becomes determined, and the velocity and pressure at any point in the flow field can be calculated. The panel method has been successfully applied not only to the flow problems involving two- and three-dimensional bodies of complex geometry, but also to the problems concerning internal flows and nonuniform and unsteady flows. The state of the art of this method can be found in the review article by Hess (1975), who is one of the pioneers responsible for the development of the panel method.

To demonstrate that a panel distributed with singularities can be used to deflect the oncoming flow and that the resultant flow can be made tangent to the panel surface by adjusting the strength of the distributed singularities, we will construct a two-dimensional source panel by using a large number of discrete line sources. It is well known that when a line source is placed in a uniform stream, it pushes the approaching flow away to form a half-body that extends to infinity in the downstream direction. When five identical sources are placed at equal intervals along a line normal to the uniform flow, the streamline pattern is qualitatively similar to that caused by a single source and is plotted in Fig. 2.8.1a. Because of symmetry, only the upper half of the entire flow is shown. These five sources effectively generate a body that encloses all of the sources and is bounded by the streamline $\psi = 0$, so that the originally uniform stream is distorted to flow around the body surface. If the same total source strength is evenly distributed among 11 equally spaced sources within the same range, the waviness exhibited by the streamline $\psi = 0$ disappears and the resulting flow pattern changes to that shown in Fig. 2.8.1b. When the same total source strength is redistributed evenly among 101 equally spaced sources, we obtain a source sheet; this causes a flow pattern from which the presence of individual sources can no longer be identified. The plot of Fig. 2.8.1c reveals that fluid ejecting from the source sheet is contained completely within the surface $\psi = 0$. The body surface described by $\psi = 0$ can be made to approach the source sheet by reducing the total source strength, as shown by the representative example in Fig. 2.8.1d. A further reduction in source strength by an appropriate amount may cause the body surface to stand at an infinitesimal distance upstream from the source sheet; that is, the sheet itself can be made approximately the surface of a body by adjusting the total strength of a large number of identical sources.

If, instead of being distributed among a number of discrete sources, the total strength is distributed uniformly throughout a finite region, a source panel is formed that can be made to be exactly tangent to the resultant flow by choosing a proper strength. For a simplified mathematical analysis, we consider for the moment a source panel of length $2L$ lying symmetrically on the y-axis (Fig. 2.8.2), on which sources of strength λ per unit length are distributed. The velocity potential at any point (x, y) induced by the sources contained within a small element dy' of the panel at $(0, y')$ is

$$\frac{\lambda \, dy'}{2\pi} \ln [x^2 + (y - y')^2]^{1/2}$$

This expression can be obtained either by taking the real part of the complex potential for a source flow described in Section 2.7, or by integrating the velocity derived from the stream function given by (2.4.2). Thus the induced

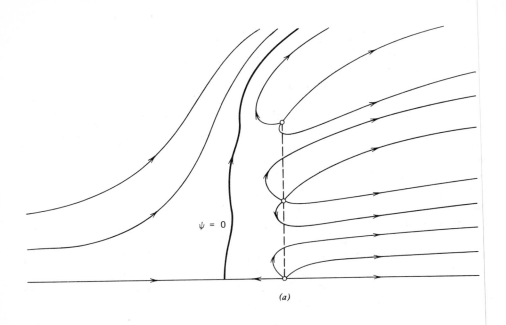

$\psi = 0$

(a)

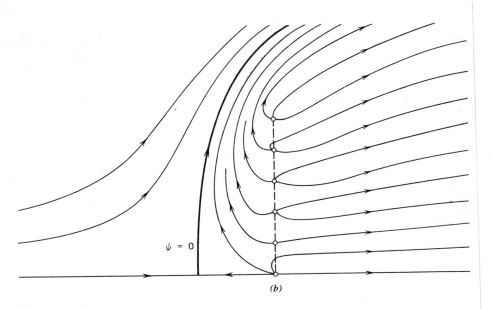

$\psi = 0$

(b)

FIGURE 2.8.1 Flow pattern resulting from distributing line sources along the dashed line in the presence of a uniform flow.

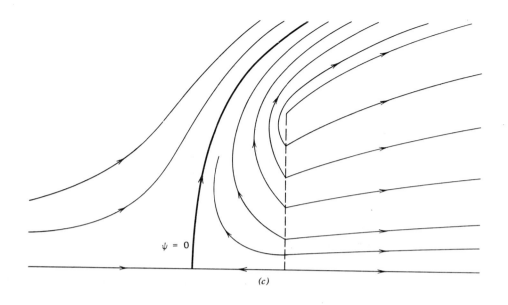

(c)

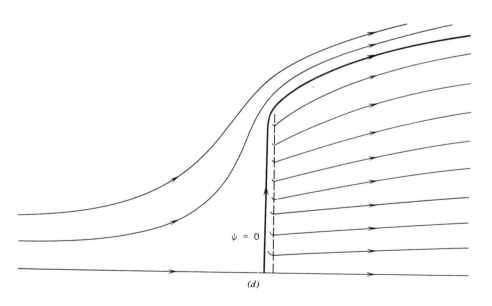

(d)

FIGURE 2.8.1 (*Continued*).

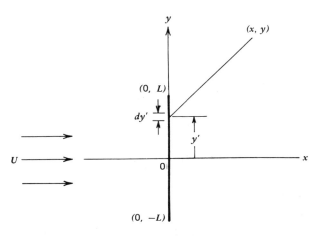

FIGURE 2.8.2 Source panel on y-axis.

velocity potential at that point caused by the entire source panel is

$$\phi(x, y) = \frac{\lambda}{4\pi} \int_{-L}^{L} \ln \left[x^2 + (y - y')^2 \right] dy' \tag{2.8.1}$$

Differentiations with respect to x and y give, respectively, the expressions for velocity components

$$u(x, y) = \frac{\lambda}{2\pi} \left[\tan^{-1} \left(\frac{y + L}{x} \right) - \tan^{-1} \left(\frac{y - L}{x} \right) \right] \tag{2.8.2}$$

and

$$v(x, y) = \frac{\lambda}{4\pi} \ln \frac{x^2 + (y + L)^2}{x^2 + (y - L)^2} \tag{2.8.3}$$

Consider a point whose coordinates satisfy the condition that $x > 0$ and $L > y > -L$. By letting $x \to 0$ this point will approach the panel surface from its right. The limiting value of (2.8.2) gives $u(0^+, y) = \lambda/2$ on the right face of the panel. On the other hand, by letting a similar point approach the panel from the left, (2.8.2) gives $u(0^-, y) = -\lambda/2$ on the left face. It shows that a source panel of strength λ generates about itself a flow having a uniform outward normal velocity of magnitude $\lambda/2$ at the surface. The tangential velocity, however, is the same on both sides of the panel according to (2.8.3). It starts from zero at the middle and accelerates along the panel surface toward the edges. The edges of a panel are singularities where the normal velocity is not defined and the tangential velocity becomes infinite.

When a source panel of strength $\lambda = 2U$ is placed normal to a uniform flow of speed U, as shown in Fig. 2.8.2, its induced normal velocity cancels exactly the oncoming flow on the left face so that the resultant flow becomes tangent to the surface there. In other words, by distributing sources of a suitable strength along a surface perpendicular to a uniform stream, that surface may become coincident with one of the streamlines of the resultant flow. To construct a surface making an angle θ with the uniform stream, sources can again be distributed along that surface, which is of such a strength that the induced outward flow can just cancel the oncoming flow in the direction normal to the surface. The panel strength is, therefore, $\lambda = 2U \sin \theta$.

The use of a set of source panels for solving the problem of flow past a two-dimensional body is illustrated in Fig. 2.8.3. A finite number of points are first selected on the surface of the body, which is represented by the dashed curve. They are designated as the "control points" in the panel method. Through these control points, lines tangent to the local body surface are drawn. The intersections of neighboring tangent lines are called the "boundary points"; the panels are defined by the line segments between a pair of neighboring boundary points. In this way the body surface is covered by a polygon of straight-line panels.

Let m be the total number of panels around the body. On each of the m panels, whose lengths are generally not the same, sources of a uniform density is distributed. Let $\lambda_1, \lambda_2, \ldots, \lambda_m$ represent the source strengths per unit length on these panels. The velocity potential at any point (x_i, y_i) in the flow caused by the source distribution on the jth panel is, generalized from (2.8.1) for a special case,

$$\frac{\lambda_j}{2\pi} \int_j \ln r_{ij} \, ds_j$$

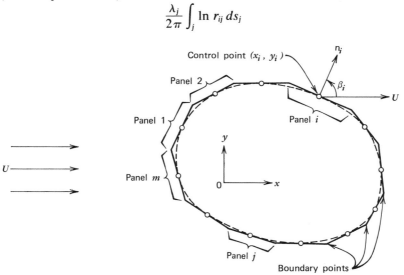

FIGURE 2.8.3 Covering the surface of a body with source panels.

Panel Method

In this expression $\lambda_j\,ds_j$ is the strength of sources contained within the element ds_j located at (x_j, y_j) on the jth panel, and $r_{ij} = \sqrt{(x_i - x_j)^2 + (y_i - y_j)^2}$ is the distance between the points (x_i, y_i) and (x_j, y_j). The subscript j following the integral sign indicates integration over the entire length S_j of the jth panel. Thus the velocity potential for the flow consisting of the uniform stream and the m source panels is given by

$$\phi(x_i, y_i) = Ux_i + \sum_{j=1}^{m} \frac{\lambda_j}{2\pi} \int_j \ln r_{ij}\, ds_j \tag{2.8.4}$$

We now let (x_i, y_i) be the coordinates of the control point on the ith panel, where the outward normal vector $\mathbf{n}_i$ makes an angle β_i with the direction of the uniform flow. The boundary condition at the body surface requires that at each control point the normal velocity of the resultant flow vanishes; that is,

$$\frac{\partial}{\partial n_i} \phi(x_i, y_i) = 0, \qquad i = 1, 2, \cdots, m \tag{2.8.5}$$

Substitution of (2.8.4), and from the property that the contribution to the normal velocity on a source panel by the source distribution on the panel itself is $\lambda/2$, the boundary conditions (2.8.5) become

$$\frac{\lambda_i}{2} + \sum_{j \neq i}^{m} \frac{\lambda_j}{2\pi} I_{ij} = - U \cos \beta_i, \qquad i = 1, 2, \cdots, m \tag{2.8.6}$$

where

$$I_{ij} \equiv \int_j \frac{\partial}{\partial n_i} (\ln r_{ij})\, ds_j \tag{2.8.7}$$

represents the contribution to the normal velocity at the control point on the ith panel by a source distribution of unit strength on the jth panel, and the summation is carried out for all values up to m except $j = i$.

For a given panel configuration both $\mathbf{n}_i$ and β_i are known at each control point. After the evaluation of the integrals I_{ij}, a set of m simultaneous algebraic equations is obtained that enables us to solve for λ_1 to λ_m. With known panel strengths, the velocity and pressure at any point in the flow can be computed by taking derivatives of ϕ expressed in (2.8.4) and using Bernoulli's equation. However, since the integrals I_{ij} in the form of (2.8.7) are not convenient for computer calculations, we need to go one step further by expressing them in terms of panel geometries.

Let us name the panels in the clockwise order as shown in Fig. 2.8.3 and, in the same order, designate the coordinates of the boundary points on the edges of the kth panel as (X_k, Y_k) and (X_{k+1}, Y_{k+1}), respectively. In this notation there are $m + 1$ boundary points for m panels, with the last boundary point coinciding with the first. The length of the kth panel can be expressed in terms of the coordinates of the two end points by the relationship

$$S_k = \sqrt{(X_{k+1} - X_k)^2 + (Y_{k+1} - Y_k)^2} \qquad (2.8.8)$$

and the angle θ_k between that panel and the x-axis, measured at the point (X_k, Y_k), is given by the actual slope of the body at the control point. Referring to Fig. 2.8.4, in which two representative panels i and j are shown, θ_i is chosen to be negative and θ_j positive. From various panel configurations a generalized expression is deduced for β_k, the angle between the outward normal on the kth panel and the x-axis, which has the form

$$\beta_k = \theta_k + \frac{\pi}{2} \qquad (2.8.9)$$

It follows that

$$\sin \beta_k = \cos \theta_k \quad \text{and} \quad \cos \beta_k = -\sin \theta_k \qquad (2.8.10)$$

Quantities involved in the integral I_{ij} are depicted in Fig. 2.8.4. (x_i, y_i) are the coordinates of the control point on the ith panel, whereas (x_j, y_j) are those of any point on the jth panel. After differentiation with respect to n_i, (2.8.7) becomes

$$I_{ij} = \int_j \frac{(x_i - x_j) \cos \beta_i + (y_i - y_j) \sin \beta_i}{(x_i - x_j)^2 + (y_i - y_j)^2} \, ds_j$$

Upon use of (2.8.10) and the geometric relations that

$$x_j = X_j + s_j \cos \theta_j \quad \text{and} \quad y_j = Y_j + s_j \sin \theta_j$$

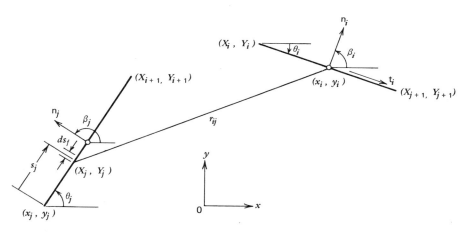

FIGURE 2.8.4 Evaluation of the integral I_{ij}.

Panel Method

the integral takes the form

$$I_{ij} = \int_0^{S_j} \frac{Cs_j + D}{s_j^2 + 2As_j + B} \, ds_j \tag{2.8.11}$$

where

$$A = -(x_i - X_j) \cos \theta_j - (y_i - Y_j) \sin \theta_j \tag{2.8.12}$$

$$B = (x_i - X_j)^2 + (y_i - Y_j)^2 \tag{2.8.13}$$

$$C = \sin (\theta_i - \theta_j) \tag{2.8.14}$$

and

$$D = -(x_i - X_j) \sin \theta_i + (y_i - Y_j) \cos \theta_i \tag{2.8.15}$$

For the denominator of the integrand, the following relationship holds.

$$4B - (2A)^2 = 4E^2 > 0$$

where

$$E = (x_i - X_j) \sin \theta_j - (y_i - Y_j) \cos \theta_j \tag{2.8.16}$$

The integral is then evaluated accordingly. After some manipulation, a final expression is obtained in the following form.

$$I_{ij} = \tfrac{1}{2} \sin (\theta_i - \theta_j) \ln \left[1 + \frac{S_j^2 + 2AS_j}{B} \right]$$

$$- \cos (\theta_i - \theta_j) \left[\tan^{-1} \left(\frac{S_j + A}{E} \right) - \tan^{-1} \left(\frac{A}{E} \right) \right] \tag{2.8.17}$$

This expression contains the orientation of the ith panel, the coordinates of the control point on its surface, the length and orientation of the jth panel and, finally, the coordinates of one of the boundary points on the jth panel.

By using (2.8.10) and with the introduction of the dimensionless variables $\lambda'_k \equiv \lambda_k/2\pi U$, (2.8.6) can be rewritten in a more convenient form as

$$\sum_{j=1}^m I_{ij}\lambda'_j = \sin \theta_i, \qquad i = 1, 2, \ldots, m \tag{2.8.18}$$

in which I_{ij} are given by (2.8.17), except $I_{ij} = \pi$ when $j = i$.

Before solving this system of equations for λ'_1 to λ'_m, we must know the coordinates of all boundary points so that the coefficients I_{ij} can be adequately computed. On two neighboring panels k and $(k + 1)$, the positions of the control points, (x_k, y_k) and (x_{k+1}, y_{k+1}), and the panel orientations, θ_k and θ_{k+1}, have already been prescribed. They are related to the intersecting boundary point (X_{k+1}, Y_{k+1}) of these two panels through the equations

$$\tan \theta_k = \frac{Y_{k+1} - y_k}{X_{k+1} - x_k} \quad \text{and} \quad \tan \theta_{k+1} = \frac{y_{k+1} - Y_{k+1}}{x_{k+1} - X_{k+1}}$$

Solving them simultaneously gives

$$X_{k+1} = \frac{x_{k+1} \tan \theta_{k+1} - x_k \tan \theta_k - y_{k+1} + y_k}{\tan \theta_{k+1} - \tan \theta_k}$$

$$Y_{k+1} = \frac{(x_{k+1} - x_k) \tan \theta_{k+1} \tan \theta_k - y_{k+1} \tan \theta_k + y_k \tan \theta_{k+1}}{\tan \theta_{k+1} - \tan \theta_k} \qquad (2.8.19)$$

However, these formulas are not appropriate for computation if one of the orientation angles becomes $\pm 90°$. When this happens, the following formulas must be used instead. For $\theta_k = \pm \pi/2$,

$$X_{k+1} = x_k$$
$$Y_{k+1} = y_{k+1} - (x_{k+1} - x_k) \tan \theta_{k+1} \qquad (2.8.20)$$

For $\theta_{k+1} = \pm \pi/2$,

$$X_{k+1} = x_{k+1}$$
$$Y_{k+1} = y_k + (x_{k+1} - x_k) \tan \theta_k \qquad (2.8.21)$$

When dealing with a body of arbitrary shape, the previously mentioned procedure for determining boundary points may become impractical, because the slopes of the tangents to the surface may not always be easily computed from a given contour shape. An alternative procedure is to specify instead a set of boundary points around the body. Panels are defined by the straight lines connecting neighboring boundary points, and their orientation angles are calculated from

$$\theta_k = \tan^{-1} \left(\frac{Y_{k+1} - Y_k}{X_{k+1} - X_k} \right), \qquad k = 1, 2, \ldots, m \qquad (2.8.22)$$

whereas control points are usually chosen to be midpoints of the panels (see, for example, the general description by Hess and Smith, 1966). The second procedure is not as accurate as the first in that the control points, which are used to represent the body, do not coincide with the actual body surface in this case. Nevertheless, the control points will approach the real surface as the number of panels increases.

It has been shown that in the presence of a single source panel, it is possible to make the resultant flow tangent to the entire panel except at the end points. When a string of panels is used to construct a body, as shown in Fig. 2.8.3, we demand that the flow be tangent to the body only at the control points. Because of the mutual influences among panels, the flow generally is not tangent to the panels at points other than the control points. Remember that the body surface is approximated by a set of discrete control points; it is not replaced by a polygon of straight-line panels. Actually, the fluid velocity becomes infinite at the edges of every panel, as shown by the result of an example concerning flow past a two-dimensional airfoil (Kuethe and Chow, 1976, p. 140), so that they are not appropriate to be included as part of a body.

Panel Method 143

There is a basic difference between von Kármán's method and the panel method. In the former singularities are distributed within a body along the axis of symmetry; in the latter they are distributed at the surface so that a nonsymmetric body also can be constructed.

We now go back to the system of simultaneous equations (2.8.18). After the dimensionless panel strengths are obtained, they are substituted into the expression for velocity potential on the right-hand side of (2.8.4). The velocity vector at the control point (x_i, y_i) on the ith panel is tangent to the panel surface by virtue of the fact that its normal component there is zero. Thus, at a control point only, the magnitude of the velocity vector is

$$V(x_i, y_i) = \frac{\partial}{\partial t_i} \phi(x_i, y_i)$$

where t_i is a vector tangent to the surface of the ith panel, as shown in Fig. 2.8.4. Following a similar procedure employed previously in carrying out the derivative of ϕ with respect to n_i, we obtain

$$\frac{V(x_i, y_i)}{U} = \cos \theta_i + \sum_{j=1}^{m} \lambda'_j I'_{ij} \qquad (2.8.23)$$

I'_{ij} represents the contribution to the tangential velocity at the control point on the ith panel by a source distribution of unit strength on the jth panel. $I'_{ij} = 0$ when $j = i$; otherwise

$$I'_{ij} = -\tfrac{1}{2} \cos (\theta_i - \theta_j) \ln \left[1 + \frac{S_j^2 + 2AS_j}{B} \right]$$
$$- \sin (\theta_i - \theta_j) \left[\tan^{-1} \left(\frac{S_j + A}{E} \right) - \tan^{-1} \left(\frac{A}{E} \right) \right] \qquad (2.8.24)$$

This expression is somewhat similar to that for I_{ij} shown in (2.8.17), with A, B, and E defined already in (2.8.12), (2.8.13), and (2.8.16). The pressure coefficient defined by (2.7.26) can still be used in the present case to represent the pressure distribution around the body surface.

To demonstrate the use of the method of source panels, the method is applied to find the velocity and pressure distributions at the surface of a circular cylinder of radius a held stationary in a uniform flow of speed U. Eight equally spaced control points are chosen on the surface of the body, which is shown by the dashed curve in Fig. 2.8.5. Eight panels are then constructed and are labeled in the clockwise order. In the present special case the coordinates of the control points can conveniently be specified by varying φ, the angle between the radial position vector of a control point and the x-axis, through the formulas

$$x_i = a \cos \varphi_i \qquad \text{and} \qquad y_i = a \sin \varphi_i \qquad (2.8.25)$$

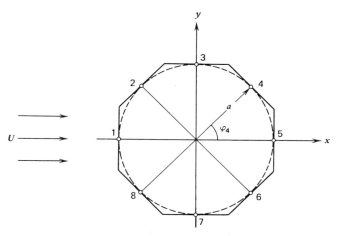

FIGURE 2.8.5 Solving the problem of uniform flow past a circular cylinder by the method of source panels.

where

$$\varphi_i = \pi - (i-1)\frac{\pi}{4} = (5-i)\frac{\pi}{4}, \qquad i = 1, 2, \cdots, 8 \tag{2.8.26}$$

The orientation angles θ_i of the panels can also be expressed in terms of φ_i by

$$\theta_i = \varphi_i - \frac{\pi}{2} \tag{2.8.27}$$

In order to be able to use the same formulas (2.8.19), (2.8.20), or (2.8.21) for computing the coordinates of all the boundary points, we tentatively call panel 1 by the name panel 9. After the ninth boundary point is determined, which is the intersection of panels 8 and 9 according to our notation described earlier, its coordinates are then reassigned to the first boundary point.

The accuracy of the numerical result can again be checked by comparing with the exact solution. The stream function of the flow is the sum of the contribution from the uniform flow, the right-hand side of (2.4.1) with $\alpha = 0$, and that from a doublet, the right-hand side of (2.4.4) with $x_0 = 0$, $y_0 = 0$, and $\kappa = 2\pi U a^2$; that is

$$\psi = Uy\left(1 - \frac{a^2}{x^2 + y^2}\right) \tag{2.8.28}$$

Velocity components are obtained by differentiations with respect to x and y. It can be shown that the speed at a surface point (x_i, y_i) is

$$V_{\text{exact}} = 2U \sin \varphi_i \tag{2.8.29}$$

Panel Method

145

It follows that the exact expression for pressure coefficient at that surface point is

$$(c_p)_{\text{exact}} = 1 - 4\sin^2\varphi_i \qquad (2.8.30)$$

The result of the panel method shows that

$$\sum_{i=1}^{8} \lambda_i' = 0$$

For the present example having panels of the same length, this equation states that the total strength of the sources and sinks distributed around the cylinder is zero, as it should be to form a closed body. Excellent agreements are obtained between the numerical result and the exact solution with only eight panels. Errors in velocity and pressure coefficient do not show up when four digits are printed after the decimal point.

A remark must be made on the sign of V shown in the computer output. The value of V is computed by differentiating the velocity potential in the direction of **t**, which is a vector tangent to the panel surface pointed in the

---

List of Principal Variables in Program 2.6

Program Symbol	Definition
(Main Program)	
A	a, radius of the circular cylinder
COSINE	Array containing values of $\cos\theta_k$
CP, CPEX	Pressure coefficient on the surface of cylinder and its exact value
INTGL1, INTGL2	Representing I_{ij} and I'_{ij}, whose values are evaluated according to (2.8.17) and (2.8.24), respectively
LAMBDA	λ', the dimensionless panel strength
PHI	φ, the angle between the radial position vector of a control point and the x-axis
S	S, panel length
SINE	Array containing values of $\sin\theta_k$
THETA	θ, panel orientation angle
U	Speed of the uniform flow
V, VEX	Flow speed on the surface of cylinder and its exact value
X, Y	x and y, coordinates of control point
XB, YB	X and Y, coordinates of boundary point

```
C                    ***** PROGRAM 2.6 *****

C         SOLUTION OF THE PROBLEM OF UNIFORM FLOW PAST A CIRCULAR
C         CYLINDER BY THE METHOD OF SOURCE PANELS

      REAL INTGL1(8,8),INTGL2(8,8),LAMBDA(8)
      DIMENSION X(9),Y(9),XB(9),YB(9),PHI(9),THETA(9),
     A          S(8),SINE(8),COSINE(8)

C    ..... LET THE RADIUS OF CYLINDER BE 1 METER AND THE SPEED OF
C            UNIFORM FLOW BE 1 METER PER SECOND .....
      A = 1.0
      U = 1.0

C    ..... SPECIFY THE COORDINATES OF CONTROL POINTS AND THE
C            PANEL ORIENTATION ANGLES .....
      PI = 3.14159265
      DO 1  I=1,9
      PHI(I) = (5.0-FLOAT(I)) * PI / 4.0
      X(I) = A * COS(PHI(I))
      Y(I) = A * SIN(PHI(I))
    1 THETA(I) = PHI(I) - PI/2.0

C    ..... COMPUTE COORDINATES OF BOUNDARY POINTS .....
      DO 4  K=1,8
      KP1 = K+1
      TANK = TAN( THETA(K) )
      TANKP1 = TAN( THETA(KP1) )
      DIFF = ABS( PI/2.0-ABS(THETA(K)) )
      IF( DIFF.LT.0.000001 )  GO TO 2
      DIFF = ABS( PI/2.0-ABS(THETA(KP1)) )
      IF( DIFF.LT.0.000001 )  GO TO 3
      XB(KP1) = ( X(KP1)*TANKP1 - X(K)*TANK - Y(KP1) + Y(K) )
     A          / ( TANKP1 - TANK )
      YB(KP1) = ( (X(KP1)-X(K))*TANKP1*TANK - Y(KP1)*TANK
     A          + Y(K)*TANKP1 ) / ( TANKP1 - TANK )
      GO TO 4
    2 XB(KP1) = X(K)
      YB(KP1) = Y(KP1) - (X(KP1)-X(K))*TANKP1
      GO TO 4
    3 XB(KP1) = X(KP1)
      YB(KP1) = Y(K) + (X(KP1)-X(K))*TANK
    4 CONTINUE
      XB(1) = XB(9)
      YB(1) = YB(9)

C    ..... CALCULATE LENGTH, SINE AND COSINE OF THETA FOR ALL PANELS...
      DO 5  K=1,8
      S(K) = SQRT( (XB(K+1)-XB(K))**2 + (YB(K+1)-YB(K))**2 )
      SINE(K) = SIN( THETA(K) )
    5 COSINE(K) = COS( THETA(K) )

C    ..... COMPUTE VALUES FOR TWO INTEGRALS ACCORDING TO
C            (2.8.17) AND (2.8.24) .....
      DO 7  I=1,8
      DO 7  J=1,8
```

```
      IF( J.EQ.I ) GO TO 6
      A = -(X(I)-XB(J))*COSINE(J) - (Y(I)-YB(J))*SINE(J)
      B = (X(I)-XB(J))**2 + (Y(I)-YB(J))**2
      C = SIN( THETA(I)-THETA(J) )
      D = COS( THETA(I)-THETA(J) )
      E = (X(I)-XB(J))*SINE(J) - (Y(I)-YB(J))*COSINE(J)
      F = ALOG( 1.0 + S(J)*(S(J)+2.*A)/B )
      G = ATAN( (S(J)+A)/E ) - ATAN( A/E )
      INTGL1(I,J) = 0.5*C*F - D*G
      INTGL2(I,J) =-0.5*D*F - C*G
      GO TO 7
    6 INTGL1(I,J) = PI
      INTGL2(I,J) = 0.0
    7 CONTINUE

C     ..... PRINT THE COEFFICIENT MATRIX ON THE LEFT AND THE COLUMN
C            MATRIX ON THE RIGHT OF (2.8.18) .....
      WRITE(6,50)
      DO 8  I=1,8
    8 WRITE(6,51)   (INTGL1(I,J),J=1,8), SINE(I)

C     ..... SOLVE SIMULTANEOUS EQUATIONS (2.8.18) FOR DIMENSIONLESS
C            PANEL STRENGTHS USING CRAMER,S RULE, THEN PRINT OUT
C            THEIR VALUES .....
      CALL CRAMER( INTGL1, SINE, LAMBDA, 8 )

C     ..... PRINT DATA ON PANEL CONFIGURATION AND STRENGTH .....
      WRITE(6,52)
      DO 9  I=1,8
    9 WRITE(6,53)   I,X(I),Y(I),XB(I),YB(I),S(I),THETA(I),LAMBDA(I)

C     ..... COMPUTE AND PRINT VELOCITY AND PRESSURE COEFFICIENT AT EACH
C            OF THE EIGHT CONTROL POINTS IN COMPARISON WITH THEIR EXACT
C            VALUES .....
      WRITE(6,54)
      DO 11  I=1,8
      VI = COSINE(I)
      DO 10  J=1,8
   10 VI = VI + LAMBDA(J)*INTGL2(I,J)
      V = U*VI
      VEX = 2.*U*SIN(PHI(I))
      CP = 1.0 - (V/U)**2
      CPEX = 1.0 - 4.*SIN(PHI(I))**2
   11 WRITE(6,55)   X(I),Y(I),V,VEX,CP,CPEX

   50 FORMAT( 1H1//////4X, 42HCOEFFICIENT MATRIX ON THE LEFT AND COLUMN .
     A        32HMATRIX ON THE RIGHT OF (2.8.18): // )
   51 FORMAT( 8F8.4, 8X, F8.4 )
   52 FORMAT( //////23X33HPANEL CONFIGURATION AND STRENGTH: // 3X1HI,
     A        7X4HX(I), 5X4HY(I), 7X5HXB(I), 4X5HYB(I), 7X4HS(I),
     B        4X8HTHETA(I), 4X9HLAMBDA(I) / 2X3H---, 4X7H-------,
     C        2X7H-------, 5X7H-------, 2X7H-------, 5X5H-----,
     D        4X8H--------, 4X9H--------- )
   53 FORMAT( I4, 3X, 2F9.4, 3X, 2F9.4, F10.4, 1X, 2F11.4 )
   54 FORMAT( //////8X46HCOMPARISON OF NUMERICAL RESULT WITH THE EXACT .
     A        19HAT 8 SURFACE POINTS // 13X1HX, 9X1HY, 11X1HV,
     B        6X6HVEXACT, 9X2HCP, 5X7HCPEXACT / 10X7H-------,
```

```
      C          3X7H-------, 5X7H---.----, 3X6H------, 6X7H-------,
      D          3X7H------- )
   55 FORMAT( 7X, 2F10.4, 3X, 2F9.4, 3X, 2F10.4 )

      END



      SUBROUTINE CRAMER( C, A, X, N )

C     ..... TAKEN FROM PROGRAM 2.4 WITH DIMENSION STATEMENT REPLACED
C             BY THE ONE SHOWN BELOW .....
      DIMENSION C(8,8),CC(8,8),A(8),X(8)



      FUNCTION DETERM( ARRAY, N )

C     ..... TAKEN FROM PROGRAM 2.4 WITH DIMENSION STATEMENT REPLACED
C             BY THE ONE SHOWN BELOW .....
      DIMENSION ARRAY(8,8),A(8,8)
```

COEFFICIENT MATRIX ON THE LEFT AND COLUMN MATRIX ON THE RIGHT OF (2.8.18):

3.1416	.3528	.4018	.4074	.4084	.4074	.4018	.3528	1.0000
.3528	3.1416	.3528	.4018	.4074	.4084	.4074	.4018	.7071
.4018	.3528	3.1416	.3528	.4018	.4074	.4084	.4074	0.0000
.4074	.4018	.3528	3.1416	.3528	.4018	.4074	.4084	-.7071
.4084	.4074	.4018	.3528	3.1416	.3528	.4018	.4074	-1.0000
.4074	.4084	.4074	.4018	.3528	3.1416	.3528	.4018	-.7071
.4018	.4074	.4084	.4074	.4018	.3528	3.1416	.3528	-.0000
.3528	.4018	.4074	.4084	.4074	.4018	.3528	3.1416	.7071

PANEL CONFIGURATION AND STRENGTH:

I	X(I)	Y(I)	XB(I)	YB(I)	S(I)	THETA(I)	LAMBDA(I)
1	-1.0000	.0000	-1.0000	-.4142	.8284	1.5708	.3765
2	-.7071	.7071	-1.0000	.4142	.8284	.7854	.2662
3	.0000	1.0000	-.4142	1.0000	.8284	0.0000	-.0000
4	.7071	.7071	.4142	1.0000	.8284	-.7854	-.2662
5	1.0000	0.0000	1.0000	.4142	.8284	-1.5708	-.3765
6	.7071	-.7071	1.0000	-.4142	.8284	-2.3562	-.2662
7	.0000	-1.0000	.4142	-1.0000	.8284	-3.1416	-.0000
8	-.7071	-.7071	-.4142	-1.0000	.8284	-3.9270	.2662

COMPARISON OF NUMERICAL RESULT WITH THE EXACT AT 8 SURFACE POINTS

X	Y	V	VEXACT	CP	CPEXACT
-1.0000	.0000	.0000	.0000	1.0000	1.0000
-.7071	.7071	1.4142	1.4142	-1.0000	-1.0000
.0000	1.0000	2.0000	2.0000	-3.0000	-3.0000
.7071	.7071	1.4142	1.4142	-1.0000	-1.0000
1.0000	0.0000	-.0000	0.0000	1.0000	1.0000
.7071	-.7071	-1.4142	-1.4142	-1.0000	-1.0000
.0000	-1.0000	-2.0000	-2.0000	-3.0000	-3.0000
-.7071	-.7071	-1.4142	-1.4142	-1.0000	-1.0000

149

clockwise direction along the surface. Thus the fluid velocity at the lower half of the cylindrical surface is in the direction opposite to that of the local t so that V has negative values there.

In Program 2.6 the panel arrangement is symmetric about the x-axis. This arrangement is not necessary, however, for a symmetric body. You may verify that, by rotating the control points of the present example through such an arbitrary angle about the cylinder axis that the panel arrangement becomes nonsymmetric, the same degree of accuracy is still obtained in the numerical result. On the other hand, if the distances between neighboring control points are chosen to be all different so that the panel lengths are unequal, you may find that the same numerical technique could give tremendous errors. A rule of thumb for the panel method is that more panels are needed in the region where the surface has a larger curvature. For example, in solving the problem of flow past an airfoil, a better result can be expected if many short panels are placed around the leading edge of the airfoil.

As mentioned earlier, panels may also be constructed by first specifying a set of boundary points on the original body. Such a procedure was followed in an example dealing with the same problem as the one presented here (Kuethe and Chow, 1976, p. 112). In that example the panel configuration is similar to the one shown in Fig. 2.8.5 except that the octagon is now confined within the circle. The body approximated by the control points, which are chosen to be the midpoints of the panels, is a circular cylinder of a radius smaller than that of the original body. It is interesting to note that despite the fact that the control points are displaced radially inward, the velocity and pressure coefficient computed there are exactly the same as those shown in the output of Program 2.6 for control points on the original cylinder. This result comes from the special form of (2.8.29) and (2.8.30) that both V and c_p evaluated at the body surface are independent of the radius of the circular cylinder.

Problem 2.11   The method of source panels is a useful tool for studying aerodynamic interactions among various components of an aircraft. To construct a simple problem, let us consider two circular cylinders of 1-m radius placed in a uniform stream flowing at a speed of 1 m/s. The centers of these cylinders are separated by a distance of 2.5 m measured in a direction perpendicular to the free stream. Eight source panels are to be used on each of the two cylinders, following the same arrangement as shown in Fig. 2.8.5. Thus the flow is determined by solving 16 simultaneous equations for the panel strengths.

Print the values of velocity and pressure coefficient at each of the 16 control points. The result will show that a low-pressure region is created in the region between the two cylinders. The effect of the resulting pressure forces is to push the cylinders toward each other.

In a two-dimensional flow, if line vortices instead of sources are distributed uniformly along a line segment, a vortex panel is obtained. The strength of a vortex panel is defined as the circulation γ around vortices contained within unit length of that panel. The flow field about a lifting body can be computed approximately by replacing the body surface with vortex panels.

Imagine that the panel on the left side of Fig. 2.8.4 is the jth vortex panel of strength γ_j on a body. Starting from (2.4.3), the expression for stream function of a line vortex, it can be shown that the velocity potential induced at the ith control point (x_i, y_i) by the vortices on the jth panel is

$$-\frac{\gamma_j}{2\pi} \int_j \tan^{-1}\left(\frac{y_i - y_j}{x_i - x_j}\right) ds_j$$

If the body is placed in a uniform flow of speed U that makes an angle α with the x-axis, and if m vortex panels are used to approximate the body, the velocity potential at the same control point will have the form

$$\phi(x_i, y_i) = U(x_i \cos\alpha + y_i \sin\alpha) - \sum_{j=1}^{m} \frac{\gamma_j}{2\pi} \int_j \tan^{-1}\left(\frac{y_i - y_j}{x_i - x_j}\right) ds_j \quad (2.8.31)$$

Problem 2.12 If the vortex panels are tangent to the body surface at the control points, the normal velocity at the ith control point must vanish; that is, $\partial\phi/\partial n_i = 0$. Verify that this boundary condition results in

$$\sum_{j=1}^{m} \frac{\gamma_j}{2\pi} J_{ij} = U \sin(\alpha - \theta_i) \quad (2.8.32)$$

in which $J_{ij} = -I'_{ij}$ and the expression for I'_{ij} has been given in (2.8.24). Verify also that the tangential velocity at the ith control point is

$$V(x_i, y_i) = U \cos(\alpha - \theta_i) + \sum_{j=1}^{m} \frac{\gamma_j}{2\pi} J'_{ij} \quad (2.8.33)$$

where J'_{ij} is exactly the same as I_{ij} defined in (2.8.17).

To demonstrate the use of vortex panels, we consider the problem of a flat-plate wing of infinite span flying at a constant speed U and at a constant angle of attack α (Fig. 2.8.6). m vortex panels of equal lengths are used to replace the airfoil of chord length c. In this simple case all the boundary

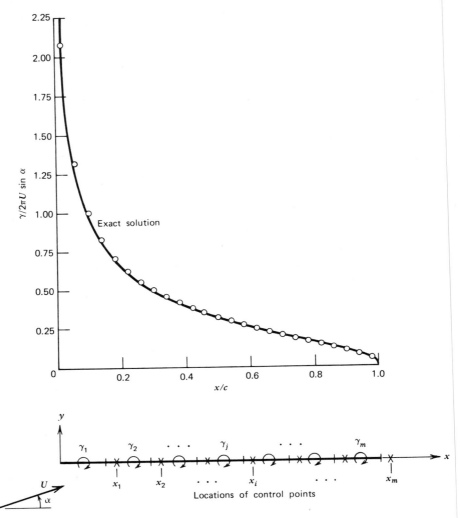

FIGURE 2.8.6 Solving the problem of uniform flow past a flat-plate wing by the method of vortex panels.

points and control points are on the x-axis and all the panel inclination angles are zero.

On a lifting body, however, there is a condition to be satisfied by the flow field in addition to those specified in (2.8.32). This is the Kutta condition of finite velocity at the sharp trailing edge of a lifting body and is described in Section 2.7. For numerical computations Kutta condition can be ap-

Inviscid Fluid Flows

proximated in various forms (Hess, 1972). In this example it may be stated as that the flow at the trailing edge must be tangent to the local airfoil surface. Because the induced velocity at an end point of a vortex panel becomes infinite and the trailing edge happens to be such a point, we will let the Kutta condition be satisfied instead at a point in the wake, which is at a small distance downstream from the trailing edge. With this additional condition, only $(m - 1)$ control points can be chosen on the airfoil in order to solve for the m unknown panel strengths. Good results are obtained by omitting the control point on the first panel, in the arrangement shown in Fig. 2.8.6. Thus the boundary conditions (2.8.32) still apply, with the understanding that the mth control point is in the wake.

A numerical computation is performed for $m = 25$ under the conditions that the last control point is $0.001c$ behind the trailing edge and the neighboring control points are separated by a distance equal to one panel length. The result is compared favorably in Fig. 2.8.6 with the exact solution (Kuethe and Chow, 1976, p. 123)

$$\gamma = 2U \sin \alpha \sqrt{\frac{c - x}{x}} \qquad (2.8.34)$$

From the Kutta-Joukowski theorem stated in Section 2.5, γ is proportional to the lift per unit area at a given location x, and therefore the plot of γ in Fig. 2.8.6 represents the lift distribution on a symmetric airfoil.

Problem 2.13 Consider a cambered airfoil whose mean camber line is represented by the parabola

$$y = 4y_m \left[\frac{x}{c} - \left(\frac{x}{c} \right)^2 \right] \qquad (2.8.35)$$

where y_m is the maximum height of the camber at the mid-chord and c is the chord. Find the lift distribution on the airfoil with $y_m = 0.1c$ flying at zero angle of attack by using the method of vortex panels, and compare it with the analytical solution (Kuethe and Chow, 1976, p. 132)

$$\gamma = 16U \frac{y_m}{c} \sqrt{\frac{x}{c} - \left(\frac{x}{c} \right)^2} \qquad (2.8.36)$$

2.9 Classification of Second-Order Partial Differential Equations

Consider a body moving through a compressible inviscid fluid. The disturbances originating from the body are propagating away in all directions at the speed of sound. If the body moves at a speed much slower than the sonic speed, these disturbances will finally reach infinity in all directions, and the flow pattern will resemble that of an incompressible potential flow past that body. In a coordinate system attached to the body, the originally uniform oncoming flow is deformed upstream as well as downstream from it. However, if the body moves at a supersonic speed, it overtakes the forward propagating disturbances so that all the disturbances created by the body are left behind. The flow pattern as seen by an observer moving with the body (e.g., a thin wedge of an infinitesimal wedge angle) is sketched in Fig. 2.9.1. The pattern is radically different from that around a body moving subsonically in that the flow upstream from the wedge is undisturbed and remains uniform. This undisturbed region is separated from the disturbed region along two straight lines called the Mach lines or, in mathematical terms, the characteristics. On the other hand, such lines do not exist in a subsonic flow.

In fact, the governing equations for these two flows are of different forms. About a thin body that produces only weak disturbances, the governing equation for a two-dimensional subsonic flow is (Kuethe and Chow, 1976, Section 11.3)

$$(1 - M^2) \frac{\partial^2 \phi}{\partial x^2} + \frac{\partial^2 \phi}{\partial y^2} = 0 \tag{2.9.1}$$

in which ϕ is the velocity potential and M, the Mach number, is the ratio of

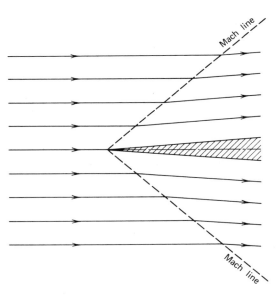

FIGURE 2.9.1 Supersonic flow past a thin wedge.

Inviscid Fluid Flows

flow speed to the speed of sound and is treated in the linearized theory as a constant. The equation takes the following from when the flow is supersonic.

$$(M^2 - 1)\frac{\partial^2 \phi}{\partial x^2} - \frac{\partial^2 \phi}{\partial y^2} = 0 \qquad (2.9.2)$$

Notice that in (2.9.1) the two coefficients have the same sign, whereas in (2.9.2) their signs are opposite. These two partial differential equations are said to be of two different types.

The type of an equation is determined by examining the existence of characteristics along which two different solutions to the same equation may be patched (as demonstrated in Fig. 2.9.1). For a general discussion let us assume that the velocity potential of a two-dimensional flow is governed by a representative equation

$$A\phi_{xx} + 2B\phi_{xy} + C\phi_{yy} = D \qquad (2.9.3)$$

in which A, B, C, and D may be functions of x, y, ϕ, ϕ_x, and ϕ_y. Here the subscripts are used to denote partial derivatives, so that ϕ_x represents $\partial\phi/\partial x$, ϕ_{xy} represents $\partial^2\phi/\partial x\,\partial y$, and so forth. If the velocity components ϕ_x and ϕ_y are both continuous functions of x and y, their changes when going from (x, y) to a neighboring point $(x + dx, y + dy)$ are, respectively,

$$dx \cdot \phi_{xx} + dy \cdot \phi_{xy} + 0 \cdot \phi_{yy} = d\phi_x \qquad (2.9.4)$$

$$0 \cdot \phi_{xx} + dx \cdot \phi_{xy} + dy \cdot \phi_{yy} = d\phi_y \qquad (2.9.5)$$

Equations (2.9.3) to (2.9.5) may be solved simultaneously for the three second-order derivatives of ϕ. For instance, by applying Cramer's rule, we find

$$\phi_{xy} = \begin{vmatrix} A & D & C \\ dx & d\phi_x & 0 \\ 0 & d\phi_y & dy \end{vmatrix} \bigg/ \begin{vmatrix} A & 2B & C \\ dx & dy & 0 \\ 0 & dx & dy \end{vmatrix} \qquad (2.9.6)$$

However, if dx and dy are the components of a displacement along a Mach line (i.e., a characteristic), the derivatives of velocity components in the direction of that line may be discontinuous because along it two different solutions may be patched. Accordingly, along a characteristic, the second-order derivatives of ϕ are indeterminate. Setting the denominator of (2.9.6) equal to zero gives

$$A\left(\frac{dy}{dx}\right)^2 - 2B\frac{dy}{dx} + C = 0 \qquad (2.9.7)$$

Classification of Second-Order Partial Differential Equations **155** where dy/dx is the slope of a characteristic. Its expression is obtained by solving the preceding equation.

$$\left(\frac{dy}{dx}\right)_{char} = \frac{B \pm \sqrt{B^2 - AC}}{A} \qquad (2.9.8)$$

Similarly, by setting the numerator of (2.9.6) equal to zero, the slope of a characteristic is determined in the hodograph plane using velocity components as the coordinates.

There are three possible results for the slope of a characteristic in the physical plane, depending on the magnitude of the expression under the radical of (2.9.8). They are stated separately as follows.

If $B^2 - AC > 0$ there exist two characteristics at a point whose slopes are represented by the two real values computed from the right-hand side of (2.9.8). The partial differential equation is then classified as being of *hyperbolic* type. The equations governing steady or unsteady wave motions usually belong to this category. One example is (2.9.2), the equation describing perturbations in a supersonic flow, in which $A = M^2 - 1$, $B = 0$, and $C = -1$, so that $B^2 - AC = M^2 - 1 > 0$. The slopes of the two characteristics are $\pm 1/\sqrt{M^2 - 1}$ according to (2.9.8), which are exactly the slopes of Mach lines in a supersonic flow (Kuethe and Chow, 1976, p. 268).

If $B^2 - AC < 0$, the right-hand side of (2.9.8) becomes complex. The existence of characteristics is impossible, and the corresponding partial differential equation is classified as being of *elliptic* type. Examples are the Laplace and Poisson equations introduced in Section 2.1 for incompressible potential flows, and the governing subsonic equation (2.9.1). In the latter case $A = 1 - M^2$, $B = 0$, and $C = 1$ and, therefore, $B^2 - AC = -(1 - M^2) < 0$.

Finally, if $B^2 - AC = 0$, there is only one real value for the slope, and such a partial differential equation is called of *parabolic* type. The following example is the equation governing the unsteady thermal conduction in a one-dimensional conductor.

$$\frac{\partial T}{\partial t} = k \frac{\partial^2 T}{\partial x^2} \qquad (2.9.9)$$

where T denotes temperature, x and t represent, respectively, spatial coordinate and time, and k is the thermal diffusivity of the conducting medium. A comparison between (2.9.9) and the standard form (2.9.3) gives $A = k$, $B = C = 0$, so that $B^2 - AC = 0$. It will be shown in the next chapter that the equation describing diffusion of vorticity in a viscous fluid also belongs to this class.

These three types of partial differential equations may be solved using different numerical techniques, which will be introduced in this and the following chapters.

2.10 Numerical Methods for Solving Elliptic Partial Differential Equations

Suppose that the Poisson equation

$$\frac{\partial^2 f}{\partial x^2} + \frac{\partial^2 f}{\partial y^2} = q(x, y) \tag{2.10.1}$$

which is an elliptic partial differential equation, is to be solved in a rectangular region subject to the condition that the values of f are prescribed on the boundary of that domain. To introduce a commonly used numerical technique, the rectangular grid system defined previously in Fig. 2.4.2 is adopted and is redrawn in Fig. 2.10.1. We use the notation (i, j) to indicate the intersecting point of the vertical grid line passing through x_i and the horizontal grid line through y_j. Our first step is to approximate the partial differential equation (2.10.1) by a finite difference equation at the representative interior point (i, j).

Let $f_{i,j}$ and $q_{i,j}$ represent $f(x_i, y_j)$ and $q(x_i, y_j)$, respectively. Applying the central-difference formula (2.2.9) for computing second-order derivatives and referring to the five neighboring points as shown in Fig. 2.10.1, we have

$$\frac{\partial^2 f}{\partial x^2} = \frac{1}{(\Delta x)^2}(f_{i+1,j} - 2f_{i,j} + f_{i-1,j}) - O\left[(\Delta x)^2 \left(\frac{\partial^4 f}{\partial x^4}\right)_{i,j}\right]$$

$$\frac{\partial^2 f}{\partial y^2} = \frac{1}{(\Delta y)^2}(f_{i,j+1} - 2f_{i,j} + f_{i,j-1}) - O\left[(\Delta y)^2 \left(\frac{\partial^4 f}{\partial y^4}\right)_{i,j}\right]$$

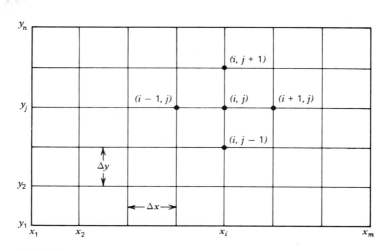

FIGURE 2.10.1 Rectangular domain.

Solving Elliptic Partial Differential Equations

The differential equation (2.10.1) may then be replaced by the finite-difference form

$$\frac{1}{(\Delta x)^2}(f_{i+1,j} - 2f_{i,j} + f_{i-1,j}) + \frac{1}{(\Delta y)^2}(f_{i,j+1} - 2f_{i,j} + f_{i,j-1}) = q_{i,j} \qquad (2.10.2)$$

with a truncation error of the order of $(\Delta x)^2 + (\Delta y)^2$.

For a square grid we let $\Delta x = \Delta y = h$, so that the preceding equation becomes

$$f_{i,j} = \tfrac{1}{4}(f_{i-1,j} + f_{i+1,j} + f_{i,j-1} + f_{i,j+1}) - \tfrac{1}{4}h^2 q_{i,j} \qquad (2.10.3)$$

Vanishing of q reduces (2.10.1) to the Laplace equation, and (2.10.3) states that the value of f at an interior point is equal to the average of the values of f at the four adjoining points. This statement is called the *mean value theorem* for harmonic functions that satisfy the Laplace equation.

When (2.10.2) or (2.10.3) is applied at all interior points bounded within the domain of Fig. 2.10.1, we obtain $(m-2)(n-2)$ algebraic equations that automatically include the boundary conditions. Solving these equations simultaneously, we can, theoretically, find the unknown values of f at the interior grid points. However, practically, this direct procedure is tedious and time-consuming, especially when the number of grid points is large. It is therefore necessary to develop some methods of solution that can conveniently be carried out on a computer.

One of these numerical techniques, using an iterative scheme, is called *Liebmann's method*. In this method values of f are first guessed at all interior points in addition to those prescribed at the boundary points on the edges of the given domain. These values are represented by $f_{i,j}^{(0)}$, with the superscript 0 indicating the zeroth iteration. In order to facilitate the following discussion, the square mesh is chosen so that the simplified difference equation (2.10.3) will be used. The values of f are computed for the next iteration by executing (2.10.3) at every interior point based on the values of f at the present iteration. The sequence of computation starts from the interior point situated at the lower left corner, proceeds upward until reaching the top, and then goes to the bottom of the next vertical line on the right. This process is repeated until the new value of f at the last interior point at the upper right corner has been obtained.

Applying (2.10.3) at the first point gives

$$f_{2,2}^{(1)} = \tfrac{1}{4}[f_{1,2}^{(0)} + f_{3,2}^{(0)} + f_{2,1}^{(0)} + f_{2,3}^{(0)}] - \tfrac{1}{4}h^2 q_{2,2} \qquad (2.10.4)$$

in which $f_{1,2}^{(0)}$ and $f_{2,1}^{(0)}$ are the constant boundary values and therefore do not vary with the number of iterations. They may be replaced, respectively, by $f_{1,2}^{(1)}$ and $f_{2,1}^{(1)}$ in (2.10.4). The computation at the next point involves $f_{2,2}^{(0)}$. Since an

Inviscid Fluid Flows

improved value $f_{2,2}^{(1)}$ is available at this time, it will be used instead. Thus

$$f_{2,3}^{(1)} = \tfrac{1}{4}[f_{1,3}^{(1)} + f_{3,3}^{(0)} + f_{2,2}^{(1)} + f_{2,4}^{(0)}] - \tfrac{1}{4}h^2 q_{2,3} \qquad (2.10.5)$$

where $f_{1,3}^{(1)}$ is used to replace the constant boundary value $f_{1,3}^{(0)}$. Repeating this argument, we deduce a general formula for computing values of f at the $(n + 1)$th iteration:

$$f_{i,j}^{(n+1)} = \tfrac{1}{4}[f_{i-1,j}^{(n+1)} + f_{i+1,j}^{(n)} + f_{i,j-1}^{(n+1)} + f_{i,j+1}^{(n)}] - \tfrac{1}{4}h^2 q_{i,j} \qquad (2.10.6)$$

This equation applies to any interior point, no matter whether it is next to some boundary points or not. It holds for the former case, as already shown by (2.10.4) and (2.10.5). If (i, j) is a deep interior point, the first and third terms on the right-hand side of (2.10.6) represent, respectively, the values of f at the grid points to the left of and below that point. These values have already been recalculated according to our sequence and, therefore, carry the superscript $(n + 1)$. (2.10.6) is named *Liebmann's iterative formula*.

We now need to prove that as $n \to \infty$, $f_{i,j}^{(n)}$ approaches $f_{i,j}$, the solution to the finite difference equation. For a simpler analysis we will examine, instead of (2.10.6), an iterative scheme called *Richardson's iterative formula*,

$$f_{i,j}^{(n+1)} = \tfrac{1}{4}[f_{i-1,j}^{(n)} + f_{i+1,j}^{(n)} + f_{i,j-1}^{(n)} + f_{i,j+1}^{(n)}] - \tfrac{1}{4}h^2 q_{i,j} \qquad (2.10.7)$$

which, as will be shown later, converges slower than (2.10.6) toward the final solution. Subtracting (2.10.3) from (2.10.7) and defining the error at (i, j) during the nth iteration as

$$e_{i,j}^{(n)} = f_{i,j}^{(n)} - f_{i,j} \qquad (2.10.8)$$

we obtain

$$e_{i,j}^{(n+1)} = \tfrac{1}{4}[e_{i-1,j}^{(n)} + e_{i+1,j}^{(n)} + e_{i,j-1}^{(n)} + e_{i,j+1}^{(n)}]$$

which implies that

$$|e_{i,j}^{(n+1)}| \leqslant \tfrac{1}{4}[|e_{i-1,j}^{(n)}| + |e_{i+1,j}^{(n)}| + |e_{i,j-1}^{(n)}| + |e_{i,j+1}^{(n)}|] \qquad (2.10.9)$$

If $E^{(n)}$ represents the highest value among the absolute errors at all grid points during the nth iteration, (2.10.9) leads to the conclusion that

$$E^{(n+1)} \leqslant E^{(n)} \qquad (2.10.10)$$

This relationship is not sufficient, however, to prove that $E^{(n)} \to 0$ as $n \to \infty$. The way in which $E^{(n)}$ shrinks can be estimated as follows.

Referring to Fig. 2.10.1, let the points adjoining to at least one boundary point be called the first-layer points, those adjoining to at least one first-layer point be called the second-layer points, and so forth. When (2.10.9) is applied at a first-layer point, at least one of the four adjoining points is a boundary point where the error is zero. Therefore,

$$|e_{i,j}^{(n+1)}| \leqslant \tfrac{3}{4}E^{(n)} = (1 - \tfrac{1}{4})E^{(n)} \qquad (2.10.11)$$

Solving Elliptic Partial Differential Equations

At a second-layer point, at least one of the adjoining points is a first-layer point. Applying (2.10.9) at such a point gives, by using (2.10.11),

$$|e_{i,j}^{(n+1)}| \leqslant \tfrac{1}{4}[3E^{(n)} + (1 - \tfrac{1}{4})E^{(n-1)}]$$

$$\leqslant \tfrac{1}{4}[3E^{(n-1)} + (1 - \tfrac{1}{4})E^{(n-1)}] = \left(1 - \frac{1}{4^2}\right)E^{(n-1)}$$

Similarly, we deduce a generalized expression for the absolute error at an mth-layer point.

$$|e_{i,j}^{(n+1)}| \leqslant \left(1 - \frac{1}{4^m}\right)E^{(n-m+1)}$$

If the total number of layers in the grid system is M, the maximum absolute error during an iteration should occur on this innermost layer according to the preceding expression. Thus we have

$$E^{(n+1)} \leqslant \left(1 - \frac{1}{4^M}\right)E^{(n-M+1)}$$

which gives, by assigning appropriate values to n,

$$E^{(M)} \leqslant \left(1 - \frac{1}{4^M}\right)E^{(0)}$$

$$E^{(2M)} \leqslant \left(1 - \frac{1}{4^M}\right)E^{(M)}$$

and so forth. In other words, if N is any positive integer, the maximum error at the (NM)th iteration is

$$E^{(NM)} \leqslant \left(1 - \frac{1}{4^M}\right)^N E^{(0)} \tag{2.10.12}$$

The right-hand side approaches zero as $N \to \infty$; that is, after a large number of iterations, the values of f computed from (2.10.7) will approach the solution to the finite-difference equation (2.10.3). Because of the condition (2.10.10), Liebmann's iterative formula (2.10.6) will converge toward the final solution faster than Richardson's iterative formula (2.10.7).

An improved numerical technique, called the *successive overrelaxation* (*S.O.R.*) *method*, gives an even faster convergence than Liebmann's method in solving the Poisson equation. This method used the following iteration scheme for a rectangular domain of square meshes.

$$f_{i,j}^{(n+1)} = f_{i,j}^{(n)} + \frac{\omega}{4}[f_{i-1,j}^{(n+1)} + f_{i+1,j}^{(n)} + f_{i,j-1}^{(n+1)} + f_{i,j+1}^{(n)} - 4f_{i,j}^{(n)} - h^2 q_{i,j}] \tag{2.10.13}$$

in which ω is a constant to be determined. The formula is equivalent to

(2.10.7) for $\omega = 1$. The iterated result converges for $1 \leq \omega < 2$, and it converges most rapidly when ω is assigned the optimum value

$$\omega_{opt} = \frac{8 - 4\sqrt{4 - \alpha^2}}{\alpha^2} \qquad (2.10.14)$$

with $\alpha = \cos(\pi/m) + \cos(\pi/n)$, where m and n are, respectively, the total number of increments into which the horizontal and vertical sides of the rectangular region are divided. The expression for ω must be modified for nonuniform grid sizes or nonrectangular domains. A more general formulation for the overrelaxation method and its discussions can be found in Chapter III of the book by Roache (1972).

As the first application of Liebmann's formula (2.10.6), we consider the flow caused by a distribution of vorticity within a rectangular domain. Vorticity vector ζ is defined as the curl of the velocity vector, or

$$\nabla \times \mathbf{V} = \zeta \qquad (2.10.15)$$

For two-dimensional fluid motion on the x-y plane, the velocity components may be derived from the stream function ψ according to (2.1.14), and the vorticity is a vector in the z direction whose magnitude is designated by ζ. Thus (2.10.15) can be written in the form of a scalar equation,

$$\frac{\partial^2 \psi}{\partial x^2} + \frac{\partial^2 \psi}{\partial y^2} = -\zeta(x, y) \qquad (2.10.16)$$

which belongs to the class of (2.10.1). Based on (2.10.6), the numerical scheme for solving (2.10.16) is

$$\psi_{i,j} = \tfrac{1}{4}(\psi_{i-1,j} + \psi_{i+1,j} + \psi_{i,j-1} + \psi_{i,j+1} + h^2 \zeta_{i,j}) \qquad (2.10.17)$$

Here the superscripts used to denote the number of iterations have been omitted, because they are automatically taken care of if this equation is applied successively at interior points in the same order as that described prior to (2.10.4).

In Program 2.7, we examine how the circular streamlines of a line vortex deform when it is confined within a finite domain defined by $-3 \leq x \leq 3$ and $-2 \leq y \leq 2$. The location of the line vortex of finite vorticity ζ_0 is fixed at (x_0, y_0). With the value 0.25 assigned to h, which is the size of increments in both x and y directions, there will be 425 grid points corresponding to $m = 25$ and $n = 17$ in the notation of Fig. 2.10.1.

Since the stream function may take on an arbitrary additive constant, the value zero will be assigned to the bounding streamline coinciding with the boundary of the rectangular region. To start the iteration, a constant value of 0.5 is guessed for the stream function at all interior points. After a new value of ψ is computed from (2.10.17) at an interior point, the absolute value of the difference between this and the previous value of ψ at the same point is calculated and is added to the value of a variable called ERROR, whose starting value at the beginning of an iteration is zero. At the end of one iteration, when the values of ψ at all interior points have been updated, the value of ERROR is compared with ERRMAX, which is the maximum allowable value of the total error. If ERROR is less than or equal to ERRMAX, the desired accuracy has been reached, so the iterating process can be stopped. Otherwise, the iteration counter ITER is increased by 1 and a new iteration is started. In our program we let ERRMAX = 0.001. In other words, on the average, the maximum error allowed at each of the 345 interior points is approximately 3×10^{-6}.

In order not to spend too much computing time on the problem, a maximum allowable number of 200 iterations is assigned to ITMAX. Iteration process stops when the iteration counter reaches this maximum value, even if the result has not yet reached the desired accuracy.

List of Principal Variables in Program 2.7

Program Symbol	Definition
(Main Program)	
CONST	A one-dimensional array containing values of stream function assigned to the plotted streamlines
ERROR, ERRMAX	Total error during one iteration and its maximum allowable value
H	Grid size in both x and y directions
ITER, ITMAX	Iteration counter and its maximum allowable value
I0, J0	Location of line vortex in index notation
M, N	Number of grid lines in x and y directions, respectively
PLTC	An array storing plotting characters
PSI, PSIPRV	Stream function at the present and at the previous iteration, respectively
X, Y	Coordinates of grid points
XMAX, YMIN, YMAX, YMIN	Limiting values on the abscissa and ordinate, respectively
X0, Y0	Coordinates of line vortex
ZETA	Vorticity
ZETA0	Vorticity at center of vortex line

```
C                      ***** PROGRAM 2.7 *****

C           SOLVING POISSON EQUATION FOR THE FLOW AROUND A VORTEX BOUNDED
C           WITHIN A RECTANGULAR REGION USING LIEBMANN,S ITERATIVE FORMULA

            DIMENSION X(25),Y(17),XX(100),YY(100),CONST(5),PLTC(5),PSI(25,17),
           1          PSIPRV(25,17),ZETA(25,17),LCHAR(3),NSCALE(4)
            COMMON M,N,H

C           ..... ASSIGN INPUT DATA .....
            DATA NSCALE / 0, 1, 0, 1 /
            DATA XMAX,XMIN,YMAX,YMIN / 3., -3., 2., -2. /
            DATA ZETA0,X0,Y0,ITMAX,ERRMAX / 100., 1., 1., 200, 0.001 /
            DATA CONST / .05, 0.2, 0.5, 1.0, 1.5 /
            DATA PLTC / 1H$, 1H., 1HX, 1H0, 1H* /
            H = 0.25

C           ..... COMPUTE COORDINATES FOR GRID POINTS .....
            M = (XMAX-XMIN)/H + 1
            N = (YMAX-YMIN)/H + 1
            X(1) = XMIN
            DO 1   I = 2,M
          1 X(I) = X(I-1) + H
            Y(1) = YMIN
            DO 2   J = 2,N
          2 Y(J) = Y(J-1) + H

C           ..... ASSIGN BOUNDARY VALUES TO PSI .....
            DO 3   I = 1,M
            PSI(I,1) = 0.
          3 PSI(I,N) = 0.
            NM1 = N-1
            DO 4   J = 2,NM1
            PSI(1,J) = 0.
          4 PSI(M,J) = 0.

C           ..... CONVERT COORDINATES (X0,Y0) OF THE VORTEX TO (I0,J0) IN
C                 INDEX NOTATION. THEN ASSIGN VALUES TO ZETA WHICH HAS A
C                 NONVANISHING VALUE OF ZETA0 AT (X0,Y0) .....
            I0 = (X0-XMIN)/H + 1
            J0 = (Y0-YMIN)/H + 1
            DO 5   I = 1,M
            DO 5   J = 1,N
          5 ZETA(I,J) = 0.
            ZETA(I0,J0) = ZETA0

C           ..... ASSIGN GUESSED VALUES TO PSI .....
            MM1 = M-1
            DO 6   I = 2,MM1
            DO 6   J = 2,NM1
          6 PSI(I,J) = 0.5

C           ..... LET ITERATION COUNTER START FROM ZERO .....
            ITER = 0
```

```
C      ..... START AN ITERATION BY INCREASING ITERATION COUNTER BY ONE
C            AND SETTING ERROR INITIALLY TO ZERO.  BEFORE APPLYING
C            LIEBMANN,S FORMULA, THE LOCAL VALUES OF PSI ARE STORED IN
C            THE ARRAY PSIPRV.  ABSOLUTE DIFFERENCES BETWEEN TWO
C            CONSECUTIVE APPROXIMATIONS TO PSI AT INDIVIDUAL INTERIOR
C            POINTS ARE SUMMED AND STORED IN ERROR .....
       DO 7  I = 1,M
       PSIPRV(I,1) = PSI(I,1)
     7 PSIPRV(I,N) = PSI(I,N)
       DO 8   J = 2,NM1
       PSIPRV(1,J) = PSI(1,J)
     8 PSIPRV(M,J) = PSI(M,J)
     9 ITER = ITER + 1
       ERROR = 0.
       DO 10   I = 2,MM1
       DO 10   J = 2,NM1
       PSIPRV(I,J) = PSI(I,J)
       PSI(I,J) = (PSI(I-1,J) + PSI(I+1,J) + PSI(I,J-1) + PSI(I,J+1)
      A            + H**2*ZETA(I,J)) / 4.0
    10 ERROR = ERROR + ABS(PSI(I,J)-PSIPRV(I,J))

C      ..... PRINT PSI IF FINAL VALUE OF ERROR IS LESS THAN OR EQUAL
C            TO ERRMAX, OR IF THE NUMBER OF ITERATIONS HAS REACHED THE
C            VALUE ITMAX, OTHERWISE GO BACK FOR ANOTHER ITERATION .....
       IF( ERROR.LE.ERRMAX )  GO TO 11
       IF( ITER-ITMAX )  9,11,11
    11 WRITE(6,21)  M,N,ITER,ERROR
       DO 12  K = 1,N,2
       L = N - K + 1
    12 WRITE(6,22)  ( PSI(I,L), I=1,M,3 )

C      ..... PRINT PSIPRV TO MAKE SURE THAT THE RESULT IS SATISFACTORY
C            AT EVERY GRID POINT, THEN SKIP TO THE NEXT PAGE .....
       WRITE(6,23)
       DO 13   K = 1,N,2
       L = N - K + 1
    13 WRITE(6,22)  ( PSIPRV(I,L), I=1,M,3 )
       WRITE(6,24)

C      ..... PLOT FIVE REPRESENTATIVE STREAMLINES .....
       CALL PLOT1( NSCALE, 3, 20, 3, 50, 1H-, 1HI )
       CALL PLOT2( XMAX, XMIN, YMAX, YMIN, .TRUE. )
       DO 14   I = 1,5
       CALL SERCH2( X, Y, PSI, CONST(I), XX, YY, KMAX )
    14 CALL PLOT3( PLTC(I), XX, YY, 1, KMAX, 1 )

C      ..... CENTER OF THE VORTEX IS MARKED BY A CROSS .....
       XX(1) = X0
       YY(1) = Y0
       CALL PLOT3( 1H+, XX, YY, 1, 1, 1 )
       LCHAR(1) = 10H
       CALL PLOT4( LCHAR, 10 )

C      ..... PRINT FIGURE NUMBER AND CAPTION .....
       WRITE(6,25)  ( (PLTC(I),CONST(I)), I=1,5 )
```

164

```
21 FORMAT( 1H1, ////// 19X, 3HM =, I3, 5X, 3HN =, I3, 9X,
  A          23HNUMBER OF ITERATIONS = , I3, 9X, 15HSUM OF ERRORS =,
  B          F8.4 // 12X, 24HTHE RESULT FOR PSI(I,J): // )
22 FORMAT( 10X, 9E12.4 / )
23 FORMAT( //// 12X, 35HTHE RESULT FOR PSI(I,J) IN PREVIOUS,
  A          11H ITERATION: // )
24 FORMAT( 1H1, ////// )
25 FORMAT( // 33X, 31HFIGURE 2.10.2   STREAMLINES OF ,
  A          39HA VORTEX BOUNDED BY A RECTANGULAR WALL. // 29X,
  B          A1, 9H FOR PSI=, F4.2, 3H  , A1, 9H FOR PSI=, F3.1,
  C          3H  , A1, 9H FOR PSI=, F3.1, 3H  , A1, 9H FOR PSI=,
  D          F3.1, 3H  , A1, 9H FOR PSI=, F3.1 )

    END




    SUBROUTINE SERCH2( X, Y, PSI, PSIA, XX, YY, KMAX )
C   ..... THIS IS A MODIFIED VERSION OF SUBROUTINE SEARCH WRITTEN
C         FOR PROGRAM 2.2, WITH THE ADDITION OF SEARCHING THROUGH
C         HORIZONTAL GRID LINES FOR POINTS ON A STREAMLINE.  THE
C         NOTATIONS USED HERE ARE THE SAME AS THOSE IN SEARCH .....

    DIMENSION X(25),Y(17),PSI(25,17),XX(100),YY(100)
    COMMON M,N,DY
    EQUIVALENCE (DX,DY)

C   ..... SEARCH THROUGH VERTICAL GRID LINES .....
    K = 0
    I = 0
  1 J = 1
    I = I + 1
    IF( I.GT.M ) GO TO 7
  2 P = PSI(I,J) - PSIA
    IF( ABS(P).LE.0.00001 )  GO TO 6
  3 J = J + 1
    IF( J.GT.N ) GO TO 1
    Q = PSI(I,J) - PSIA
    IF( P*Q ) 5,5,4
  4 P = Q
    GO TO 3
  5 K = K + 1
    XX(K) = X(I)
    YY(K) = Y(J) - DY*ABS(Q)/(ABS(P)+ABS(Q))
    P = Q
    GO TO 3
  6 K = K + 1
    XX(K) = X(I)
    YY(K) = Y(J)
    J = J + 1
    IF( J.GT.N ) GO TO 1
    GO TO 2
```

```
C      ..... SEARCH THROUGH HORIZONTAL GRID LINES .....
    7 J = 0
    8 I = 1
      J = J + 1
      IF( J.GT.N )  GO TO 14
    9 P = PSI(I,J) - PSIA
      IF( ABS(P).LE.0.00001 )  GO TO 13
   10 I = I + 1
      IF( I.GT.M )  GO TO 8
      Q = PSI(I,J) - PSIA
      IF( P*Q )  12,12,11
   11 P = Q
      GO TO 10
   12 K = K + 1
      XX(K) = X(I) - DX*ABS(Q)/(ABS(P)+ABS(Q))
      YY(K) = Y(J)
      P = Q
      GO TO 10
   13 K = K + 1
      XX(K) = X(I)
      YY(K) = Y(J)
      I = I + 1
      IF( I.GT.M )  GO TO 8
      GO TO 9
   14 KMAX = K
      RETURN
      END
```

      M = 25   N = 17   NUMBER OF ITERATIONS = 153   SUM OF ERRORS =   .0010

THE RESULT FOR PSI(I,J):

0.	0.	0.	0.	0.	0.	0.	0.	0.
0.	3.2182E-02	8.1210E-02	1.7963E-01	4.1169E-01	9.1304E-01	7.0016E-01	2.4871E-01	0.
0.	5.7460E-02	1.4250E-01	3.0533E-01	6.8182E-01	1.9776E+00	1.2621E+00	4.0653E-01	0.
0.	7.1310E-02	1.7216E-01	3.4847E-01	6.8826E-01	1.2779E+00	1.0196E+00	4.1581E-01	0.
0.	7.2657E-02	1.7007E-01	3.2226E-01	5.5623E-01	7.9920E-01	6.8755E-01	3.3603E-01	0.
0.	6.3227E-02	1.4376E-01	2.5695E-01	4.0088E-01	5.0805E-01	4.4402E-01	2.3764E-01	0.
0.	4.6029E-02	1.0230E-01	1.7515E-01	2.5606E-01	3.0435E-01	2.6575E-01	1.4808E-01	0.
0.	2.4101E-02	5.2804E-02	8.8118E-02	1.2432E-01	1.4319E-01	1.2457E-01	7.0536E-02	0.
0.	0.	0.	0.	0.	0.	0.	0.	0.

THE RESULT FOR PSI(I,J) IN PREVIOUS ITERATION:

0.	0.	0.	0.	0.	0.	0.	0.	0.
0.	3.2184E-02	8.1213E-02	1.7964E-01	4.1170E-01	9.1304E-01	7.0016E-01	2.4871E-01	0.
0.	5.7464E-02	1.4251E-01	3.0533E-01	6.8182E-01	1.9776E+00	1.2621E+00	4.0653E-01	0.
0.	7.1316E-02	1.7217E-01	3.4848E-01	6.8826E-01	1.2779E+00	1.0196E+00	4.1581E-01	0.
0.	7.2663E-02	1.7007E-01	3.2226E-01	5.5624E-01	7.9921E-01	6.8755E-01	3.3602E-01	0.
0.	6.3233E-02	1.4376E-01	2.5696E-01	4.0089E-01	5.0805E-01	4.4401E-01	2.3764E-01	0.
0.	4.6034E-02	1.0230E-01	1.7516E-01	2.5606E-01	3.0435E-01	2.6575E-01	1.4808E-01	0.
0.	2.4104E-02	5.2808E-02	8.8122E-02	1.2432E-01	1.4319E-01	1.2457E-01	7.0535E-02	0.
0.	0.	0.	0.	0.	0.	0.	0.	0.

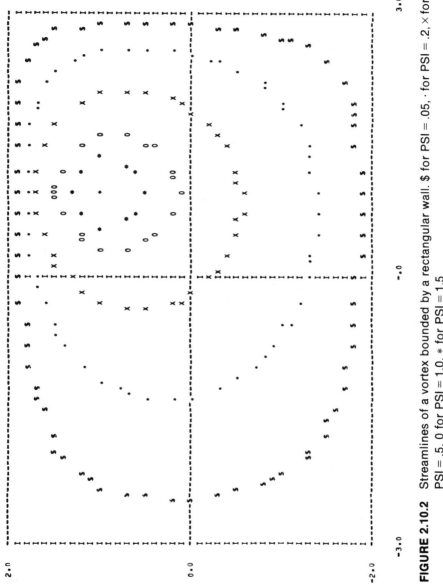

FIGURE 2.10.2 Streamlines of a vortex bounded by a rectangular wall. $ for PSI = .05, · for PSI = .2, × for PSI = .5, 0 for PSI = 1.0, * for PSI = 1.5

167

The final values of ψ are printed only at some selected grid points because of the limited space for the output. The values in the previous iteration are also printed for comparison. Points on several $\psi = $ constant curves are then located by searching through both vertical and horizontal grid lines. This is achieved by calling a subroutine named SERCH2, a modified version of subroutine SEARCH used in Program 2.2 that searches only through the vertical grid lines for such points. Finally, the flow pattern shown in Fig. 2.10.2 is printed by use of subroutines PLOTN.

Problem 2.14 Rerun Program 2.7 after replacing Liebmann's iterative formula (2.10.17) by the successive overrelaxation formula (2.10.13). Compare the efficiency of these two methods in view of the number of iterations required to obtain a solution of the same accuracy.

Problem 2.15 Run Program 2.7 with another concentrated vorticity added in the rectangular domain. If the two vorticities are of opposite signs, the result will show that the two vortices are confined within two regions separated by a branch of the $\psi = 0$ streamline. However, if the vorticities are of the same sign, the two regions containing the vortices are connected by a streamline having the shape of an "8," which is encircled by other streamlines.

2.11 Potential Flows in Ducts or Around Bodies—Irregular and Derivative Boundary Conditions

The flow considered in Section 2.10 is bounded by a rectangular wall along which the stream function assumes a constant value. In many flow problems, however, the surfaces of a body or wall may be of complicated geometries, or there may exist surfaces along which the derivatives of stream function instead of the function itself are known. Two examples are shown in this section to demonstrate the techniques used for handling irregular and derivative boundary conditions. In the meantime, an alternative way of plotting the flow pattern is presented.

The first example is concerned with a rectangular chamber whose upper right corner is blocked off by a plate, making a 45° angle with the x-axis. A two-dimensional flow field is established when an inviscid incompressible fluid is flowing steadily through the chamber between an inlet and an outlet, both situated on the bottom wall. Dimensions of the chamber are specified in Fig.

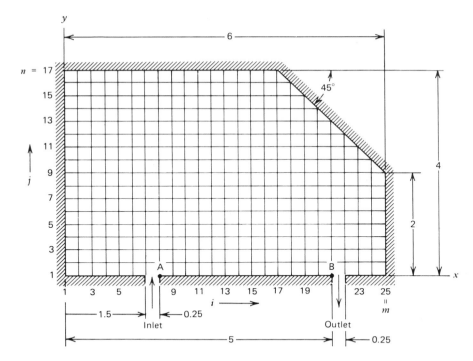

FIGURE 2.11.1 A chamber with irregular boundaries.

2.11.1. When a square mesh of size $h = 0.25$ is laid, as shown in the figure, the slant surface intersects grid lines exactly at some grid points, which are naturally the boundary points on that surface. The more general case, in which boundary points do not coincide with grid points, will be taken up in the next example.

In the absence of vorticity, the governing equation is deduced from (2.10.16).

$$\frac{\partial^2 \psi}{\partial x^2} + \frac{\partial^2 \psi}{\partial y^2} = 0 \tag{2.11.1}$$

To solve this equation the successive overrelaxation scheme (2.10.13) will be employed in the present example. For computer calculations the iteration formula can be written as

$$\psi_{i,j} = (1 - \omega)\psi_{i,j} + \frac{\omega}{4}[\psi_{i-1,j} + \psi_{i+1,j} + \psi_{i,j-1} + \psi_{i,j+1}] \tag{2.11.2}$$

in which the $\psi_{i,j}$ on the right-hand side represents its value at the nth iteration

Potential Flows in Ducts or Around Bodies   **169**

and that on the left-hand side represents the value at the next iteration. The optimum value of ω given by (2.10.14) will be used in the computation.

It is known that stream function represents the volume flow rate passing between a point and a reference streamline. Let it be normalized so that the net flow rate through the chamber is unity. In Fig. 2.11.1 the line AB between the inlet and outlet is chosen to be the base streamline along which $\psi = 0$. Then $\psi = 1$ is the streamline described by the remaining part of the chamber wall. Having specified the boundary conditions, we can now apply (2.11.2) at interior points following exactly the same iterative procedure as that adopted for executing (2.10.17) in Program 2.7, except that the interior domain is now redefined to accommodate the slant surface. The numerical solution is considered satisfactory when the sum of errors at all interior points is less than or equal to 0.001.

The result will be presented graphically in a different fashion. The fluid region outlined in Fig. 2.11.1 is to be printed by filling it with an array of symbols, each signifying the range within which the local value of the stream function lies. Such a plot is often used to display a rough view of a solution.

Suppose the dimensions of a given domain are XRANGE and YRANGE, respectively. The entire region is to be covered by NY printed horizontal lines, each containing NX symbols. The printer of most computers prints 10 symbols per inch horizontally, but only 6 lines per inch vertically. Thus if NX is prescribed, NY cannot be arbitrary and is equal to the truncated integer value of 0.6(NX)(YRANGE/XRANGE). The dimensions represented by a printed symbol are therefore DELX = XRANGE/NX and DELY = YRANGE/NY, respectively.

Consider the case in which a printed symbol is smaller in size than a grid cell, whose area is $h \times h$ in the present problem, as sketched in Fig. 2.11.2. The grid point at the lower left corner of the cell is designated (i, j) in index notation. Lines passing through the center of the symbol divide the cell into four rectangular regions of areas A_1, A_2, A_3, and A_4 as marked in the figure following the clockwise direction.

Let the origin of the coordinate system be fixed at the lower left corner of the chamber in Fig. 2.11.1. For a symbol printed in line number JSYMBL counting from the top and in position number ISYMBL on that line counting from the left, the coordinates of its center are, respectively,

$$X = (\text{ISYMBL} - \tfrac{1}{2})(\text{DELX}) \tag{2.11.3}$$
$$Y = \text{YRANGE} - (\text{JSYMBL} - \tfrac{1}{2})(\text{DELY}) \tag{2.11.4}$$

The location of the cell containing (X, Y) can readily be found by assigning the expressions $1 + X/h$ and $1 + Y/h$, respectively, to the integer variables i and j. The value of stream function at the center of the printed symbol can be

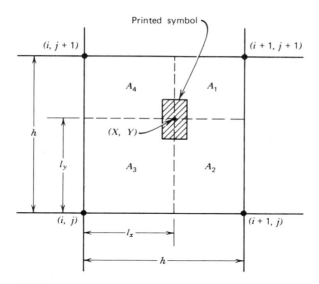

FIGURE 2.11.2 A symbol printed in a grid cell.

interpolated from the values at the four corners of the containing cell, which are already known from the solution to the Laplace equation (2.11.1). It can easily be verified that, when linear interpolation is used, the stream function at (X, Y) is

$$\psi_0 = [A_1\psi_{i,j} + A_2\psi_{i,j+1} + A_3\psi_{i+1,j+1} + A_4\psi_{i+1,j}]/h^2 \qquad (2.11.5)$$

The four areas are the weighting factors, and they are computed from the lengths $l_x = X - (i - 1)h$ and $l_y = Y - (j - 1)h$ according to Fig. 2.11.2.

After ψ_0 has been found, the remaining task is to convert this value to a particular symbol. In this case it is known that the stream function varies from 0 to 1 in the flow field. If this whole range is equally divided into nine subranges, the size of each subrange is then $\delta = \frac{1}{9}$. In the computer program we assign symbols according to the following table.

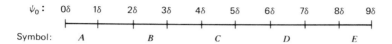

The subranges are numbered from left to right in an ascending order beginning with 1. The table shows that blanks are assigned to the even-numbered subranges. For a given value of ψ_0, the corresponding subrange number is obtained by first calculating the numerical value of $(1 + \psi_0/\delta)$ and then

Potential Flows in Ducts or Around Bodies

171

truncating it after the decimal point. This number determines the symbol to be printed from an array named SYMBOL, whose nine alphanumeric elements are the alphabets and blanks arranged in the same order as that shown in the table. By analogy, a graph can be constructed having the whole stream function range subdivided into any desired number of intervals.

To print the flow pattern, we start from the top printing line with JSYMBL = 1, compute the stream function at the center of each of the NX symbols in that line according to (2.11.3) to (2.11.5), and then convert them into symbols that are stored as elements of an array named GRAPH. After it is printed, GRAPH will be used to store symbols for the second printing line. The procedure is repeated until the last line is printed.

Care must be taken when a printing line crosses the slant surface of the chamber wall, since some symbols in that line are contained by triangular instead of rectangular cells (see Fig. 2.11.1). We use the notation (ILAST(J), J), for $J = 9, 10, \ldots, 16$, to designate the grid points at the lower left corners of these triangular cells. Instead of deriving an interpolation formula for a triangular region, we instruct the computer to print the symbol E at any location inside a triangular cell, assuming that these cells are very close to the streamline $\psi = 1$. This method is simple, but may cause errors in the plot when coarse meshes are used for the grid system. Blanks are printed in the space to the right of the slant surface.

Figure 2.11.3, the output of Program 2.8, is actually a plot of stream tubes in the flow field. Thus a wider cross section along a stream tube indicates a region of slower flow speed. In this flow fast speeds are found at the inlet and outlet openings. To obtain the final solution, 41 iterations were performed. However, tests show that if Liebmann's method is used instead, the number of interactions is increased to 256 for a solution having the same accuracy! The saving in computing time by using the successive overrelaxation method is amazing.

As a second example of dealing with various boundary conditions, we consider a channel flow past a circular cylinder, described in Fig. 2.11.4. The flow enters the channel from a small central opening and leaves it at a section downstream from the cylinder. Assume that buffers are installed at the exit so that the flow there becomes horizontal. In this problem it is more convenient to place the origin of the coordinate system at the center of the circular cylinder. Because of its symmetry about the x-axis, only the upper half of the flow will be computed. Again, a square mesh of size $h = 0.25$ is used to cover the region of interest.

The governing equation (2.11.1) will be solved numerically using Liebmann's iterative formula.

$$\psi_{i,j} = \tfrac{1}{4}(\psi_{i-1,j} + \psi_{i+1,j} + \psi_{i,j-1} + \psi_{i,j+1}) \tag{2.11.6}$$

The boundary conditions are: (1) $\psi = 0$ on the cylinder surface and on the

List of Principal Variables in Program 2.8

Program Symbol	Definition
ALPHA	The parameter α in (2.10.14)
A1, A2, A3, A4	The four areas defined in Fig. 2.11.2
DELTA	Size of a subrange within the whole range of stream function
DELX, DELY	Lengths in x and y directions represented by a printed symbol
ERROR	The same as that defined in Program 2.7
GRAPH	An array used to store symbols for a printed line
H	h, size of square grid mesh
I, J	position of a grid point in index notation
ILAST	Value of i at the last interior point on a horizontal grid line cutting the slant surface of the chamber
ISYMBL, JSYMBL	Position of a printed symbol in the matrix containing all such symbols, counting from the upper left corner of the matrix
ITER	Iteration counter
JLAST	Value of j at the last grid point on a vertical grid line
LX, LY	Lengths l_x and l_y defined in Fig. 2.11.2
M, N	Total numbers of vertical and horizontal lines, respectively, in the grid system
NX, NY	Total number of symbols in a line and total number of lines in the matrix containing plotting symbols
OMEGA	The parameter ω in (2.10.13), computed according to (2.10.14)
PSI, PSIPRV	Steam function at a grid point and its value in the previous iteration
PSIO	Stream function at center of a printed symbol
SYMBOL	An array containing plotting symbols
X, Y	Coordinates at the center of a printed symbol
XLNGTH	Chamber length in x direction, which is determined by the slant surface
XRANGE, YRANGE	Dimensions of rectangular domain containing the chamber

Potential Flows in Ducts or Around Bodies

```
C                    ***** PROGRAM 2.8 *****
C          NUMERICAL SOLUTION OF FLOW THROUGH A CHAMBER HAVING AN
C          IRREGULAR BOUNDARY USING SUCCESSIVE OVERRELAXATION METHOD
      REAL LX,LY
      DIMENSION SYMBOL(9),ILAST(17),JLAST(25),PSI(25,17),GRAPH(100)
C     ..... ASSIGN INPUT DATA .....
      DATA XRANGE,YRANGE,M,N,H,NX / 6.0, 4.0, 25, 17, 0.25, 100 /
      DATA SYMBOL / 1HA, 1H , 1HB, 1H , 1HC, 1H , 1HD, 1H , 1HE /
      MM1 = M-1
      NM1 = N-1
C     ..... DEFINE THE LAST VALUE OF J ALONG EACH VERTICAL GRID LINE ...
      DO 1  I = 1,17
    1 JLAST(I) = 17
      DO 2  I = 18,M
    2 JLAST(I) = N - I + 17
C     ..... ASSIGN BOUNDARY VALUES TO PSI .....
      DO 3  J = 1,N
    3 PSI(1,J) = 1.0
      DO 4  J = 1,9
    4 PSI(M,J) = 1.0
      DO 5  I = 2,MM1
      PSI(I,1) = 1.0
      IF( I.GT.7 .AND. I.LT.22 )  PSI(I,1) = 0.0
      J = JLAST(I)
    5 PSI(I,J) = 1.0
C     ..... ASSIGN GUESSED VALUES TO PSI .....
      DO 6  I = 2,MM1
         JEND = JLAST(I) - 1
      DO 6  J = 2,JEND
    6 PSI(I,J) = 0.5
C     ..... ITERATE THE SUCCESSIVE OVERRELAXATION FORMULA UNTIL
C           THE SUM OF ERRORS IS LESS THAN OR EQUAL TO 0.001 .....
      PI = 4.0 * ATAN(1.0)
      ALPHA = COS(PI/M) + COS(PI/N)
      OMEGA = (8.-4.*SQRT(4.-ALPHA**2)) / ALPHA**2
      ITER = 0
    7 ITER = ITER + 1
      ERROR = 0.0
      DO 8  I = 2,MM1
         JEND = JLAST(I) - 1
      DO 8  J = 2,JEND
      PSIPRV = PSI(I,J)
      PSI(I,J) = (1.0-OMEGA)*PSI(I,J) + OMEGA/4.0*
     A          ( PSI(I-1,J) + PSI(I+1,J) + PSI(I,J-1) + PSI(I,J+1) )
    8 ERROR = ERROR + ABS(PSI(I,J)-PSIPRV)
      IF( ERROR.GT.0.001 )  GO TO 7
      WRITE(6,9)  OMEGA,ITER
    9 FORMAT( 1H1, /// 45X, 8HOMEGA = , F6.4, 10X,
     A        23HNUMBER OF ITERATIONS = , I2 /// )
C     ..... PREPARE FOR PLOTTING.  0.001 ADDED IN THE STATEMENT BELOW
C           IS USED TO AVOID TRUNCATION ERROR .....
      NY = 0.6*NX*YRANGE/XRANGE + 0.001
      DELX = XRANGE / NX
      DELY = YRANGE / NY
      DELTA = 1.0 / 9.0
      DO 10  J = 9,16
   10 ILAST(J) = M - J + 8
```

174

```
C         ..... PRINT THE UPPER HALF OF THE GRAPH CONTAINING
C                THE SLANT SURFACE OF CHAMBER WALL .....
          NYHALF = NY / 2
          DO 14  JSYMBL = 1,NYHALF
          Y = YRANGE - (JSYMBL-0.5)*DELY
          J = 1 + Y/H
              LY = Y - (J-1)*H
              HMLY = H - LY
              XLNGTH = 4.0 + (YRANGE-Y)
          DO 13  ISYMBL = 1,NX
          X = (ISYMBL-0.5) * DELX
          IF( X.GE.XLNGTH )  GO TO 11
          I = 1 + X/H
          IF( I.EQ.ILAST(J) )  GO TO 12
              LX = X - (I-1)*H
              HMLX = H - LX
              A1 = HMLX * HMLY
              A2 = HMLX * LY
              A3 = LX * LY
              A4 = LX * HMLY
          PSIO = ( A1*PSI(I,J) + A2*PSI(I,J+1) + A3*PSI(I+1,J+1)
         A         + A4*PSI(I+1,J) ) / H**2
          NRANGE = 1 + PSIO/DELTA
          GRAPH(ISYMBL) = SYMBOL(NRANGE)
          GO TO 13
       11 GRAPH(ISYMBL) = 1H
          GO TO 13
       12 GRAPH(ISYMBL) = 1HE
       13 CONTINUE
       14 WRITE(6,15)  GRAPH
       15 FORMAT( 20X, 100A1 )

C         ..... PRINT THE LOWER HALF OF THE GRAPH .....
          NYNEXT = NYHALF + 1
          DO 17  JSYMBL = NYNEXT,NY
          Y = YRANGE - (JSYMBL-0.5)*DELY
          J = 1 + Y/H
              LY = Y - (J-1)*H
              HMLY = H - LY
          DO 16  ISYMBL = 1,NX
          X = (ISYMBL-0.5)*DELX
          I = 1 + X/H
              LX = X - (I-1)*H
              HMLX = H - LX
              A1 = HMLX * HMLY
              A2 = HMLX * LY
              A3 = LX * LY

              A4 = LX * HMLY
          PSIO = ( A1*PSI(I,J) + A2*PSI(I,J+1) + A3*PSI(I+1,J+1)
         A         + A4*PSI(I+1,J) ) / H**2
          NRANGE = 1 + PSIO/DELTA
       16 GRAPH(ISYMBL) = SYMBOL(NRANGE)
       17 WRITE(6,15)  GRAPH

C         ..... PRINT FIGURE CAPTION .....
          WRITE(6,18)
       18 FORMAT( ///// 45X, 30HFIGURE 2.11.3   PATTERN OF FLO,
         A         20HW THROUGH A CHAMBER. )
          END
```

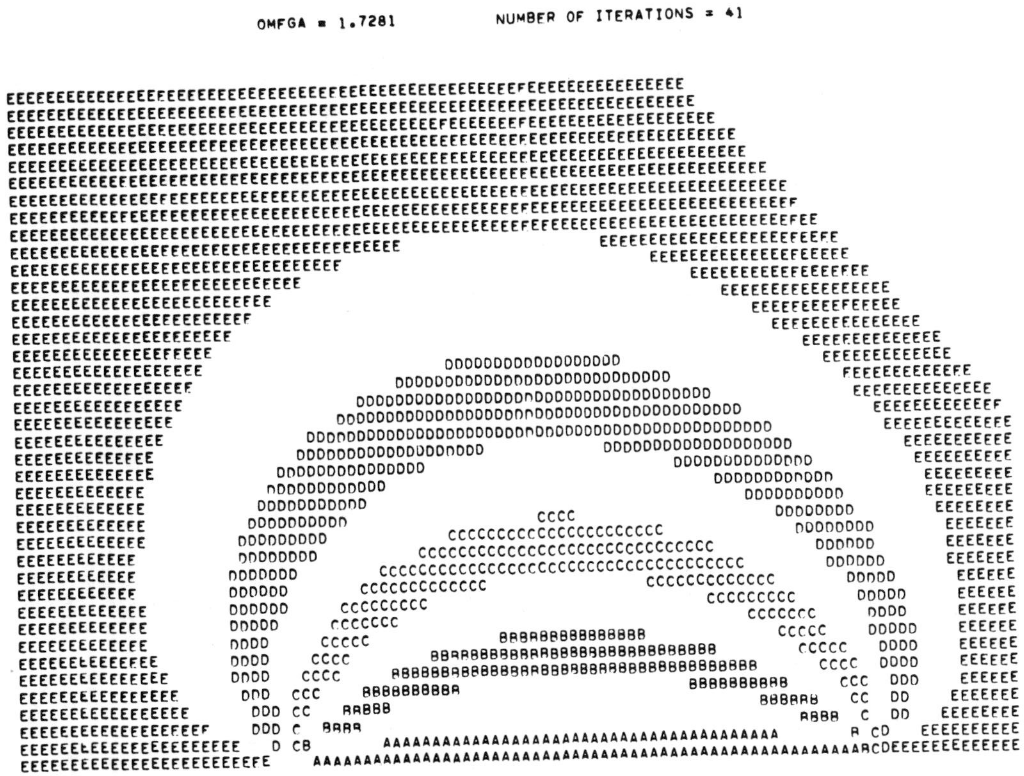

FIGURE 2.11.3 Pattern of flow through a chamber.

176

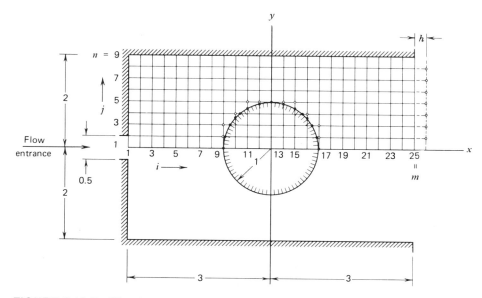

FIGURE 2.11.4 Circular cylinder inside a channel.

x-axis outside the body; (2) ψ is a constant, say 1, on the upper wall of the channel; and (3) $v = -\partial\psi/\partial x = 0$ at the exit section.

Two problems arise when specifying these boundary conditions. Boundary points on the cylinder are defined as the intersections of the body surface and the horizontal and vertical grid lines. These points, as marked by small solid circles in Fig. 2.11.4, do not usually coincide with the grid points. Thus the formula (2.11.6) cannot be used at some grid points in the immediate neighborhood of the cylinder. On the other hand, a numerical technique is needed to incorporate the derivative boundary condition at the exit.

Equation (2.11.6) is a finite difference approximation of the Laplace equation (2.11.1), based on the assumption that point (i, j) is at equal distances from its four neighboring grid points. To solve the first problem, the difference equation (2.11.6) has to be generalized for an arbitrary situation, shown in Fig. 2.11.5, in which the four neighboring points are at different distances a, b, c, and d from point (i, j). We let the stream functions evaluated at these neighboring points be denoted ψ_a, ψ_b, ψ_c, and ψ_d, respectively, and then approximate the left-hand side of (2.11.1) at (i, j) by the following linear function:

$$\left(\frac{\partial^2\psi}{\partial x^2} + \frac{\partial^2\psi}{\partial y^2}\right)_{i,j} = \alpha_0\psi_{i,j} + \alpha_a\psi_a + \alpha_b\psi_b + \alpha_c\psi_c + \alpha_d\psi_d \qquad (2.11.7)$$

Potential Flows in Ducts or Around Bodies

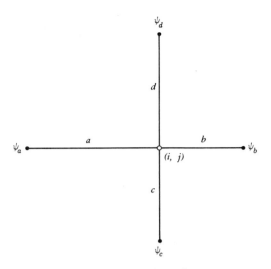

FIGURE 2.11.5 Evaluation of $\psi_{i,j}$.

where the α's are coefficients to be determined. Each of the four stream functions can be expanded in Taylor's series about (i, j). For example,

$$\psi_a = \psi_{i,j} - a\left(\frac{\partial \psi}{\partial x}\right)_{i,j} + \tfrac{1}{2}a^2\left(\frac{\partial^2 \psi}{\partial x^2}\right)_{i,j} - O(a^3)$$

$$\psi_d = \psi_{i,j} + d\left(\frac{\partial \psi}{\partial y}\right)_{i,j} + \tfrac{1}{2}d^2\left(\frac{\partial^2 \psi}{\partial y^2}\right)_{i,j} + O(d^3)$$

and so forth. Substitution of these equations into (2.11.7) gives, after rearrangement and dropping the cubic and higher-order terms,

$$\left(\frac{\partial^2 \psi}{\partial x^2} + \frac{\partial^2 \psi}{\partial y^2}\right)_{i,j} = (\alpha_0 + \alpha_a + \alpha_b + \alpha_c + \alpha_d)\psi_{i,j} + (b\alpha_b - a\alpha_a)\left(\frac{\partial \psi}{\partial x}\right)_{i,j} + (d\alpha_d - c\alpha_c)$$

$$\times \left(\frac{\partial \psi}{\partial y}\right)_{i,j} + \tfrac{1}{2}(a^2\alpha_a + b^2\alpha_b)\left(\frac{\partial^2 \psi}{\partial x^2}\right)_{i,j} + \tfrac{1}{2}(c^2\alpha_c + d^2\alpha_d)\left(\frac{\partial^2 \psi}{\partial y^2}\right)_{i,j}$$

Equating corresponding coefficients on the two sides results in five simultaneous algebraic equations whose solution is

$$\alpha_0 = -2\left(\frac{1}{ab} + \frac{1}{cd}\right), \qquad \alpha_a = \frac{2}{a(a+b)}, \qquad \alpha_b = \frac{2}{b(a+b)},$$

$$\alpha_c = \frac{2}{c(c+d)}, \qquad \alpha_d = \frac{2}{d(c+d)}$$

Vanishing of the right-hand side of (2.11.7) gives the desired difference approximation of the Laplace equation.

$$\psi_{i,j} = \left[\frac{\psi_a}{a(a+b)} + \frac{\psi_b}{b(a+b)} + \frac{\psi_c}{c(c+d)} + \frac{\psi_d}{d(c+d)}\right]\Big/\left(\frac{1}{ab} + \frac{1}{cd}\right) \quad (2.11.8)$$

It can easily be shown that (2.11.8) is equivalent to (2.11.6) when $a = b = c = d = h$.

We now return to Fig. 2.11.4. At the boundary points on the cylinder the value $\psi = 0$ is assigned. By inspection we have picked out 10 grid points, marked by hollow circles in the neighborhood of the cylinder, at which the generalized formula (2.11.8) must apply. These grid points can be divided into two groups, depending on whether they are on the left or on the right side of the cylinder. A representative point (i, j) on the left side of the cylinder is sketched in Fig. 2.11.6. At such a point $\psi_a = \psi_{i-1,j}$, $\psi_d = \psi_{i,j+1}$ and $a = d = h$. If within a distance h to the right there is a boundary point on the cylinder, we let

$$b = (13 - i)h - \sqrt{1 - [(j-1)h]^2} \quad \text{and} \quad \psi_b = 0 \quad (2.11.9)$$

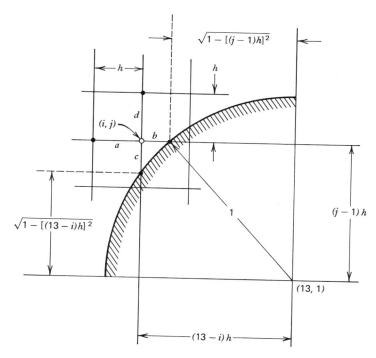

FIGURE 2.11.6   A grid point (i,j) near the left surface of the circular cylinder.

Potential Flows in Ducts or Around Bodies

Otherwise we let $b = h$ and $\psi_b = \psi_{i+1,j}$. However, in order to simplify the algorithm, the second equation in (2.11.9) will still be written as $\psi_b = \psi_{i+1,j}$, but the value zero will be assigned to stream function at all grid points inside the cylinder. Thus, if a boundary point appears below the point (i, j) within a distance h, we need only to set

$$c = (j - 1)h - \sqrt{1 - [(13 - i)h]^2} \tag{2.11.10}$$

while $\psi_c = \psi_{i,j-1}$ is automatically equal to zero.

Similarly, at a point on the right side of the cylinder, $b = d = h$, the modification for c is still (2.11.10), and that for a is

$$a = (i - 13)h - \sqrt{1 - [(j - 1)h]^2} \tag{2.11.11}$$

In this way the curved boundary is taken care of by defining the four lengths a, b, c, and d at each of the points where the formula (2.11.8) is to be used. ψ_a, ψ_b, ψ_c, and ψ_d are the values of ψ at the four neighboring grid points of (i, j) and therefore need not be redefined in the computer program.

Next we consider the derivative boundary condition that $\partial \psi / \partial x = 0$ at the exit. By using the central-difference formula (2.2.8), the condition becomes

$$\psi_{i+1,j} = \psi_{i-1,j} \qquad \text{for } i = 25 \text{ and } j = 2, \cdots, 8 \tag{2.11.12}$$

It suggests that an additional column of fictitious grid points are needed at a distance h to the right of the exit section, as indicated in Fig. 2.11.4. These fictitious points do not show specifically in the computation because, with substitution of (2.11.12) the iterative formula (2.11.6) at $i = 25$ becomes

$$\psi_{i,j} = \tfrac{1}{4}(2\psi_{i-1,j} + \psi_{i,j-1} + \psi_{i,j+1}) \tag{2.11.13}$$

In summary, there are three types of grid points besides the boundary points. They are: (1) the points in the close neighborhood of the cylinder, (2) those at the exit section of the channel, and (3) the remaining grid points inside the fluid. To classify the type, we introduce an identification number ID at each of the interior points and assign to it the integer value of 1, -1, or 0 according to the location of that point as specified in the order just mentioned. Proper actions can then be directed when ID is inserted in an arithmetic IF statement.

After the numerical solution for ψ has been obtained, the flow pattern is plotted following the same procedure as that described in Program 2.8. The flow pattern below the centerline of the channel is the mirror image of that above. It is displayed by printing a two-dimensional array named GRALOW whose elements are the plotting symbols taken properly from the upper half of the figure.

Most of the variable names used in Program 2.9 are exactly the same as those used in Program 2.8. New names are listed as follows: ID, GRALOW, A,

```
C                    ***** PROGRAM 2.9 *****

C        CHANNEL FLOW PAST A CIRCULAR CYLINDER, AN EXAMPLE ON
C        CURVED SURFACE AND THE DERIVATIVE BOUNDARY CONDITION

         REAL LX,LY
         DIMENSION SYMBOL(9),JFIRST(25),ID(25,9),PSI(25,9),A(25,9),
        A         B(25,9),C(25,9),D(25,9),GRAPH(100),GRALOW(100,20)

C        ..... ASSIGN INPUT DATA .....
         DATA XRANGE,YRANGE,M,N,H,NX / 6.0, 2.0, 25, 9, 0.25, 100 /
         DATA SYMBOL / 1H0, 1H , 1H1, 1H , 1H2, 1H , 1H3, 1H , 1H4 /
         NM1 = N-1

C        ..... DEFINE THE VALUE OF J AT THE LOWEST INTERIOR GRID
C              POINT ALONG EACH VERTICAL GRID LINE .....
         DO 1  I = 1, M
         JFIRST(I) = 2
         IF( I.GE.11 .AND. I.LE.15 )  JFIRST(I) = 5
       1 CONTINUE
         JFIRST(10) = 4
         JFIRST(13) = 6
         JFIRST(16) = 4

C        ..... ASSIGN PSI AT GRID POINTS ON OR INSIDE SOLID BOUNDARIES ....
         DO 2  I = 1, M
              J1 = JFIRST(I) - 1
         DO 2  J = 1, J1
       2 PSI(I,J) = 0.0
         DO 3  I = 1, M
       3 PSI(I,N) = 1.0
         DO 4  J = 2, 8
       4 PSI(1,J) = 1.0

C        ..... ASSIGN ID NUMBERS TO ALL INTERIOR POINTS .....
         DO 6  I = 2, M
              JA = JFIRST(I)
         DO 6  J = JA, NM1
         ID(I,J) = 0
       6 IF( I.EQ.M )  ID(I,J) = -1
         DO 7  I = 9, 17
         J = JFIRST(I)
       7 ID(I,J) = 1
         ID( 9,3) = 1
         ID(13,6) = 0
         ID(17,3) = 1

C        ..... COMPUTE A, B, C, AND D FOR THE POINTS
C              AT WHICH (2.11.A) WILL BE USED .....
         DO 9  I = 9, 17
         DO 9  J = 2, 5
         IF( ID(I,J) )  9,9,8
       8 A(I,J) = H
         B(I,J) = H
         C(I,J) = H
         D(I,J) = H
```

```
      IF( I.EQ.9 .OR. I.EQ.10 )  B(I,J) = (13-I)*H-SQRT(1.-((J-1)*H)**2)
      IF( I.EQ.16 .OR. I.EQ.17 )  A(I,J) = (I-13)*H-SQRT(1.-((J-1)*H)**2)
      IF( I.GE.10.AND.I.LE.16 )  C(I,J) = (J-1)*H-SQRT(1.-((13-I)*H)**2)
    9 CONTINUE

C     ..... ASSIGN GUESSED VALUES TO PSI .....
      DO 10  I = 2, M
            JA = JFIRST(I)
      DO 10  J = JA, NM1
   10 PSI(I,J) = 0.5


C     ..... ITERATE UNTIL SUM OF ERRORS IS ≤ 0.0005 .....
      ITER = 0
   11 ITER = ITER + 1
      ERROR = 0.0
      DO 15  I = 2, M
            JA = JFIRST(I)
      DO 15  J = JA, NM1
      PSIPRV = PSI(I,J)
→     IF( ID(I,J) )  12,13,14
C        (2.11.13) IS USED TO INCORPORATE DERIVATIVE
C        BOUNDARY CONDITION AT THE EXIT!
   12 PSI(I,J) = 0.25 * ( 2.*PSI(I-1,J) + PSI(I,J-1) + PSI(I,J+1) )
      GO TO 15
C        AT GRID POINTS AWAY FROM CURVED BOUNDARY (2.11.6) IS USED!
   13 PSI(I,J) = 0.25 * ( PSI(I-1,J) + PSI(I+1,J) + PSI(I,J-1)
     A                     + PSI(I,J+1) )
      GO TO 15
C        AT POINTS NEIGHBORING CURVED BOUNDARY (2.11.8) IS USED!
   14 APB = A(I,J) + B(I,J)
      CPD = C(I,J) + D(I,J)
      DENOM = 1.0/(A(I,J)*B(I,J)) + 1.0/(C(I,J)*D(I,J))
      PSI(I,J) = ( PSI(I-1,J)/(A(I,J)*APB) + PSI(I+1,J)/(B(I,J)*APB)
     A    + PSI(I,J-1)/(C(I,J)*CPD) + PSI(I,J+1)/(D(I,J)*CPD) ) / DENOM
   15 ERROR = ERROR + ABS(PSI(I,J)-PSIPRV)
      IF( ERROR.GT.0.0005 )  GO TO 11
      WRITE(6,16)  ITER
   16 FORMAT( 1H1, /// 57X, 23HNUMBER OF ITERATIONS = , I2 /// )


C     ..... PRINT THE UPPER HALF OF THE GRAPH .....
      NY = 0.6*NX*YRANGE/XRANGE + 0.001
      DELX = XRANGE / NX
      DELY = YRANGE / NY
      DELTA = 1.0/9.0
      DO 22  JSYMBL = 1, NY
      Y = YRANGE - (JSYMBL-0.5)*DELY
      J = 1 + Y/H
            LY = Y - (J-1)*H
            HMLY = H - LY
      DO 19  ISYMBL = 1, NX
      XX = (ISYMBL-0.5) * DELX
      X = XX - 3.0
      IF( (Y.GT.1.0) .OR. (Y.LE.1.0 .AND. ABS(X).GT.1.0) )  GO TO 17
      R = SQRT( X**2 + Y**2 )
      IF( R.LT.1.0 )  GO TO 18
```

182

```
  17 I = 1 + XX/H
        LX = XX - (I-1)*H
        HMLX = H - LX
        A1 = HMLX * HMLY
        A2 = HMLX * LY
        A3 = LX * LY
        A4 = LX * HMLY
     PSIO = ( A1*PSI(I,J) + A2*PSI(I,J+1) + A3*PSI(I+1,J+1)
    A         + A4*PSI(I+1,J) ) / H**2
        NRANGE = 1 + PSIO/DELTA
        GRAPH(ISYMBL) = SYMBOL(NRANGE)
        GO TO 19
  18 GRAPH(ISYMBL) = 1H
  19 CONTINUE
C        STORE THE IMAGE IN ARRAY GRALOW TO COMPOSE
C        THE LOWER HALF OF THE GRAPH:
        JBELOW = NY - JSYMBL + 1
        DO 21   K = 1, NX
  21        GRALOW(K,JBELOW) = GRAPH(K)
  22 WRITE(6,23)   GRAPH
  23 FORMAT( 20X, 100A1 )

C      ..... PRINT THE LOWER HALF OF THE GRAPH .....
        DO 24   J = 1, NY
  24 WRITE(6,23)   ( GRALOW(I,J), I=1,NX )

C      ..... PRINT FIGURE CAPTION .....
        WRITE(6,25)
  25 FORMAT( ///// 43X, 29HFIGURE 2.11.7   CHANNEL FLOW ,
    A          25HPAST A CIRCULAR CYLINDER. )
        END
```

B, C, and D have just been defined, JFIRST(I) is the value of j at the lowest interior grid point on the ith grid line, and R is the radial distance of the center of a printed symbol from the origin.

In this program the ID numbers are assigned to grid points according to Fig. 2.11.4. This is not a tedious process in the present case, since the number of grid points near the curved boundary is not large. An elaborate algorithm could actually be developed to let the computer examine and decide the type of each of the grid points.

The output of Program 2.9, Fig. 2.11.7, shows that the flow speed on the right side of the cylinder is slower than that on the left. The pressure difference has a tendency to push the body upstream.

Problem 2.16 Compute and plot the channel flow past a cylinder of rectangular cross section shown in Fig. 2.11.8. The potential flow pattern is assumed to be symmetric about both x- and y-axes, so that only the upper left quarter of the flow needs to be computed. It is further assumed that the flow entering the section AB is approximately uniform and the speed there is unity.

Potential Flows in Ducts or Around Bodies

183

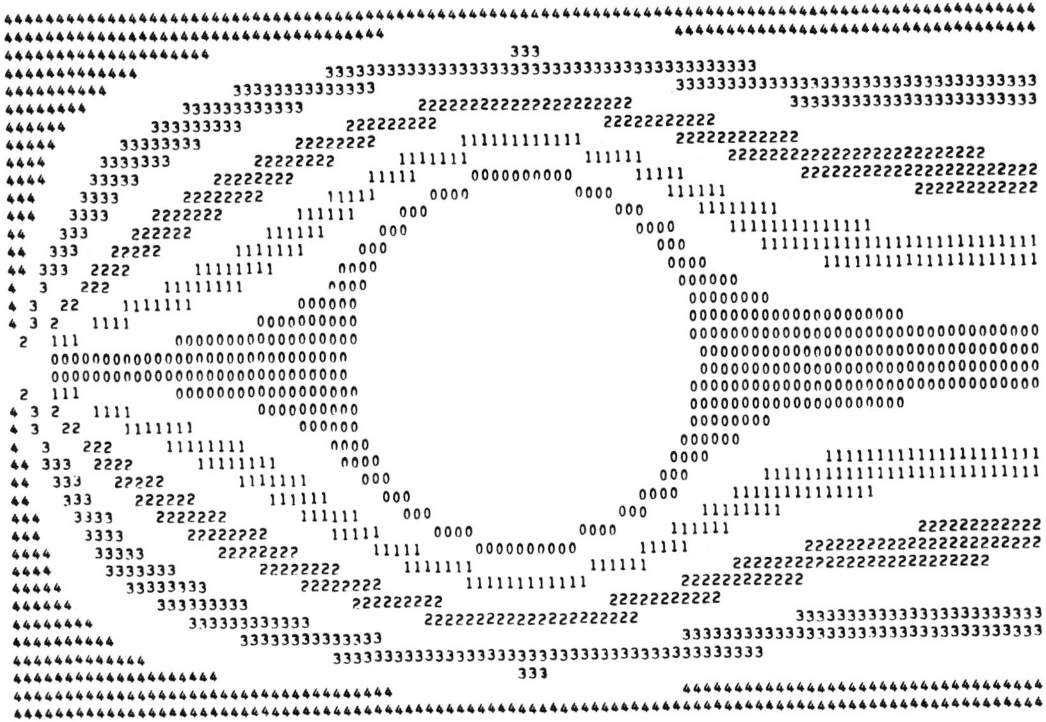

FIGURE 2.11.7 Channel flow past a circular cylinder.

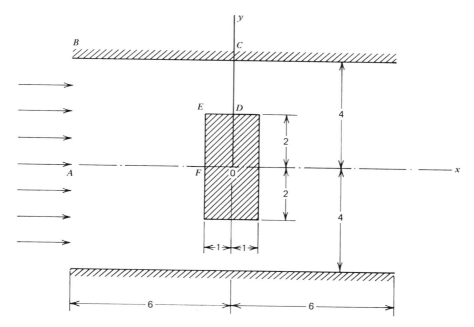

FIGURE 2.11.8 Channel flow past a rectangular cylinder.

The boundary conditions are that $\psi = 0$ along $AFED$, $\psi = 1$ along BC, $\psi = y/4$ along AB, and $\partial\psi/\partial x = 0$ along CD. In the computation let $h = 0.25$.

Problem 2.17 The equation governing the stream function of an axisymmetric potential flow was derived previously as (2.6.2) having the form

$$\frac{\partial^2\psi}{\partial r^2} - \frac{1}{r}\frac{\partial\psi}{\partial r} + \frac{\partial^2\psi}{\partial z^2} = 0$$

Show that for a square mesh of size h, this partial differential equation may be solved numerically by using the following iterative formula:

$$\psi_{i,j} = \frac{1}{4}\left[\psi_{i-1,j} + \psi_{i+1,j} + \left(1 + \frac{h}{2r_j}\right)\psi_{i,j-1} + \left(1 - \frac{h}{2r_j}\right)\psi_{i,j+1}\right]$$

$$(2.11.14)$$

where r_j is the radial distance of point (i, j) from the z-axis.

Using this numerical scheme, compute and then plot the flow in the region bounded by $OABCDO$ within an axisym-

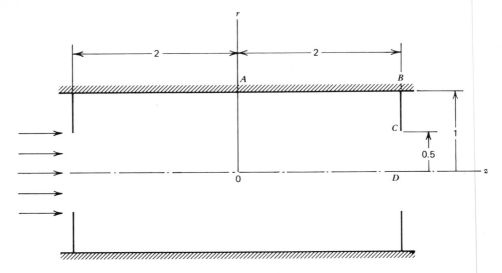

FIGURE 2.11.9 Axisymmetric flow through a tube containing repeated partitions.

metric tube containing repeated partitions (see Fig. 2.11.9). Choose a square mesh of size $h = 0.1$ for numerical computation.

Around the boundary of this region, $\psi = 0$ along OD, $\psi = 1$ along ABC, and $\partial\psi/\partial x = 0$ along both OA and CD. The derivative boundary conditions are the result of symmetry of stream function about these two sections.

2.12 Numerical Solution of Hyperbolic Partial Differential Equations

Problems concerning wave motions in fluid mechanics are governed by hyperbolic partial differential equations. One example mentioned in Section 2.9 is the supersonic flow past a thin body whose governing equation is (2.9.2). Another commonly cited example is the propagation of a one-dimensional sound wave of small amplitude, described by (see Liepmann and Roshko, 1957, p. 68)

$$\frac{\partial^2 u}{\partial t^2} = a^2 \frac{\partial^2 u}{\partial x^2} \tag{2.12.1}$$

in which t is time, x is the coordinate in the direction of wave propagation, a is the speed of sound treated as constant in the linearized analysis, and u is

Inviscid Fluid Flows

the fluid speed. It can be shown that density, pressure, and temperature are all governed by equations of the same form.

In this section a numerical technique is developed for solving (2.12.1) to find u at any time $t > 0$ in the spatial domain $0 \leqslant x \leqslant L$, provided that the initial conditions of u are given at $t = 0$ and are expressed in the following form, with functions f and g to be specified for a particular problem.

$$u(x, 0) = f(x) \tag{2.12.2}$$

$$\frac{\partial u}{\partial t}(x, 0) = g(x) \tag{2.13.3}$$

Boundary conditions are to be specified at both ends of the gaseous domain, say within a channel of constant cross-sectional area. If one end of the channel is enclosed by a rigid wall, then u must be zero there at all times. On the other hand, at an end that opens to the atmosphere, the pressure there must be a constant or, alternatively, $\partial u / \partial x$ must vanish at that section.

To solve this mixed initial-boundary-value problem numerically, we divide the spatial range of the domain into small intervals of length h and the time axis into intervals of size τ. The total number of vertical grid lines is m, whereas that of horizontal grid lines can be as many as needed in a particular computation. Lines and points in the grid system are named according to Fig. 2.12.1.

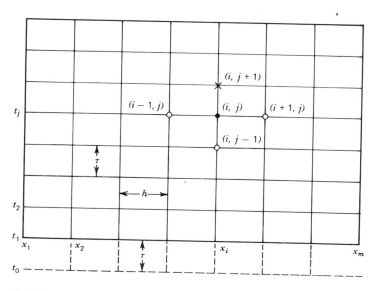

FIGURE 2.12.1 Grid system for numerical computation.

Numerical Solution of Hyperbolic Partial Differential Equations

A difference equation can be derived following exactly the same procedure as that used to obtain the numerical scheme (2.10.2) for solving the Poisson equation. Using the central-difference formula to approximate the derivatives in (2.12.1), we obtain, after regrouping,

$$u_{i,j+1} = 2u_{i,j} + C^2(u_{i-1,j} - 2u_{i,j} + u_{i+1,j}) - u_{i,j-1} \qquad (2.12.4)$$

where C is a dimensionless parameter called the *Courant number*, defined by

$$C \equiv \frac{a\tau}{h} \qquad (2.12.5)$$

(2.12.4) computes the solution at a certain time level based on the solutions at two previous time levels.

Actually, various numerical schemes can be constructed for solving the same partial differential equation by using different finite-difference approximations. The applicability of a numerical scheme is determined by whether it is stable, that is, whether the numerical solution grows and becomes unbounded after repeatedly applying the scheme. In a way we have proved in Section 2.10 that Richardson's iterative formula is stable, and so is Liebmann's. Here we will use a different approach to find the conditions under which (2.12.4) is computationally stable. It turns out that the stability of this numerical scheme is determined by the magnitude of C.

Following *von Neumann's stability analysis*, we assume that time and space variables are separable and that the solution to (2.12.4) can be expanded in the form of a Fourier series. A representative Fourier component may be written

$$u_{i,j} = U_j \, e^{Iikh} \qquad (2.12.6)$$

where U_j is the amplitude at t_j of the wave component whose wave number is k, and $I = \sqrt{-1}$. Similarly,

$$u_{i,j\pm1} = U_{j\pm1} \, e^{Iikh}, \qquad u_{i\pm1,j} = U_j \, e^{I(i\pm1)kh}$$

Substituting this into (2.12.4) gives, after canceling the common factor e^{Iikh},

$$U_{j+1} = 2U_j + C^2 U_j(e^{-Ikh} + e^{Ikh} - 2) - U_{j-1}$$

By using the identity that $(e^{I\theta} + e^{-I\theta})/2 = \cos\theta$, it becomes

$$U_{j+1} = AU_j - U_{j-1} \qquad (2.12.7)$$

where $A \equiv 2[1 - C^2(1 - \cos kh)]$. By introduction of an *amplification factor* λ such that

$$U_j = \lambda U_{j-1} \qquad \text{and} \qquad U_{j+1} = \lambda U_j = \lambda^2 U_{j-1} \qquad (2.12.8)$$

(2.12.7) is reduced to

$$\lambda^2 - A\lambda + 1 = 0 \qquad (2.12.9)$$

whose roots are

$$\lambda = \frac{A}{2} \pm \sqrt{\left(\frac{A}{2}\right)^2 - 1} \qquad (2.12.10)$$

For $|A| \geqslant 2$, the roots are real, but their magnitudes are $|\lambda| \geqslant 1$; for $|A| < 2$ the magnitudes of the complex roots are less than 1. An inspection of (2.12.8) concludes that the amplitude grows indefinitely with increasing time unless $|\lambda| \leqslant 1$. Thus the inequality that $|A| \leqslant 2$ or, $A^2 \leqslant 4$, determines the condition for stability; that is, for stability,

$$[1 - C^2(1 - \cos kh)]^2 \leqslant 1$$

After expanding the left side and rearranging, we obtain

$$C^2 \leqslant \frac{2}{1 - \cos kh}$$

When $\cos kh$ varies from -1 to $+1$, the function on the right-hand side varies from 1 to infinity, of which the lowest value is chosen to insure stability. Therefore the stability criterion for the numerical scheme (2.12.4) is $C^2 \leqslant 1$, or

$$\frac{a\tau}{h} \leqslant 1 \qquad (2.12.11)$$

To arrive at this expression, we have used the fact that each of the three variables on its left is positive. This relationship implies that τ and h cannot be chosen independently.

If the Courant number is chosen to be

$$\frac{a\tau}{h} = 1 \qquad (2.12.12)$$

(2.12.4) takes an especially simple form.

$$u_{i,j+1} = u_{i-1,j} + u_{i+1,j} - u_{i,j-1} \qquad (2.12.13)$$

As shown in Fig. 2.12.1, this equation states that the value of u at a grid point marked by a cross is computed from the values already computed at three circled grid points at two previous time steps. The numerical scheme (2.12.13) is commonly referred to as the *leapfrog method.* It will now be proved that this numerical method actually gives the exact solution to the differential equation (2.12.1).

It is well known that the solution to (2.12.1) satisfying initial conditions (2.12.2) and (2.12.3) is

$$u(x, t) = \tfrac{1}{2}[f(x + at) + f(x - at)] + \frac{1}{2a} \int_{x-at}^{x+at} g(v) \, dv \qquad (2.12.14)$$

which can be verified by substituting it back into each of those equations. The

Numerical Solution of Hyperbolic Partial Differential Equations

solution may be written in the simpler functional form

$$u(x, t) = F(x - at) + G(x + at) \qquad (2.12.15)$$

where the functions F and G represent simple waves propagating without changing shape along the positive and negative x directions at constant speed a. The lines of slope $dx/dt = \pm a$ in the x-t plane, which trace the progress of the waves, are called the characteristics of the wave equation (see Liepmann and Roshko, 1957, p. 69).

When applied at a grid point (x_i, t_j), (2.12.15) becomes

$$u_{i,j} = F(x_i - at_j) + G(x_i + at_j)$$

From Fig. 2.12.1 we have

$$x_i = x_1 + (i - 1)h \qquad \text{and} \qquad t_j = t_1 + (j - 1)\tau$$

so that

$$u_{i,j} = F(\alpha + ih - ja\tau) + G(\beta + ih + ja\tau)$$

in which $\alpha \equiv (x_1 - h) - a(t_1 - \tau)$ and $\beta \equiv (x_1 - h) + a(t_1 - \tau)$. With $a\tau = h$, obtained from the condition that $C = 1$, it reduces to

$$u_{i,j} = F[\alpha + (i - j)h] + G[\beta + (i + j)h]$$

According to this relation, the right-hand side of (2.12.13) is rewritten

$$
\begin{aligned}
u_{i-1,j} + u_{i+1,j} - u_{i,j-1} &= F[\alpha + (i - j - 1)h] + G[\beta + (i + j - 1)h] \\
&\quad + F[\alpha + (i - j + 1)h] + G[\beta + (i + j + 1)h] \\
&\quad - F[\alpha + (i - j + 1)h] - G[\beta + (i + j - 1)h] \\
&= F[\alpha + (i - j - 1)h] + G[\beta + (i + j + 1)h]
\end{aligned}
$$

which is exactly $u_{i,j+1}$ or the left-hand side of (2.12.13). It follows that exact solution is computed for (2.12.1) by the leapfrog scheme (2.12.13).

Having introduced the concept of characteristics, we are now in a position to interpret the physical meaning of the stability criterion (2.12.11) by use of Fig. 2.12.2. An examination of (2.12.4) reveals that the solution at the grid point P is influenced by the solution at each of the grid points at previous time steps contained within two diagonals PQ and PR of slope $(dx/dt)_n = \pm h/\tau$. Thus the region $PQRP$ is the domain of dependence of point P in the numerical computation. If Pq and Pr are the backward characteristics of slope $(dx/dt)_c = \pm a$ passing through P, and if $a < h/\tau$ or, equivalently, if $|(dx/dt)_c| < |(dx/dt)_n|$, these lines will lie between PQ and PR, as shown in the figure. However, from the theory of characteristics, it is known that point P can receive signals only from the region $PqrP$, which is its physical domain of dependence. In the present case of $a\tau/h < 1$, in which the computational domain of dependence contains the physical domain of dependence, all the

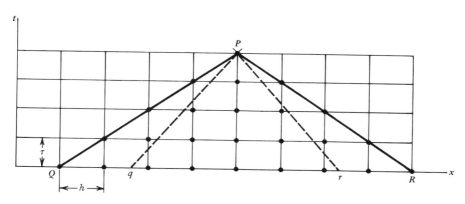

FIGURE 2.12.2 Physical interpretation of stability criterion (2.12.11).

information required to determine the condition at P is included in the computation so that the numerical scheme is stable. The result is inaccurate because of the inclusion of some unnecessary information originating from the region between PQ and Pq and the region between PR and Pr. If $a\tau/h > 1$, the characteristics Pq and Pr would be drawn outside of PQ and PR. In this case only a part of the needed information is used to determine the solution at P, and the computation is unstable. It becomes obvious that when $a\tau/h = 1$ (i.e. when the computational and physical domains of dependence coincide), the numerical solution is the exact solution.

In using the formula (2.12.13) information is needed at two previous time steps. It cannot be used directly at the initial stage to compute the solution at t_2, since conditions are specified only at the initial instant $t_1 = 0$. To help start the numerical procedure, we construct in Fig. 2.12.1 a row of fictitious grid points at $t_0 = t_1 - \tau$, and then rewrite the initial conditions (2.12.2) and (2.12.3) in index notation:

$$u_{i,1} = f_i \qquad (2.12.16)$$

$$u_{i,0} = u_{i,2} - 2\tau g_i \qquad (2.12.17)$$

in which f_i and g_i represent, respectively, $f(x_i)$ and $g(x_i)$. In obtaining the second expression we have approximated $\partial u/\partial t$ by the central-difference form (2.2.8). For $j = 1$ and with substitution from the preceding equations, (2.12.13) becomes

$$u_{i,2} = u_{i-1,1} + u_{i+1,1} - u_{i,0}$$
$$= f_{i-1} + f_{i+1} - u_{i,2} + 2\tau g_i$$

or

$$u_{i,2} = \tfrac{1}{2}(f_{i-1} + f_{i+1}) + \tau g_i, \qquad i = 2, \cdots, m-1 \qquad (2.12.18)$$

This is called the starting formula for (2.12.13).

Numerical Solution of Hyperbolic Partial Differential Equations

2.13 Propagation and Reflection of a Small-Amplitude Wave

Although exact solutions of the linear partial differential equation (2.12.1) can be written out in the form of (2.12.14), the work is still tedious when waves interact or reflect from boundaries. In this section these phenomena are examined numerically by using the leapfrog method just derived.

We consider a one-dimensional tube 1 m in length, with the left end closed and the right end open (Fig. 2.13.1). At $t = 0$ a sinusoidal wavelet of 0.4-m wavelength is somehow generated in the tube at a distance 0.2 m from the left end. Let the amplitude of the wavelet be 1 m/s, and let the functions f and g used in (2.12.2) and (2.12.3) to describe the initial conditions be defined as follows.

$$f(x) = \sin\left(2\pi \frac{x - 0.2}{0.4}\right) \qquad \text{for } 0.2 \leqslant x \leqslant 0.6$$

$$= 0 \text{ elsewhere} \tag{2.13.1}$$

$$g(x) = 0 \qquad \text{for } 0 \leqslant x \leqslant 1 \tag{2.13.2}$$

The boundary conditions are:

$$u(0, t) = 0 \tag{2.13.3}$$

$$\frac{\partial u}{\partial x}(1, t) = 0 \tag{2.13.4}$$

Under sea level conditions, the speed of sound is $a = 340$ m/s.

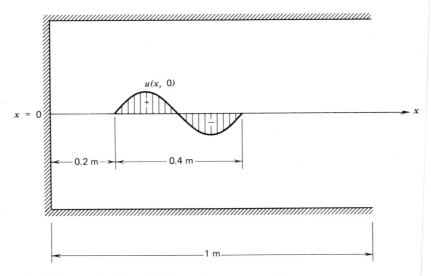

FIGURE 2.13.1 Sinusoidal wavelet in a one-dimensional tube.

For numerical computation we choose $h = 0.02$ m, so that the number of vertical grid lines is $m = 51$ and the space occupied initially by the wavelet is between grid lines $11 \leq i \leq 31$. Thus the nonvanishing f_i's for these values of i are given by

$$f_i = \sin\{5\pi[(i-1)h - 0.2]\} \tag{2.13.5}$$

and $g_i = 0$ for all values of i.

The boundary condition (2.13.3) when written in index notation is

$$u_{1,j} = 0 \qquad \text{for all } j \tag{2.13.6}$$

To handle the condition (2.13.4), which contains a derivative with respect to x, we construct a column of fictitious grid points in the x-t plane along the grid line $i = m + 1$, in the same fashion as shown in Fig. 2.12.4. Consequently, we obtain

$$u_{m+1,j} = u_{m-1,j} \qquad \text{for all } j \tag{2.13.7}$$

Under this condition the iterative formula (2.12.13) is modified at $i = m$ by

$$u_{m,j+1} = 2u_{m-1,j} - u_{m,j-1} \qquad \text{for } j > 2 \tag{2.13.8}$$

while the starting formula (2.12.18) at the same location becomes

$$u_{m,2} = f_{m-1} + \tau g_m \tag{2.13.9}$$

Finally, the size of time steps is computed from $\tau = h/a$. This is the time required for the wave to travel through the distance h. Thus it takes a time period of 10τ for the left-propagating wave to reach the closed end starting from its initial position.

After the solution $u_{i,j}$ has been obtained for $j \leq j_{max}$, instead of printing the numerical data, the result is plotted in the form of time sequential pictures. The wave shape at one instant is shown in a frame, with the abscissa representing the spatial coordinate of the tube and the ordinate the value of u. One frame is plotted every NT time steps, with NF frames on each page of the output paper.

To facilitate the plotting, a rectangular frame is first constructed consisting of 51 printed symbols on the upper and lower boundaries and 13 symbols on the left and the right boundaries. A horizontal line and a vertical line are added passing through the center of the region, and the remaining space is left blank. All the symbols are stored as elements of a two-dimensional array FRAME of dimensions 51×13. Thus the distance between two neighboring symbols on a horizontal line represents the distance h, and the ith column in FRAME can be used to describe the condition at x_i. Here we use the whole range of the ordinate to cover values of u between -1 and $+1$ m/s, with the horizontal line at the middle denoting $u = 0$. It follows that the distance between two neighboring symbols on a vertical line represents the value $\frac{1}{6}$ m/s.

Propagation and Reflection of a Small-Amplitude Wave

To plot the solution at a certain instant, we add proper symbols onto the array FRAME and then store the resultant in another two-dimensional array GRAPH. In this way the elements in GRAPH will be changed and printed from time to time while those in FRAME are kept the same. At a fixed time step j, $u_{i,j}$ at x_i is represented by a symbol located at (i, j_{wave}) in the array GRAPH. The value of j_{wave} is obtained by first truncating $[(u_{i,j}/\frac{1}{6}) \pm 0.5]$ to an integer, where 0.5 is added or subtracted, depending on whether $u_{i,j}$ is positive or negative to round off the number, and then adding to it the integer value of 7. For a meaningful presentation, we construct the ith column of GRAPH by taking the ith column from FRAME and adding to it the symbol "+" in the space between rows 7 and j_{wave} if $u_{i,j}$ is positive, or adding the symbol "−" in the space between rows j_{wave} and 7 if $u_{i,j}$ is negative. If $u_{i,j}$ vanishes, no extra symbols will be added. The wave shape is thus roughly outlined for that time step after all values of i have been examined.

Variable names used in Program 2.10 are consistent with those described in this and the previous sections, and therefore are not specifically listed in a table.

The value 5 has been assigned to NT in Program 2.10 in order to cut down the number of printed pages in the output. Suppose we let NT = 1 and let the images of GRAPH at all time steps be displayed one after another on a film strip; a movie would be produced showing the detailed development of the wave system in time. Nevertheless, important features of the solution have already been retained in the output of Program 2.10.

```
C                      ***** PROGRAM 2.10 *****

C              LEAPFROG METHOD IS USED TO STUDY THE PROPAGATION
C              OF A SINUSOIDAL WAVELET IN A TUBE WITH LEFT END
C              CLOSED AND RIGHT END OPEN TO THE ATMOSHERE

        DIMENSION F(51),G(51),U(51,116),FRAME(51,13),GRAPH(51,13)

C          ..... ASSIGN INPUT DATA .....
        DATA A,H,M,JMAX,NT,NF / 340., 0.02, 51, 116, 5, 4 /
        MM1 = M - 1
        JMAXM1 = JMAX - 1
        TAU = H / A

C          ..... DEFINE FUNCTIONS F AND G .....
        PI = 4.0 * ATAN(1.0)
        DO 1   I = 1, M
        F(I) = 0.0
        G(I) = 0.0
      1 IF( I.GE.11 .AND. I.LE.31 )  F(I) = SIN( 5.*PI*((I-1)*H-0.2) )

C          ..... ASSIGN INITIAL CONDITION TO U .....
        DO 2   I = 1, M
      2 U(I,1) = F(I)
```

```
C       ..... ASSIGN BOUNDARY CONDITION AT THE CLOSED LEFT END .....
        DO 3  J = 2, JMAX
      3 U(1,J) = 0.0

C       ..... COMPUTE SOLUTION AT J=2 USING STARTING
C             FORMULAE (2.12.18) AND (2.13.9) .....
        DO 4  I = 2, MM1
      4 U(I,2) = (F(I-1)+F(I+1))/2.0 + TAU*G(I)
        U(M,2) = F(M-1) + TAU*G(M)

C       ..... THEN COMPUTE SOLUTION AT SUBSEQUENT TIME STEPS USING
C             (2.12.13) AND (2.13.8).  DERIVATIVE BOUNDARY CONDITION
C             IS INCORPORATED IN THE LATTER FORMULA .....
        DO 6  J = 2, JMAXM1
        DO 5  I = 2, MM1
      5 U(I,J+1) = U(I-1,J) + U(I+1,J) - U(I,J-1)
        U(M,J+1) = 2.0*U(M-1,J) - U(M,J-1)
      6 CONTINUE

C       ..... CONSTRUCT THE FRAME FOR PLOTTING WAVE SHAPE .....
        DO 7  I = 1, 51
        FRAME(I,1)  = 1H=
      7 FRAME(I,13) = 1H=
        DO 8  I = 1, 51
        DO 8  JJ = 2, 12
        FRAME(I,JJ) = 1H
        IF( I.EQ.26 .OR. JJ.EQ.7 )  FRAME(I,JJ) = 1H.
      8 IF( I.EQ.1 .OR. I.EQ.51 )  FRAME(I,JJ) = 1H1

C       ..... PLOT SOLUTION EVERY NT TIME STEPS, NF FRAMES PER PAGE .....
        DO 22  J = 1, JMAX, NT
        JM1 = J - 1
        IF( JM1/NF*NF .EQ. JM1 )  GO TO 9
        GO TO 11
      9 WRITE(6,10)
     10 FORMAT( 1H1 )
     11 DO 18  I = 1, 51
        INTGER = U(I,J)*6.0 + 0.5
        IF( U(I,J).LT.0. )  INTGER = U(I,J)*6.0 - 0.5
        JWAVE = INTGER + 7
        IF( JWAVE-7 )  12, 14, 16
     12    DO 13 JJ = 1, 13
          GRAPH(I,JJ) = FRAME(I,JJ)
     13    IF( JJ.GE.JWAVE .AND. JJ.LE.7 )  GRAPH(I,JJ) = 1H-
        GO TO 18
     14    DO 15 JJ = 1, 13
     15    GRAPH(I,JJ) = FRAME(I,JJ)
        GO TO 18
     16    DO 17 JJ = 1, 13
          GRAPH(I,JJ) = FRAME(I,JJ)
     17    IF( JJ.GE.7 .AND. JJ.LE.JWAVE )  GRAPH(I,JJ) = 1H+
     18 CONTINUE
        T = JM1 * TAU
        WRITE(6,19) JM1, T
     19 FORMAT( / 19X, 6HTIME =, I3, 7H*TAU = , 1PE9.3, 8H SECONDS )
        DO 20 JINVRT = 1, 13
        JJ = 14 - JINVRT
     20 WRITE(6,21) ( GRAPH(I,JJ), I=1,51 )
     21 FORMAT( 10X, 51A1 )
     22 CONTINUE
        END
```

TIME =   0*TAU = 0.   SECONDS

TIME =   5*TAU = 2.941E-04 SECONDS

TIME = 10*TAU = 5.882E-04 SECONDS

TIME = 15*TAU = 8.824E-04 SECONDS

196

```
                TIME = 20*TAU = 1.176E-03 SECONDS
       ========================================================
       1                            .                          1
       1                            .                          1
       1                            .            +++           1
       1                            .         ++++++++         1
       1                            .         +++++++++        1
       1-------------                .........++++++++++.------1
       1---------  ---                        +++++++++++------1
       1---------  ---                             ------------1
       1 -------                                       ---     1
       1 -------                                               1
       1 -----                      .                          1
       ========--=================================================
```

```
                TIME = 25*TAU = 1.471E-03 SECONDS
       ========================================================
       1                            .                          1
       1                            .                          1
       1                            .              +++         1
       1                            .           ++++++++       1
       1                            .           +++++++++      1
       1.....----------              .........++++++++++.-----1
       1    ----------               .                 -----   1
       1    --------                 .                 -----   1
       1      ---                    .                 ----    1
       1                             .                 ----    1
       1                             .                 ---     1
       ===================================================---
```

```
                TIME = 30*TAU = 1.765E-03 SECONDS
       ========================================================
       1                            .                          1
       1                            .                          1
       1    +++                     .                          1
       1 +++++++                    .                          1
       1+++++++++                   .                          1
       1+++++++++.----------........................-------....1
       1          --------          .                          1
       1          -------           .                          1
       1            ---             .                          1
       1                            .                          1
       1                            .                          1
       =====3=================================================
```

```
                TIME = 35*TAU = 2.059E-03 SECONDS
       ====================================================++
       1                            .                      +++
       1                            .                     ++++
       1    +++                     .                     ++++
       1  +++++++                   .                     +++++
       1 +++++++++                  .                     +++++
       1....+++++++++.---------.........----------.+++++
       1              --------          ---------       1
       1              -------  .        -------         1
       1                ---    .          ---           1
       1                       .                        1
       1                       .                        1
       =====================================================
```

TIME = 40*TAU = 2.353E-03 SECONDS

TIME = 45*TAU = 2.647E-03 SECONDS

TIME = 50*TAU = 2.941E-03 SECONDS

TIME = 55*TAU = 3.235E-03 SECONDS

198

```
          TIME = 60*TAU = 3.529E-03 SECONDS
===================================================
1                                                 1
1                          .                       1
1                        + + +        + + +         1
1                      +++++++      +++++++         1
1                      +++++++++  +++++++++         1
1.......  ---------    .++++++++..++++++++.  ------------1
1          -------                  .         -------1
1          -------            .           -------1
1            ---                            ---1
1                                               1
1                                               1
===================================================
```

```
          TIME = 65*TAU = 3.824E-03 SECONDS
===================================================
1                          .                       1
1                                                 1
1                    + + +         + + +           1
1                   +++++++       +++++++           1
1                   +++++++++     +++++++++         1
1....  ---------    +++++++++. . .+++++++++.  -----1
1          -------                      -----1
1          --------                     ----1
1          ---                          ----1
1                                       ---1
1                                       ---1
===================================================
```

```
          TIME = 70*TAU = 4.118E-03 SECONDS
===================================================
1                          .                       1
1                                                 1
1                    + + +                         1
1                   +++++++                         1
1                   +++++++++                       1
1----------     .++++++++. . . . . . . . . . . . . . . 1
1---------                                         1
1  --------                                        1
1    ---                   .                       1
1                          .                       1
1                          .                       1
===================================================
```

```
          TIME = 75*TAU = 4.412E-03 SECONDS
=======================================================++
1                          .                        +++1
1                                                  ++++1
1        + + +             .                       ++++1
1       +++++++                                    ++++1
1       +++++++++                                  ++++1
1. . . . . +++++++++ . . . . . . . . . . . . . ---------.++++1
1                                 ---------         1
1                          .      -------           1
1                          .        ---             1
1                          .                        1
1                          .                        1
=======================================================
```

199

```
              TIME = 80*TAU = 4.706E-03 SECONDS
====+.+==========================================
1   +++++                   .                          1
1  +++++++                  .                          1
1  +++++++                  .                    +++   1
1 +++++++++                 .                 ++++++   1
1 +++++++++                 .                ++++++++  1
1 ++++++++++.................  .........--------.++++++++ 1
1                            .         ---------        1
1                            .          ------          1
1                            .           ---            1
1                            .                          1
1                            .                          1
==================================================
```

```
              TIME = 85*TAU = 5.000E-03 SECONDS
==================================================
1                           .                          1
1                           .                          1
1        +++                .              +++          1
1      +++++++              .            +++++++        1
1     +++++++++             .           +++++++++       1
1.....++++++++++..........--------.++++++++++.....1
1                          .--------                  1
1                          .--------                  1
1                          .   ---                    1
1                          .                          1
1                          .                          1
==================================================
```

```
              TIME = 90*TAU = 5.294E-03 SECONDS
==================================================
1                           .                          1
1                           .                          1
1        +++                .              +++          1
1      +++++++              .            +++++++        1
1     +++++++++             .           +++++++++       1
1---------.++++++++++.---------.++++++++++.........1
1---------            ---------                         1
1 -------             -------                           1
1   ---                 ---                             1
1                       .                              1
1                       .                              1
==================================================
```

```
              TIME = 95*TAU = 5.588E-03 SECONDS
==================================================
1                           .                          1
1                           .                          1
1                           .        +++               1
1                           .      +++++++             1
1                           .     +++++++++            1
1.....--------.........+++++++++++.............1
1       --------                .                       1
1        -------                .                       1
1          ---                  .                       1
1                               .                       1
1                               .                       1
==================================================
```

200

TIME =100*TAU = 5.882E-03 SECONDS

TIME =105*TAU = 6.176E-03 SECONDS

TIME =110*TAU = 6.471E-03 SECONDS

TIME =115*TAU = 6.765E-03 SECONDS

The first picture shows the initial state of the wavelet having an amplitude of 1 m/s and a wavelength of 0.4 m. It decomposes into two identical wavelets of amplitude 0.5 m/s, but of the original wavelength propagating in opposite directions. Let us call the one traveling toward the left the L wave and that toward the right the R wave. At $t = 5\tau$, when each of the two waves have traveled through a distance equal to one-quarter of the wavelength, the trailing halves of them coincide to cancel each other so that only the leading halves are shown. At $t = 10\tau$, these two waves have just come out of each other, but the L wave starts to impinge on the closed end. Numerical computations show that after being reflected from the rigid wall, the L wave reverses not only its direction of propagation, but also the sign of u. As can be observed from the figures plotted respectively at 10τ and 30τ, the incidental wave is led by positive u, whereas the reflected one is led by negative u. The three figures in between describe the interaction of the reflected portion with the oncoming part of the L wave.

To find the effect of the open end on the R wave, let us examine the figures plotted between 20τ and 40τ. The conclusion is that after being reflected from an open end, the wave reverses only its direction of propagation, but not the sign of u. Thus the reflected R wave is still led by negative u, as it was before reaching the open end. The four plots after 40τ show the interaction of the two reflected waves when passing each other. At $t = 50\tau$ they just cancel each other, and the tube becomes quiet momentarily. Further reflections and interactions are described in the remaining figures, and the process may go on forever—in an inviscid fluid.

Problem 2.18 When a compressible fluid of density ρ_0 is disturbed in such a way that its density becomes $\rho_0 + \rho'(x, t)$, where $\rho' \ll \rho_0$, the density fluctuation, similar to the fluid speed, is also governed by a wave equation

$$\frac{\partial^2 \rho'}{\partial t^2} = a^2 \frac{\partial^2 \rho'}{\partial x^2} \qquad (2.13.10)$$

This can readily be proved by combining the Euler equation (2.1.5) and the continuity equation

$$\frac{\partial \rho}{\partial t} + \nabla \cdot (\rho \mathbf{V}) = 0 \qquad (2.13.11)$$

linearizing them for one-dimensional flow, and then introducing the speed of sound $a = \sqrt{dp/d\rho}$ under the isentropic condition. For details consult, for example, Section 3.4 of Liepmann and Roshko (1957).

Inviscid Fluid Flows

Consider a 1 m-long tube of uniform cross section with both ends closed. The tube is divided into two chambers by a diaphragm at the middle section across which the densities are slightly different. Suppose ρ' is normalized so that it is 1 at and to the right of the diaphragm and 0 to the left. At $t = 0$ when $\partial\rho'/\partial t = 0$, the diaphragm is suddenly removed and a wave system is set up in the tube. Solve this linearized shock tube problem by finding $\rho'(x, t)$ numerically with leapfrog method. Density distributions along the tube at some representative time levels are to be plotted.

Boundary conditions for density fluctuation in this problem are determined as follows. At a closed end u vanishes regardless of time; thus, $\partial u/\partial t = 0$ there. It implies that $\partial p/\partial x = 0$ and, therefore, $\partial\rho'/\partial x = 0$ at the closed end by considering the linearized Euler equation. On the other hand, if there were an end open to the atmosphere, ρ' should vanish there because the density at that location is always the same as that outside.

Computer results should show that two waves are generated at the center after the eruption of the diaphragm: an expansion wave propagating toward the right and a compression wave propagating toward the left. A wave is classified as expansive or compressive, depending on whether the wave is traveling into a region of higher or lower density. The result should also show that at a rigid wall, an incidental expansion wave reflects as an expansion wave and an incidental compression wave reflects as a compression wave. It can also be shown that the type of a wave is changed after being reflected from an open end.

The conclusion on wave reflection from a closed or open end seems to contradict the conclusion obtained from the output of Program 2.10. However, when the type of a wave is defined in terms of the sign of u in combination with the direction of propagation, as shown in Fig. 3.3, p. 72 of Liepmann and Roshko (1957), the contradiction is automatically resolved.

2.14 Propagation of a Finite-Amplitude Wave: Formation of a Shock

The numerical solution obtained for the preceding example shows that a sound wave propagates at a constant velocity without changing its shape. This property of the wave is the result of linearization of the governing equations, assuming small perturbations about an equilibrium state of the gas. For a wave whose amplitude is not small, its properties are expected to change (see Zucrow and Hoffman, 1977, Section 19.5). This section offers a numerical study of such a wave.

To derive the equations governing nonlinear wave motions, we rewrite the continuity equation (2.13.11) and Euler equation (2.1.5) for unsteady one-dimensional flow.

$$\frac{\partial \rho}{\partial t} + u \frac{\partial \rho}{\partial x} + \rho \frac{\partial u}{\partial x} = 0 \qquad (2.14.1)$$

$$\frac{\partial u}{\partial t} + u \frac{\partial u}{\partial x} + \frac{a^2}{\rho} \frac{\partial \rho}{\partial x} = 0 \qquad (2.14.2)$$

In (2.14.2) the expression $a^2 \, \partial \rho / \partial x$ has been used to replace $\partial p / \partial x$, where the speed of sound, a, is also a function of x and t. We assume that the particle speed, u, is less than the sonic speed. Under isentropic conditions one of the three dependent variables, ρ, can be eliminated from the above equations. Let the subscript 0 indicate the undisturbed conditions and γ be the ratio of specific heats of the gas; then

$$\frac{p}{p_0} = \left(\frac{\rho}{\rho_0}\right)^\gamma$$

$$a^2 = \frac{dp}{d\rho} = \gamma \frac{p}{\rho} = \gamma \frac{p_0}{\rho_0} \left(\frac{\rho}{\rho_0}\right)^{\gamma-1} = a_0^2 \left(\frac{\rho}{\rho_0}\right)^{\gamma-1}$$

or

$$\rho = \rho_0 \left(\frac{a}{a_0}\right)^{2/(\gamma-1)} \qquad (2.14.3)$$

Substituting (2.14.3) into (2.14.1) and (2.14.2), we obtain

$$\frac{2}{\gamma-1}\left(\frac{\partial a}{\partial t} + u \frac{\partial a}{\partial x}\right) + a \frac{\partial u}{\partial x} = 0$$

$$\frac{\partial u}{\partial t} + u \frac{\partial u}{\partial x} + \frac{2a}{\gamma-1} \frac{\partial a}{\partial x} = 0$$

Adding and subtracting, respectively, give the following equations

$$\left[\frac{\partial}{\partial t} + (u+a)\frac{\partial}{\partial x}\right]\left(u + \frac{2a}{\gamma-1}\right) = 0 \qquad (2.14.4)$$

$$\left[\frac{\partial}{\partial t} + (u-a)\frac{\partial}{\partial x}\right]\left(u - \frac{2a}{\gamma-1}\right) = 0 \qquad (2.14.5)$$

Equation (2.14.4) states that the quantity $P = u + 2a/(\gamma-1)$ is constant along a curve in the x-t plane. On this curve $dP = (\partial P/\partial t)\, dt + (\partial P/\partial x)\, dx =$

204   Inviscid Fluid Flows

0, or, equivalently,

$$\left(\frac{\partial}{\partial t} + \frac{dx}{dt}\frac{\partial}{\partial x}\right)P = 0$$

Comparing this equation with (2.14.4), we obtain $dx/dt = u + a$, which is the expression for the slope of that curve. (2.14.5) can be similarly interpreted. Thus (2.14.4) and (2.14.5) exhibit a salient property that the quantities P and $Q = u - 2a/(\gamma - 1)$ are constant on curves that have slopes $dx/dt = u + a$ and $dx/dt = u - a$, respectively. These curves are called the *characteristics*; P and Q are called the *Riemann invariants*. Since both u and a vary with x and t, the characteristics are generally curved lines in the x-t plane.

The *method of characteristics* is developed based on the previously mentioned property. Referring to Fig. 2.14.1, suppose the initial data at $t = 0$ are given and the conditions at an arbitrary point C at $t_C > 0$ are to be computed. Through this point there are two characteristics, one of slope $u + a$ and the other of slope $u - a$, which intersect the x-axis at points A and B, respectively. Since $P_C = P_A$ and $Q_C = Q_B$ or, specifically,

$$u_C + \frac{2a_C}{\gamma - 1} = u_A + \frac{2a_A}{\gamma - 1}$$

$$u_C - \frac{2a_C}{\gamma - 1} = u_B - \frac{2a_B}{\gamma - 1}$$

u_C and a_C are computed immediately by adding and subtracting the above equations. Thus

$$u_C = \tfrac{1}{2}(u_A + u_B) + \frac{1}{\gamma - 1}(a_A - a_B) \tag{2.14.6}$$

$$a_C = \frac{\gamma - 1}{4}(u_A - u_B) + \tfrac{1}{2}(a_A + a_B) \tag{2.14.7}$$

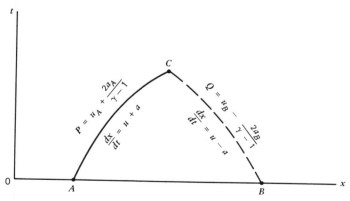

FIGURE 2.14.1 Method of characteristics.

Propagation of a Finite-Amplitude Wave

Although the exact shape of the characteristics cannot be determined unless u and a are both known in the region below point C, the solution may be computed approximately if the distance between points A and B is small. Under this condition the curved characteristics can be treated as two straight lines of slopes $u_A + a_A$ and $u_B - a_B$, respectively, and the conditions at point C, which is the intersection of these two straight lines, are described approximately by (2.14.6) and (2.14.7).

Thus, when a row of points is selected at small distances apart on the x-axis at $t = 0$, the solution can be computed step by step at later times by constructing a network using the local slopes of the characteristics, as shown in Fig. 2.14.2. This method, however, is not practical for machine calculation, because the locations of data points in the network are not known a priori. Many finite-difference techniques have been developed for solving the same problem on a predetermined rectangular mesh. The one chosen to be used here, suggested by Courant, Isaacson, and Rees (1952), is convenient for programming and yet is physically interpretable.

Let us adopt a grid system, exactly the same as that sketched in Fig. 2.12.1, for numerical solution of the simultaneous nonlinear equations (2.14.4) and (2.14.5). Suppose u and a have already been computed at all grid points on time level t_j; we now proceed to find the conditions at the grid points on the next time level t_{j+1}, for example, at point C, or $(i, j + 1)$, as shown in Fig. 2.14.3. If the characteristics through this point intersect with the back horizontal grid line at points A and B, respectively, then, according to the

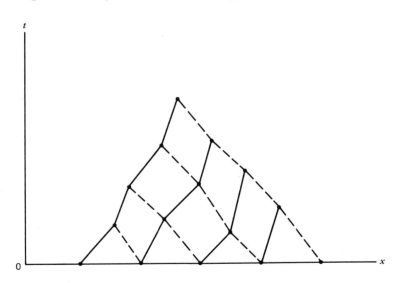

FIGURE 2.14.2 Characteristic network.

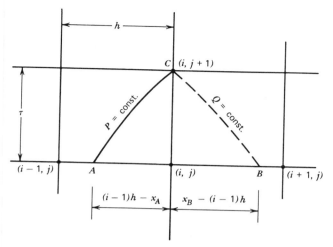

FIGURE 2.14.3 Finite-difference scheme for rectangular mesh.

method of characteristics described in Fig. 2.14.1, the condition at C is determined approximately from those at A and B by using formulas (2.14.6) and (2.14.7). The location and condition at points A and B are to be interpolated from the known information at grid points on that same time level.

Treating the characteristics as straight lines and approximating their slopes by $u_{i,j} + a_{i,j}$ and $u_{i,j} - a_{i,j}$, respectively, we find, from Fig. 2.14.3,

$$x_A = (i - 1)h - \tau(u_{i,j} + a_{i,j}) \tag{2.14.8}$$

$$x_B = (i - 1)h - \tau(u_{i,j} - a_{i,j}) \tag{2.14.9}$$

After points A and B have been located, linear interpolations are then used to find u_A and a_A within the interval between x_{i-1} and x_i. Thus

$$u_A = u_{i,j} + \frac{\tau}{h}(u_{i,j} + a_{i,j})(u_{i-1,j} - u_{i,j}) \tag{2.14.10}$$

$$a_A = a_{i,j} + \frac{\tau}{h}(u_{i,j} + a_{i,j})(a_{i-1,j} - a_{i,j}) \tag{2.14.11}$$

Similarly, at B,

$$u_B = u_{i,j} - \frac{\tau}{h}(u_{i,j} - a_{i,j})(u_{i+1,j} - u_{i,j}) \tag{2.14.12}$$

$$a_B = a_{i,j} - \frac{\tau}{h}(u_{i,j} - a_{i,j})(a_{i+1,j} - a_{i,j}) \tag{2.14.13}$$

Propagation of a Finite-Amplitude Wave

u_C and a_C or, in fact, $u_{i,j+1}$ and $a_{i,j+1}$, are computed immediately following the substitution of (2.14.10) to (2.14.13) into (2.14.6) and (2.14.7).

Detailed analysis of this numerical method has been made by Courant et al. for a more general system of equations. The result is accurate to the first order of h. For computational stability it requires that the ratio τ/h be so chosen that the numerical domain of dependence of any point in the mesh is not less than the physical domain of dependence determined by the characteristics. In other words, the stability condition is that both

$$\frac{\tau}{h}|u + a| \leqslant 1 \quad \text{and} \quad \frac{\tau}{h}|u - a| \leqslant 1 \tag{2.14.14}$$

are satisfied everywhere in the domain of computation.

An improved numerical scheme with second-order accuracy was outlined by Hartree (1958). In his method the curved characteristics through C are approximated by two straight lines whose slopes are, respectively, the average of the values at C and A and the average of those at C and B instead of the values at (i, j). The improved accuracy is paid for by the increased effort in computing the conditions at points A and B by iteration. In the following example the simpler method of Courant et al. will be used for the purpose of illustration.

A 2 m-long tube is considered whose left end is closed and right end is open, similar to the tube described in Fig. 2.13.1. The length of the tube is again divided into 0.02-m intervals, resulting in 101 vertical grid lines for the numerical work. Let a_0 be the speed of sound in the tube in an undisturbed state. At $t = 0$ a rightward-propagating wavelet of 1-m wavelength is produced in the left half of the tube. To exaggerate the effect of finite amplitude and to show an easily recognized distortion in wave shape, we assume that the initial amplitude of the wavelet is $a_0/2$ and that its initial shape is described by a set of broken straight lines defined by

$$u_{i,1} = \begin{cases} \dfrac{a_0}{2} \dfrac{i - 1}{12} & \text{for } 1 \leqslant i \leqslant 13 \\[2mm] \dfrac{a_0}{2} \dfrac{26 - i}{13} & \text{for } 13 < i \leqslant 39 \\[2mm] \dfrac{a_0}{2} \dfrac{i - 51}{12} & \text{for } 39 < i \leqslant 51 \\[2mm] 0 & \text{elsewhere} \end{cases}$$

The initial condition for a is determined by that for u through

$$a = a_0 \pm \frac{\gamma - 1}{2} u \tag{2.14.15}$$

which relates the local speed of sound to the local particle velocity (same sign as x) in an isentropic wave. The negative sign is to be taken only if the wave is traveling in the negative x direction. The derivation of (2.14.15) can be found in Section 3.9 of Liepmann and Roshko (1957) and therefore is not repeated here.

The boundary conditions at the ends can be derived with the help of Fig. 2.14.3. At the left end of the tube where $i = 1$, only one backward characteristic can be drawn through a grid point along which Q = constant. Combining this with the condition that particle velocity vanishes at a closed end gives

$$u_{1,j+1} = 0 \qquad\qquad (2.14.16)$$

$$a_{1,j+1} = a_B - \frac{\gamma - 1}{2} u_B \qquad\qquad (2.14.17)$$

where u_B and a_B are computed for $i = 1$ from (2.14.12) and (2.14.13). At a grid point at the right end of the tube open to the atmosphere, the density and therefore the sonic speed is the same as that outside. Furthermore, through such a point there is only one backward characteristic along which P = constant. Thus the boundary conditions there are

$$u_{m,j+1} = u_A + \frac{2}{\gamma - 1} (a_A - a_0) \qquad\qquad (2.14.18)$$

$$a_{m,j+1} = a_0 \qquad\qquad (2.14.19)$$

where m is the maximum value of i and u_A, a_A are computed from (2.14.10) and (2.14.11) for $i = m$.

The variables in Program 2.11 are named according to their original form. For example, UA, UB, AA, and AB represent, respectively, u_A, u_B, a_A, and a_B. We use AMPLTD for the initial amplitude of the velocity distribution, RATIO for τ/h, and COEFF for $(\gamma - 1)/2$. The plotting part is almost the same as that in Program 2.10, except that in this case the ordinate of GRAPH is normalized by the initial amplitude.

γ is chosen to be 1.4 for air at the sea level. For the fixed value of $h = 0.02$ m, the size of time steps is determined from the stability condition (2.14.14). It involves u and a, which are both unknown before the problem is solved. In the program a trial value $\tau = 0.5h/a_0$ is used. Before computing the solution at a grid point, we check (2.14.14) and discontinue the computation whenever this condition is violated. The result shows that this trial value is satisfactory for the present problem at all grid points.

The output of Program 2.11 displays the velocity distributions at increasing time steps. It is now examined to find the effects of keeping the nonlinear terms in the governing equations.

In the central part of the wave $\partial u/\partial x$ is negative; that is, the right-moving

Propagation of a Finite-Amplitude Wave

```
C                      ***** PROGRAM 2.11 *****

C            PROPAGATION OF A FINITE-AMPLITUDE WAVE IN A TUBE
C            HAVING A CLOSED LEFT END AND AN OPEN RIGHT END

       DIMENSION A(101,56),U(101,56),FRAME(101,13),GRAPH(101,13)

C        ..... ASSIGN INITIAL DATA .....
         DATA A0,H,GAMMA,M,JMAX,NT,NF / 340., 0.02, 1.4, 101, 55, 5, 4 /
         MM1 = M - 1
         TAU = 0.5 * H/A0
         RATIO = TAU / H

C        ..... ASSIGN INITIAL CONDITION TO U.  FOR A RIGHTWARD-TRAVELING
C              WAVE, CHOOSE THE UPPER SIGN IN (2.14.15) TO COMPUTE A .....
         AMPLTD = A0 / 2.0
         COEFF = (GAMMA-1.0) / 2.0
         DO 1  I = 1, M
         U(I,1) = 0.
         IF( I.GE. 1 .AND. I.LE.13 )  U(I,1) = AMPLTD*( I-1)/12.
         IF( I.GT.13 .AND. I.LE.39 )  U(I,1) = AMPLTD*(26-I)/13.
         IF( I.GT.39 .AND. I.LE.51 )  U(I,1) = AMPLTD*(I-51)/12.
       1 A(I,1) = A0 + COEFF*U(I,1)

C        ..... ASSIGN BOUNDARY CONDITION FOR U ON THE CLOSED LEFT END
C              AND THAT FOR A ON THE OPEN RIGHT END .....
         DO 2  J = 1, JMAX
         U(1,J+1) = 0.
       2 A(M,J+1) = A0

C        ..... COMPUTE SOLUTION USING THE METHOD OF COURANT, ISAACSON,
C              AND REES.  SPECIAL FORMULAE ARE USED AT THE BOUNDARIES
C              WHERE I=1 AND I=M.  STABILITY CONDITION FOR THE NUMERICAL
C              SCHEME IS CHECKED AT EVERY GRID POINT .....
         DO 4  J = 1, JMAX
            UPA = RATIO * (U(1,J)+A(1,J))
            UMA = RATIO * (U(1,J)-A(1,J))
            IF( ABS(UPA).GT.1. .OR. ABS(UMA).GT.1. )  GO TO 5
            UB = U(1,J) - UMA*(U(2,J)-U(1,J))
            AB = A(1,J) - UMA*(A(2,J)-A(1,J))
            A(1,J+1) = AB - COEFF*UB
         DO 3  I = 2, MM1
         UPA = RATIO * (U(I,J)+A(I,J))
         UMA = RATIO * (U(I,J)-A(I,J))
         IF( ABS(UPA).GT.1. .OR. ABS(UMA).GT.1. )  GO TO 5
         UA = U(I,J) + UPA*(U(I-1,J)-U(I,J))
         AA = A(I,J) + UPA*(A(I-1,J)-A(I,J))
         UB = U(I,J) - UMA*(U(I+1,J)-U(I,J))
         AB = A(I,J) - UMA*(A(I+1,J)-A(I,J))
         U(I,J+1) = 0.5 * ( (UA+UB) + (AA-AB)/COEFF )
       3 A(I,J+1) = 0.5 * ( COEFF*(UA-UB) + (AA+AB) )
            UPA = RATIO * (U(M,J)+A(M,J))
            UMA = RATIO * (U(M,J)-A(M,J))
            IF( ABS(UPA).GT.1. .OR. ABS(UMA).GT.1. )  GO TO 5
            UA = U(M,J) + UPA*(U(MM1,J)-U(M,J))
            AA = A(M,J) + UPA*(A(MM1,J)-A(M,J))
```

210

```
          U(M,J+1) = UA + (AA-A0)/COEFF
    4 CONTINUE

C     ..... COMPUTE THE LAST VALUE OF J UP TO WHICH
C             THE COMPUTATION IS STABLE .....
    5 JLAST = J

C     ..... CONSTRUCT THE FRAME FOR PLOTTING WAVE SHAPE .....
      DO 6   I = 2, 100
      DO 6   JJ = 2, 12
      FRAME(I,JJ) = 1H
    6 IF( I.EQ.26 .OR. I.EQ.51 .OR. I.EQ.76 )  FRAME(I,JJ) = 1H.
      DO 7   I = 1, 101
      FRAME(I,1)  = 1H=
      FRAME(I,7)  = 1H.
    7 FRAME(I,13) = 1H=
      DO 8   JJ = 2, 12
      FRAME(  1,JJ) = 1H1
    8 FRAME(101,JJ) = 1H1

C     ..... PLOT U EVERY NT TIME STEPS, NF FRAMES PER PAGE .....
      DO 22 J = 1, JLAST, NT
      JM1 = J - 1
      IF( JM1/NF*NF .EQ. JM1 )  GO TO 9
      GO TO 11
    9 WRITE(6,10)
   10 FORMAT( 1H1 )
   11 DO 18  I = 1, 101
      INTGER = 6.0*U(I,J)/AMPLTD + 0.5
      IF( U(I,J).LT.0. )  INTGER = 6.0*U(I,J)/AMPLTD - 0.5
      JWAVE = INTGER + 7
      IF( JWAVE-7 )  12, 14, 16
   12     DO 13  JJ = 1, 13
          GRAPH(I,JJ) = FRAME(I,JJ)
   13     IF( JJ.GE.JWAVE .AND. JJ.LE.7 )  GRAPH(I,JJ) = 1H-
      GO TO 18
   14     DO 15  JJ = 1, 13
   15     GRAPH(I,JJ) = FRAME(I,JJ)
      GO TO 18
   16     DO 17  JJ = 1, 13
          GRAPH(I,JJ) = FRAME(I,JJ)
   17     IF( JJ.GE.7 .AND. JJ.LE.JWAVE )  GRAPH(I,JJ) = 1H+
   18 CONTINUE
      T = JM1 * TAU
      WRITE(6,19)  JM1, T
   19 FORMAT( / 44X, 6HTIME =, I3, 7H*TAU = , 1PE9.3, 8H SECONDS )
      DO 20  JINVRT = 1, 13
      JJ = 14 - JINVRT
   20 WRITE(6,21)  ( GRAPH(I,JJ), I=1,101 )
   21 FORMAT( 10X, 101A1 )
   22 CONTINUE
      END
```

TIME = 0*TAU = 0. SECONDS

TIME = 5*TAU = 1.471E-04 SECONDS

TIME = 10*TAU = 2.941E-04 SECONDS

TIME = 15*TAU = 4.412E-04 SECONDS

TIME = 20*TAU = 5.882E-04 SECONDS

TIME = 25*TAU = 7.353E-04 SECONDS

TIME = 30*TAU = 8.824E-04 SECONDS

TIME = 35*TAU = 1.029E-03 SECONDS

TIME = 40*TAU = 1.176E-03 SECONDS

TIME = 45*TAU = 1.324E-03 SECONDS

TIME = 50*TAU = 1.471E-03 SECONDS

TIME = 55*TAU = 1.618E-03 SECONDS

213

velocity of a gas particle is faster than that of a particle ahead of it. Thus in this region gas particles are catching up with those ahead; it is called a compression region. The time sequence plots show that the width of this region decreases with time, so that the wave profile is steepening in the central portion. On the other hand, in the leading and trailing parts of the wave where $\partial u/\partial x$ is positive, gas particles ahead are moving faster than those behind. In each of these expansion regions the width increases and the profile flattens with increasing time.

As the wave propagates its profile is no longer antisymmetric about the node. The left portion, within which u is positive, becomes shorter than that on the right portion, within which u is negative, while the total length of the wave is kept constant. In addition, the amplitude of the wave on the left side of the node diminishes faster in time than that on the right.

When $t = 40\tau$, the wave profile in the compression region becomes almost a vertical line. In other words, at this time a drastic change in u can be found across a very short distance. This is, in fact, a shock wave.

Actually, as explained by Liepmann and Roshko (1957, Section 3.10), before this situation is reached, the velocity gradient becomes so great that viscous and heat transfer effects can no longer be neglected and the governing equations (2.14.4) and (2.14.5) break down. Even if it is numerically feasible, the computation should be stopped before any part of the wave becomes too steep.

The development of a finite-amplitude wave into a shock is analogous to the breaking of a water wave to form a bore. A description of the latter phenomenon can be found in Section 10.10 of Stoker (1957). In the next chapter the structure of a shock wave in a real gas is studied numerically.

Problem 2.19 In the computation of Program 2.11, the wave has not yet been affected by the boundary conditions unless more time steps are added. To study the nonlinear effects on wave reflection, run Program 2.11, which is modified for the following two waves.

1. A left-propagating wave having the same initial velocity distribution as that specified in Program 2.11.

2. A right-propagating wave of the same initial velocity profile, but shifted one wavelength to the right.

chapter three
VISCOUS FLUID FLOWS

The viscous effects ignored in the previous chapter are examined in this chapter. The governing equations for viscous flows are first summarized in Section 3.1. The numerical techniques, devised previously for solving ordinary initial-value and boundary-value problems, are then used or modified to be used in the next three sections to study the structures of a shock wave, a velocity boundary layer, and a thermal boundary layer. In Section 3.5, an open-channel flow problem is solved utilizing a numerical method constructed in Section 2.10 for solving elliptic partial differential equations.

There are two different approaches in solving parabolic equations. The explicit methods and their computational stabilities are discussed in Section 3.6, with an example given on the unsteady flow caused by a suddenly accelerated plane wall. The computationally stable implicit methods are introduced in Section 3.7. The illustrative example used there is the starting flow in a channel caused by the application of a constant pressure gradient.

The last section deals with Stokes' flows, which are governed by a biharmonic equation. Such an equation is solved by first rewriting it in the form of two coupled elliptic equations and then by applying twice an appropriate iterative method derived in Section 2.10. This numerical scheme is used to solve for the cavity flow caused by a moving surface.

3.1 Governing Equations for Viscous Flows

Before solving viscous flow problems, the equations governing the unsteady motion of a viscous compressible fluid having variable physical properties are summarized. Derivation of these equations can be found in many texts on fluid dynamics, for example, in Schlichting (1968). The equations are expressed here in vector notation. Their specific forms in Cartesian, cylindrical, and spherical coordinates are shown in the book by Hughes and Gaylord (1964).

No matter whether the fluid is viscous or inviscid, the continuity equation

(2.13.11) remains the same form, so that

$$\frac{\partial \rho}{\partial t} + \nabla \cdot (\rho \mathbf{V}) = 0 \qquad (3.1.1)$$

However, viscous forces appear in a real fluid, as discussed in Section 2.1. The equation of motion, constructed by adding to Euler's equation (2.1.5) the forces per unit volume caused by viscosity, is of the form

$$\rho \frac{D\mathbf{V}}{Dt} = -\nabla p - \nabla \times [\mu(\nabla \times \mathbf{V})] + \nabla[(2\mu + \lambda)\nabla \cdot \mathbf{V}] \qquad (3.1.2)$$

The operator on the left side, defined previously in (2.1.6), is the substantial derivative. μ is the coefficient of viscosity, and λ is the second coefficient of viscosity of the fluid; both are functions of temperature. For a monatomic gas $\lambda = -\frac{2}{3}\mu$. (3.1.2) is generally referred to as the *Navier-Stokes equation*. The addition of viscous forces increases the order of the differential equation by one. The extra constant of integration to appear in the solution is determined from the additional boundary condition required for a viscous fluid that the velocity component tangent to a stationary rigid surface must also be zero. The boundary condition for an inviscid fluid is that the component normal to such a surface vanishes.

The energy equation may be written in several alternative forms. The following form is often used for an ideal gas in terms of specific enthalpy $h(=c_p T)$:

$$\rho \frac{Dh}{Dt} = \frac{Dp}{Dt} + \nabla \cdot (k\nabla T) + \Phi \qquad (3.1.3)$$

in which c_p and k are, respectively, the constant-pressure specific heat and the thermal conductivity of the gas, T is the absolute temperature, and Φ is the *dissipation function* defined in Cartesian coordinates as

$$\Phi = 2\mu \left[\left(\frac{\partial u}{\partial x}\right)^2 + \left(\frac{\partial v}{\partial y}\right)^2 + \left(\frac{\partial w}{\partial z}\right)^2 \right] + \mu \left(\frac{\partial u}{\partial y} + \frac{\partial v}{\partial x}\right)^2$$
$$+ \mu \left(\frac{\partial v}{\partial z} + \frac{\partial w}{\partial y}\right)^2 + \mu \left(\frac{\partial w}{\partial x} + \frac{\partial u}{\partial z}\right)^2 + \lambda(\nabla \cdot \mathbf{V})^2 \qquad (3.1.4)$$

It represents the time rate at which energy of the ordered fluid motion per unit volume is dissipated into heat through the action of viscosity.

Finally, the equation of state for an ideal gas is

$$p = \rho R T \qquad (3.1.5)$$

where R is the gas constant.

We now have six scalar equations, (3.1.1) to (3.1.3) and (3.1.5), with (3.1.2) written in component form, which form a complete set of equations for the six unknowns ρ, p, T, u, v, and w. To find solutions to this system is extremely difficult because not only the equations are nonlinear, but the unknowns are related in such a way that all six equations must be solved simultaneously.

Great simplifications are obtained by assuming that the fluid is incompressible and the temperature variation is not too large so that fluid properties are constant. In this case ρ becomes a constant, the equation of state is not needed, and the energy equation is uncoupled from the continuity equation and the equation of motion. The latter two are simplified in Cartesian coordinates to

$$\nabla \cdot \mathbf{V} = 0 \tag{3.1.6}$$

$$\rho \frac{D\mathbf{V}}{Dt} = -\nabla p + \mu \nabla^2 \mathbf{V} \tag{3.1.7}$$

The procedure for solving the problem is first to obtain p and $\mathbf{V}$ from simultaneous equations (3.1.6) and (3.1.7), and then to find T after substituting the result into the simplified version of (3.1.3).

It is still possible to introduce the stream function defined in Section 2.1; the velocity potential can no longer be used because the flow is generally rotational in the presence of rotational viscous forces.

3.2 Structure of a Plane Shock Wave

The result of Section 2.14 shows that as an isentropic wave of large amplitude propagates, the wave profile in the compression region steepens and a shock wave is ultimately formed. In a real fluid the diffusive action of viscosity and conductivity counteracts the steepening tendency. A stationary state will finally be reached at which the shock wave propagates without further distortion. Thus, in reality, all flow variables (velocity, pressure, etc.) vary smoothly through a finite but thin shock region instead of jumping abruptly across the shock front, as described in the inviscid theory.

The theoretical structure of a shock in the presence of viscosity and conductivity was first thoroughly investigated by Becker (1922). Various numerical methods have been developed for computing unsteady, one-dimensional shock profiles. Discussions on many of these numerical methods can be found in Chapter 12 of Richtmyer and Morton (1967) and Chapter 5 of Roache (1972). In this section we consider a simplified problem to find the steady-state structure of a normal shock.

Suppose the shock propagates at a constant supersonic speed u_1 along the negative x direction. It is convenient to let the coordinate system move at the wave speed, so that the shock becomes stationary with respect to the coordinates and the flow enters the shock at the speed u_1. The subscripts 1 and 2 are used to denote, respectively, the given quantities far upstream and the quantities far downstream. The conditions far downstream are determined if one of the quantities there is prescribed.

For steady one-dimensional flow the governing equations (3.1.1) to (3.1.3) reduce to

$$\frac{d}{dx}(\rho u) = 0 \tag{3.2.1}$$

$$\rho u \frac{du}{dx} = -\frac{dp}{dx} + \frac{d}{dx}\left(\mu' \frac{du}{dx}\right) \tag{3.2.2}$$

$$\rho u \frac{d}{dx}(c_p T + \tfrac{1}{2}u^2) = \frac{d}{dx}\left(u\mu' \frac{du}{dx} + k \frac{dT}{dx}\right) \tag{3.2.3}$$

where μ' denotes $2\mu + \lambda$. Use has been made of (3.2.2) in obtaining (3.2.3) from the energy equation (3.1.3). Integrating once yields

$$\rho u = \rho_1 u_1 = m \tag{3.2.4}$$

$$\mu' \frac{du}{dx} - mu - p = -mu_1 - p_1 \tag{3.2.5}$$

$$u\mu' \frac{du}{dx} + k \frac{dT}{dx} - m(c_p T + \tfrac{1}{2}u^2) = -m(c_p T_1 + \tfrac{1}{2}u_1^2) \tag{3.2.6}$$

where m represents the mass flux through the shock and is a constant throughout the flow field. Before further integrating the equations, let us evaluate their left sides in the region far downstream where both velocity and temperature become uniform. Thus we obtain

$$\rho_2 u_2 = \rho_1 u_1 \tag{3.2.7}$$

$$m(u_1 - u_2) = p_2 - p_1 \tag{3.2.8}$$

$$c_p T_2 + \tfrac{1}{2}u_2^2 = c_p T_1 + \tfrac{1}{2}u_1^2 \tag{3.2.9}$$

These equations simply state the laws of conservation of mass, momentum and energy across a steady normal shock. When written in terms of ratios,

$$\frac{u_1}{u_2} = \frac{p_2}{\rho_1} = \left(1 + \frac{\gamma+1}{\gamma-1}\frac{p_2}{p_1}\right) \Big/ \left(\frac{\gamma+1}{\gamma-1} + \frac{p_2}{p_1}\right) = \frac{p_2}{p_1}\frac{T_1}{T_2} \tag{3.2.10}$$

they are called the *Rankine-Hugoniot relations*, which relate the conditions upstream to those downstream from a shock. The proof of (3.2.10) is referred to in Section 10.6 of Kuethe and Chow (1976), although a more general case of an oblique shock is considered there. It is interesting to point out that Rankine-Hugoniot relations are not affected by viscosity or conductivity. Their effects show in the thickness and the detailed structure of the shock wave.

Now we return to (3.2.5) and (3.2.6). These equations can be integrated again to obtain analytical solutions under special conditions, for example, for constant gas properties and Prandtl number unity (see Becker, 1922). But solutions generally are to be obtained by numerical integration. (3.2.5) and (3.2.6) in their present form are not convenient to work with. As the first step of simplification, pressure in (3.2.5) is replaced by the following expression, obtained from the equation of state (3.1.5).

$$p = \rho R T = m R \frac{T}{u}$$

Then the energy relation (3.2.6) is simplified by substituting $\mu' du/dx$ from (3.2.5). After introducing dimensionless variables

$$U = \frac{m}{mu_1 + p_1} u, \qquad T' = \frac{m^2 R}{(mu_1 + p_1)^2} T \tag{3.2.11}$$

and using the fact that $c_p/R = \gamma/(\gamma - 1)$, the resulting equations can be written as

$$\frac{dU}{dx} = \frac{m}{\mu'} \left(U + \frac{T'}{U} - 1 \right) \tag{3.2.12}$$

$$\frac{dT'}{dx} = \frac{\gamma - 1}{\gamma} \frac{m}{\mu} Pr \left(\frac{1}{\gamma - 1} T' - \tfrac{1}{2} U^2 + U - \tfrac{1}{2}\alpha \right) \tag{3.2.13}$$

in which $Pr = \mu c_p/k$ is the *Prandtl number* and α is a dimensionless parameter determined by the upstream conditions,

$$\alpha = \frac{2m^2(c_p T_1 + \tfrac{1}{2} u_1^2)}{(mu_1 + p_1)^2}. \tag{3.2.14}$$

(3.2.12) and (3.2.13) hold for any gaseous continuum medium. But, to simplify our analysis, let us assume a monatomic gas for which $\lambda = -\tfrac{2}{3}\mu$, $\mu' = \tfrac{4}{3}\mu$, and $\gamma = \tfrac{5}{3}$. These equations are then reduced to

$$\frac{dU}{dx} = \frac{3m}{4\mu} \left(U + \frac{T'}{U} - 1 \right) \tag{3.2.15}$$

$$\frac{dT'}{dx} = \frac{m}{5\mu} Pr(3T' + 2U - U^2 - \alpha) \tag{3.2.16}$$

The boundary conditions at the end of the shock are

$$\frac{dU}{dx} = \frac{dT'}{dx} = 0 \qquad \text{at } x = \pm \infty \tag{3.2.17}$$

Imposition of these conditions upon (3.2.15) and (3.2.16) yields two algebraic equations that are solved simultaneously to get

$$U = \tfrac{1}{8}(5 \pm \epsilon) \tag{3.2.18}$$

$$T' = \tfrac{1}{64}(15 - \epsilon^2 \mp 2\epsilon) \tag{3.2.19}$$

in terms of the new shock-strength parameter

$$\epsilon = \sqrt{25 - 16\alpha} \tag{3.2.20}$$

(3.2.18) and (3.2.19) are, in fact, the Rankine-Hugoniot relations expressed in terms of the present dimensionless variables. The upper and lower signs pertain to the upstream and downstream conditions, respectively.

Following Grad (1952), we rewrite the differential equations by introducing the new variables w and t through the relations

$$U = \tfrac{1}{8}(5 + \epsilon w) \tag{3.2.21}$$

$$T' = \tfrac{1}{64}(15 - \epsilon^2 + 2\epsilon t) \tag{3.2.22}$$

so that (3.2.15) and (3.2.16) take the form

$$\frac{dw}{dx} = \frac{3m}{4\mu} \frac{2(t + w) - \epsilon(1 - w^2)}{5 + \epsilon w} \tag{3.2.23}$$

$$\frac{dt}{dx} = \frac{mPr}{10\mu} [6(t + w) + \epsilon(1 - w^2)] \tag{3.2.24}$$

Dividing the second equation by the first gives the single differential equation in w and t.

$$\frac{dt}{dw} = \frac{2}{15} Pr(5 + \epsilon w) \frac{6(t + w) + \epsilon(1 - w^2)}{2(t + w) - \epsilon(1 - w^2)}. \tag{3.2.25}$$

The boundary conditions (3.2.18) and (3.2.19) are represented by two points in the w-t plane:

$$P_1: \quad w = +1, \qquad t = -1 \tag{3.2.26}$$

$$P_2: \quad w = -1, \qquad t = +1 \tag{3.2.27}$$

with P_1 representing the upstream, and P_2, the downstream conditions. As pointed out by Grad (1952) and by Gilbarg and Paolucci (1953), P_1 is an unstable node and P_2 is a saddle point in the w-t plane. It is concluded that integration of (3.2.25) must be from P_2 to P_1; that is, the path of integration is from downstream to upstream in the direction of decreasing x.

Numerical integration of (3.2.25) can be performed by using the fourth-order Runge-Kutta method described in Section 1.1. However, because of the indeterminate nature of (3.2.25) at the starting point, the derivative at the first step must be evaluated using an alternative expression. According to Grad (1952), the solution of (3.2.25) can be put in the form of the following expansion.

$$t = -w + \epsilon(1 - w^2)(t_1 + \epsilon t_2 + \epsilon^2 t_3 + \cdots) \tag{3.2.28}$$

The first three coefficients are, with $\theta = 2Pr/15$,

$$t_1 = \frac{1 - 5\theta}{2(1 + 15\theta)}$$

$$t_2 = \frac{8\theta(1 - 10\theta)}{(1 + 15\theta)^3} w$$

$$t_3 = -\frac{8\theta(1 - 10\theta)}{(1 + 15\theta)^5} [10\theta + (1 - 32\theta + 45\theta^2)w^2]$$

and succeeding coefficients can be obtained from the recurrence relation

$$(1 + 15\theta)t_{n+1} = -\tfrac{1}{2}T_n - 3\theta w t_n + \sum_{r=1}^{n} t_r T_{n-r+1}$$

where

$$T_n = (1 - w^2)\frac{dt_n}{dw} - 2wt_n$$

When only three coefficients are used in the expansion, the derivative at point P_2 is approximated by

$$\left(\frac{dt}{dw}\right)_{w=-1} = -1 + \epsilon\frac{1 - \tfrac{2}{3}Pr}{1 + 2Pr} - \epsilon^2\frac{\tfrac{32}{15}Pr(1 - \tfrac{4}{3}Pr)}{(1 + 2Pr)^3}$$

$$- \epsilon^3\frac{\tfrac{32}{15}Pr(1 - \tfrac{4}{3}Pr)(1 - \tfrac{44}{15}Pr + \tfrac{4}{5}Pr^2)}{(1 + 2Pr)^5}. \tag{3.2.29}$$

Having found t as a function of w within the shock, w may be computed as a function of x by numerical integration of (3.2.23). Because our numerical result will finally be compared with that measured by Sherman (1955), it is more convenient at this time to introduce a dimensionless distance adopted in his report defined by

$$X = \frac{\tfrac{2}{5}Pr}{1 + 2Pr}\frac{m\epsilon}{\mu_*}x, \tag{3.2.30}$$

in which the reference viscosity coefficient μ_* is to be evaluated at the temperature $T_* = 3T_0/4$, T_0 being the stagnation temperature of the upstream flow. With the approximation of temperature variation of μ by the power law

$$\frac{\mu}{\mu_*} = \left(\frac{T}{T_*}\right)^\omega \tag{3.2.31}$$

Structure of a Plane Shock Wave

and in terms of X, (3.2.23) may be rewritten as

$$dX = \frac{\frac{8}{15}\epsilon Pr}{1 + 2Pr} \left(\frac{T}{T_*}\right)^\omega \frac{5 + \epsilon w}{2(t + w) - \epsilon(1 - w^2)} \, dw \qquad (3.2.32)$$

The absolute temperature T is related to the dimensionless temperature t through (3.2.11) and (3.2.22), or,

$$T = \frac{(mu_1 + p_1)^2}{64m^2 R} (15 - \epsilon^2 + 2\epsilon t) \qquad (3.2.33)$$

As shown by Sherman, the upstream Mach number for a monatomic gas is related to the shock-strength parameter by the formula

$$\epsilon = \frac{15(M_1^2 - 1)}{3 + 5M_1^2} \qquad (3.2.34)$$

In Program 3.1 we consider specifically the structure of a shock wave of $M_1 = 1.82$ in helium with $Pr = 2/3$ and $\omega = 0.647$. The upstream stagnation temperature is $T_0 = 294°K$. From (3.2.34) and (3.2.20), we obtain $\epsilon = 1.77$ and $\alpha = 1.37$, respectively. Since $c_p T_0 = c_p T_1 + \frac{1}{2}u_1^2$, (3.2.14) gives

$$\frac{(mu_1 + p_1)^2}{m^2 R} = \frac{2c_p}{\alpha R} T_0 = 3.65 T_0$$

Substituting into (3.2.33) and knowing $T_* = 3T_0/4$, we have

$$\frac{T}{T_*} = 0.076(15 - \epsilon^2 + 2\epsilon t) \qquad (3.2.35)$$

For this particular experimental case, Sherman calculated the reference length x/X defined in (3.2.30) and obtained the value 0.0013 m, which is about three times the mean free path of gas molecules ahead of the shock. This condition barely justifies the use of Navier-Stokes equations, which describe the motion of a continuum.

The numerical procedure to be followed in Program 3.1 is summarized as follows. Starting from $w = -1$ and $t = +1$, (3.2.25) is integrated to obtain t as a function of w by application of the fourth-order Runge-Kutta formulas (1.1.8) and (1.1.9); however, at the first step, Euler's method (described in Section 1.1) is used based on a slope approximated by (3.2.29). Increments of a constant step size of $\Delta w = 0.02$ are used in the numerical method. The dimensionless coordinate X is then computed as a function of w by integrating (3.2.32) after substitution of T/T_* from (3.2.35). Finally, the variations of w and t with respect to X are plotted and the temperature profile is compared with Sherman's experimental data.

Care must be taken in integrating (3.2.32), which is written symbolically as

$$dX = f(w)\, dw \tag{3.2.36}$$

If the range of w is divided into m (200 in Program 3.1) equally spaced intervals of size Δw, and if the values of f and X evaluated at the end points $w_1, w_2, \ldots, w_{m+1}$ of these intervals are called $f_1, f_2, \ldots, f_{m+1}$ and $X_1, X_2, \ldots, X_{m+1}$, respectively, then (3.2.36) may be replaced at w_i by

$$X_{i+1} = X_i + \frac{\Delta w}{2}(f_i + f_{i+1}) \tag{3.2.37}$$

Here we have approximated the right-hand side of (3.2.36) within the interval between w_i and w_{i+1} by the area of a trapezoid. It is well known that the error in using the trapezoidal rule is proportional to $(\Delta w)^3$, which is not significant for small Δw. However, for the present problem (3.2.37) cannot be used at the end points w_1 and w_{m+1} where the function becomes unbounded. To avoid the singularities there we start the integration at an arbitrary station in between, say at w_{116}, to correspond to Sherman's coordinate system and set $X_{116} = 0$, since the solution to the differential equation (3.2.36) may take an arbitrary constant. We first integrate forward using (3.2.37) until $i = m - 1$, and then

---

List of Principal Variables in Program 3.1

Program Symbol	Definition
(Main Program)	
DW	Step size Δw
E	Shock-strength parameter ϵ
F	The function $f(w)$ on right side of (3.2.32)
M	Total number of intervals in the range of w
OMEGA	The exponent ω in (3.2.31)
PR	Prandtl number, Pr
SLOPE	The initial slope defined in (3.2.29)
T	Dimensionless temperature t defined in (3.2.22)
TE	Measured dimensionless temperature t_e
W	Dimensionless speed w defined in (3.2.21)
X	Dimensionless distance X defined in (3.2.30)
XE	Dimensionless distance X_e at which the measured temperature is t_e

---

Structure of a Plane Shock Wave

```
C                        ***** PROGRAM 3.1 *****

C          STRUCTURE OF A SHOCK WAVE AT MACH NUMBER 1.82 IN HELIUM

      EXTERNAL F1
      DIMENSION T(201),W(201),X(201),F(201),XE(19),TE(19),
     A             LCHAR(3),NSCALE(4)
      COMMON PR,E

C     ..... SPECIFY INPUT DATA .....
      PR = 2./3.
      E  = 1.77
      OMEGA = 0.647
      M = 200
      MP1 = M + 1
      MM1 = M - 1
      DW = 0.01

C     ..... INTEGRATE (3.2.25) FROM THE POINT WHERE W=-1 AND T=+1 USING
C           FOURTH-ORDER RUNGE-KUTTA METHOD WITH DW=0.01, EXCEPT AT THE
C           FIRST STEP EULER.S METHOD IS USED WITH A SLOPE APPROXIMATED
C           BY THE EXPRESSION SHOWN IN (3.2.29) .....
      W(1) = -1.0
      T(1) =  1.0
      A = 1.0 + 2.*PR
      B = 32./15. * PR * (1.-4./3.*PR) / A**3
      C = B * (1. - 44./15.*PR + 0.8*PR**2) / A**2
      SLOPE = -1.0 + E*(1.-2./3.*PR)/A - E**2*B - E**3*C
      W(2) = W(1) + DW
      T(2) = T(1) + SLOPE*DW
      WW = W(2)
      TT = T(2)
      DO 1  I = 3, MP1
      CALL RK4( TT, WW, DW, F1 )
      W(I) = WW
    1 T(I) = TT

C     ..... COMPUTE THE FUNCTION F ON THE RIGHT-HAND SIDE OF (3.2.36).
C           LET X EQUAL ZERO AT W(116), THEN INTEGRATE FORWARD AND
C           BACKWARD FROM THERE TO FIND X USING (3.2.37) AND (3.2.38),
C           RESPECTIVELY.  X(1) AND X(MP1) ARE COMPUTED FROM (3.2.39)
C           AND (3.2.40) .....
      D = 8./15. * E * PR / A
      DO 2  I = 2, M
      F(I) = D * (0.076*(15.-E*E+2.*E*T(I)))**OMEGA
     A     * (5.+E*W(I)) / (2.*(T(I)+W(I))-E*(1.-W(I)**2))
    2 CONTINUE
      X(141) = 0.0
      DO 3  I = 141, MM1
    3 X(I+1) = X(I) + DW/2.*(F(I)+F(I+1))
      DO 4  II = 3, 141
      I = 144 - II
    4 X(I-1) = X(I) - DW/2.*(F(I-1)+F(I))
```

```
      X(1) = X(2) - DW*F(2)
      X(MP1) = X(M) + DW*F(M)

C     ..... PRINT X, W, AND T .....
      WRITE(6,5)
    5 FORMAT( 1H1 /// 17X, 39HVARIATIONS OF W AND T WITH RESPECT TO X
     A      // 16X, 1HX, 19X, 1HW, 19X, 1HT / 3( 13X, 7H------ ) / )
      WRITE(6,6)  ( (X(I),W(I),T(I)), I = 1, MP1, 5 )
    6 FORMAT( 3F20.4 )

C     ..... ASSIGN DATA FOR TEMPERATURE DISTRIBUTION
C           BASED ON SHERMAN,S EXPERIMENT .....
      DATA XE/ -2.20, -1.85, -1.45, -1.25, -1.05, -0.85, -0.65, -0.45,
     A         -0.25, -0.05,  0.15,  0.35,  0.55,  0.75,  0.95,  1.10,
     B          1.50,  1.95,  2.30 /
      DATA TE/ -0.99, -0.96, -0.94, -0.90, -0.85, -0.80, -0.70, -0.60,
     A         -0.45, -0.30, -0.10,  0.12,  0.30,  0.50,  0.62,  0.75,
     B          0.92,  0.98,  0.99 /

C     ..... PLOT W AND T AS FUNCTIONS OF X.  THE LATTER
C           IS COMPARED WITH EXPERIMENTAL RESULT .....
      DATA NSCALE/ 0, 1, 0, 0 /
      CALL PLOT1( NSCALE, 5, 10, 7, 17, 1H-, 1HI )
      CALL PLOT2( 3., -3., 1., -1., .TRUE. )
      CALL PLOT3( 1HW, X, W, 1, MP1, 4 )
      CALL PLOT3( 1HT, X, T, 1, MP1, 4 )
      CALL PLOT3( 1HO, XE, TE, 1, 19, 1 )
      WRITE(6,7)
    7 FORMAT( 1H1 )
      LCHAR(1) = 7HW AND T
      CALL PLOT4( LCHAR, 7 )
      WRITE(6,8)
    8 FORMAT( / 69X, 1HX /// 12X, 35HFIGURE 3.2.1    DIMENSIONLESS SPEED
     A        51H(W) AND TEMPERATURE (T) PROFILES IN A SHOCK WAVE AT,
     B        29H MACH NUMBER 1.82 IN HELIUM. / 39X, 15H(O) REPRESENTS ,
     C        47HTEMPERATURE PROFILE MEASURED BY SHERMAN (1955).  )
      END




      FUNCTION F1( T, W )
      COMMON PR,E
      F1 = 2./15. * PR * (5.+E*W) * ( 6.*(T+W)+E*(1.-W**2) )
     A      / ( 2.*(T+W)-E*(1.-W**2) )
      RETURN
      END




      SUBROUTINE RK4( X, T, H, F )
C     ..... FOURTH-ORDER RUNGE-KUTTA FORMULAE (1.1.8) AND (1.1.9) ARE
C           PROGRAMMED FOR SOLVING A FIRST-ORDER DIFFERENTIAL EQUATION
C           OF THE FORM  DX/DT=F(X,T).  H IS THE STEP SIZE IN T .....

      D1X = H * F( X, T )
      D2X = H * F( X+D1X/2., T+H/2. )
      D3X = H * F( X+D2X/2., T+H/2. )
      D4X = H * F( X+D3X, T+H )
      T = T + H
      X = X + (D1X + 2.*D2X + 2.*D3X + D4X)/6.0
      RETURN
      END
```

x	w	T
2.8575	-1.0000	1.0000
1.9018	-.9500	.9690
1.6412	-.9000	.9371
1.4815	-.8500	.9044
1.3626	-.8000	.8707
1.2659	-.7500	.8362
1.1829	-.7000	.8007
1.1093	-.6500	.7643
1.0424	-.6000	.7270
.9805	-.5500	.6888
.9223	-.5000	.6497
.8670	-.4500	.6096
.8139	-.4000	.5687
.7626	-.3500	.5267
.7125	-.3000	.4838
.6635	-.2500	.4400
.6151	-.2000	.3952
.5672	-.1500	.3494
.5194	-.1000	.3027
.4716	-.0500	.2549
.4235	.0000	.2062
.3749	.0500	.1565
.3256	.1000	.1057
.2754	.1500	.0540
.2240	.2000	.0012
.1712	.2500	-.0527
.1165	.3000	-.1076
.0596	.3500	-.1636
0.0000	.4000	-.2207
-.0628	.4500	-.2789
-.1296	.5000	-.3382
-.2012	.5500	-.3986
-.2791	.6000	-.4603
-.3648	.6500	-.5231
-.4611	.7000	-.5872
-.5717	.7500	-.6525
-.7035	.8000	-.7191
-.8687	.8500	-.7871
-1.0953	.9000	-.8565
-1.4727	.9500	-.9275
-2.8865	1.0000	-.9945

integrate backward from w_{116} using the formula

$$X_{i-1} = X_i - \frac{\Delta w}{2} (f_{i-1} + f_i) \qquad (3.2.38)$$

until $i = 3$. The end values are computed, using Euler's method, from the formulas

$$X_1 = X_2 - f_2 \Delta w \qquad (3.2.39)$$

$$X_{m+1} = X_m + f_m \Delta w \qquad (3.2.40)$$

Nineteen temperature data points are read off from Fig. 15 of Sherman's report, measured with a 0.00025-in. wire in a low-density wind tunnel. These

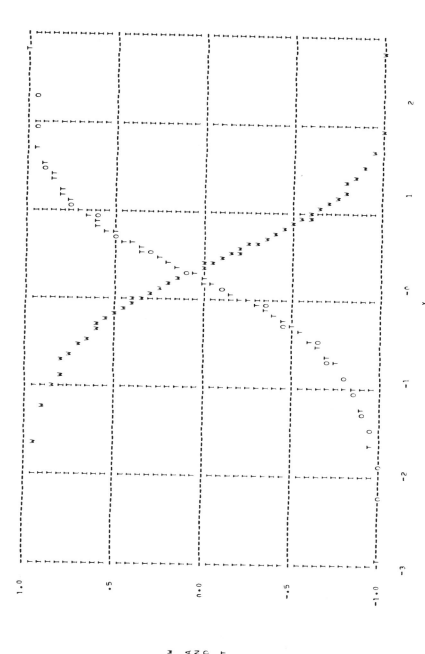

FIGURE 3.2.1 Dimensionless speed (W) and temperature (T) profiles in a shock wave at mach number 1.82 in helium. (O) represents temperature profile measured by Sherman (1955).

experimental data points, designated (X_e, t_e), are used to check the accuracy of the numerical result.

The result of numerical integration of (3.2.25) is shown in the second and third columns of the table printed by Program 3.1. We started from $w = -1$ and $t = +1$, the downstream state, and after a correct integration, should end up in the upstream state in which $w = +1$ and $t = -1$. The 0.55% error displayed in the printed value of t at the end of our numerical integration is caused by the error in using (3.2.29) to approximate the starting slope. The accuracy can be improved either by taking more terms in the expansion (3.2.28) or by reducing the size of Δw.

Figure 3.2.1, the second part of the output of Program 3.1, shows an excellent agreement between the computed and the measured temperature profiles. The variations of both w and t from their respectively uniform upstream to the uniform downstream conditions are completed within a dimensionless distance of approximately six units, or roughly 0.78 cm, for this weak shock in a low-density gas.

Problem 3.1 Study numerically the shock-wave structure by varying the Mach number, viscosity coefficient, and thermal conductivity.

3.3 Self-Similar Laminar Boundary-Layer Flows

It has been demonstrated in the preceding example that real-gas effects are important only within a short distance in a shock wave, and viscosity and conductivity can be ignored outside of this thin shock layer. Another example showing a similar property is the high-Reynolds number flow past a streamline body discussed in Section 2.1. In this case effects of viscosity and conductivity are confined within a thin boundary layer next to the body surface.

The governing equations for boundary-layer flows are deduced from Navier-Stokes equations under the assumption that the boundary-layer thickness δ is small compared with the characteristic length L of the body. Considering a thin, two-dimensional boundary layer around a body having its surface parallel to the x-axis and normal to the y-axis, Kuethe and Chow (1976, Appendix B), using an order-of-magnitude analysis, found that $v \ll u$ and $\partial/\partial x \ll \partial/\partial y$ when operating on either velocity or temperature. If the x component of the equation of motion is of order unity, then the y component is of order δ/L. When the latter is ignored completely, with $\partial p/\partial y$ approximated by zero, the resulting Navier-Stokes and energy equations are, respectively,

$$\rho \left(\frac{\partial u}{\partial t} + u \frac{\partial u}{\partial x} + v \frac{\partial u}{\partial y} \right) = -\frac{\partial p}{\partial x} + \frac{\partial}{\partial y} \left(\mu \frac{\partial u}{\partial y} \right) \tag{3.3.1}$$

$$\rho c_p \left(\frac{\partial T}{\partial t} + u \frac{\partial T}{\partial x} + v \frac{\partial T}{\partial y} \right) = \frac{\partial p}{\partial t} + u \frac{\partial p}{\partial x} + \frac{\partial}{\partial y} \left(k \frac{\partial T}{\partial y} \right)$$

They are called the *boundary-layer equations* for two
pressible flow. These, together with the continuity equai
equation of state (3.1.5), form a system of four scalar equ
unknowns ρ, T, u, and v. The pressure p in a thin boundary
treated as an unknown. Instead, it is considered to be c
boundary layer, while its x dependence is obtained by solving the problem of
an inviscid flow past the same body. The previously mentioned approximations form the basis of Prandtl's boundary-layer theory.

If we consider a simple problem of a semiinfinite flat plate aligned with a
uniform flow of constant speed U and of constant physical properties
including density ρ, as sketched in Fig. 3.3.1, the governing equations for a
steady flow are further simplified to

$$u \frac{\partial u}{\partial x} + v \frac{\partial u}{\partial y} = \nu \frac{\partial^2 u}{\partial y^2} \qquad (3.3.3)$$

$$\frac{\partial u}{\partial x} + \frac{\partial v}{\partial y} = 0 \qquad (3.3.4)$$

where $\nu = \mu/\rho$ is the kinematic viscosity of the fluid. These two equations are
sufficient for computing the velocity field. The use of the decoupled energy
equation will be demonstrated in Section 3.4. This formulation closely describes the thin laminar boundary layer on the surface of a two-dimensional
streamlined body that moves in an incompressible fluid, or in a compressible
fluid but at a speed much slower than the speed of sound.

At any fixed x on the plate, three boundary conditions are needed, two for
the first equation and one for the second. They are the no-slip condition at the

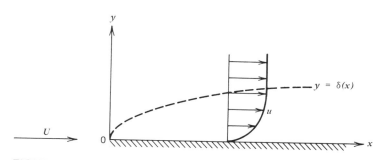

FIGURE 3.3.1 Boundary layer on a flat plate.

Self-Similar Laminar Boundary-Layer Flows

and the condition of uniform flow at infinity; that is,

$$u = v = 0 \quad \text{at} \quad y = 0 \tag{3.3.5}$$

$$u \to U \quad \text{as} \quad y \to \infty \tag{3.3.6}$$

Finding a solution to the system consisting of (3.3.3) to (3.3.6) is called the *Blasius problem* (Blasius, 1908). In terms of stream function ψ defined in (2.1.14), that is, $u = \partial\psi/\partial y$ and $v = -\partial\psi/\partial x$, (3.3.4) is satisfied automatically and (3.3.3) becomes

$$\frac{\partial\psi}{\partial y}\frac{\partial^2\psi}{\partial x \partial y} - \frac{\partial\psi}{\partial x}\frac{\partial^2\psi}{\partial y^2} = \nu \frac{\partial^3\psi}{\partial y^3} \tag{3.3.7}$$

Experiments show that by stretching the vertical coordinate according to the law $y/\sqrt{x}$, the dimensionless velocity profiles u/U measured in a laminar boundary layer at different distances x from the leading edge collapse into one. In other words, these velocity profiles are similar to one another, and the boundary layer flow is said to be *self-similar*. This evidence suggests a great simplification for our mathematical analysis of the Blasius problem. If the independent variables x and y were combined according to the stretching law just mentioned to form a new single independent variable η, we would expect that the governing partial differential equation (3.3.7) could be transformed into an ordinary differential equation and that the boundary conditions (3.3.5) and (3.3.6) would contain neither x nor y explicitly.

Let us try the following transformations, in which η is nondimensionalized and f is a dimensionless function.

$$\eta = y\left(\frac{U}{\nu x}\right)^{1/2} \tag{3.3.8}$$

$$\psi = (\nu U x)^{1/2} f(\eta) \tag{3.3.9}$$

In terms of the new variables, and with a prime denoting differentiation with respect to η, the velocity components become

$$u = U f' \tag{3.3.10}$$

$$v = \tfrac{1}{2}\left(\frac{\nu U}{x}\right)^{1/2}(\eta f' - f) \tag{3.3.11}$$

and the governing equation becomes

$$f''' + \tfrac{1}{2}ff'' = 0 \tag{3.3.12}$$

with boundary conditions

$$f = f' = 0 \quad \text{at} \quad \eta = 0 \tag{3.3.13}$$

$$f' \to 1 \quad \text{as} \quad \eta \to \infty \tag{3.3.14}$$

230

Sure enough we have reduced our problem to solving an ordinary differential equation, although it is still a nonlinear equation. η is called the *similarity variable*, and the solution to the transformed system is called the *similarity solution* of the original formulation.

It should be pointed out that the transformations (3.3.8) and (3.3.9) are not valid at the leading edge where $x = 0$, and that x (or y) must appear on the right-hand side of (3.3.9) if u/U is to be a function of η alone. Actually, these two transformations would also be derived if we started with the general form that $\eta = Ax^m y$ and $\psi = Bx^n f(\eta)$, and if we then required that after transformation x and y could not appear explicitly.

The solution of (3.3.12) satisfying accompanying boundary conditions was obtained by Blasius (1908) in the form of a power series expansion about $\eta = 0$ and an asymptotic expansion for $\eta \to \infty$, the two being matched at an intermediate point. Here we will find the solution using a numerical approach.

If f, f', and f'' are all known at a certain dimensionless height η_i, the fourth-order Runge-Kutta method may be utilized to find the solution at $\eta_{i+1} = \eta_i + h$ and at stations thereafter step by step. To prepare for using the method, the third-order equation (3.3.12) is first written as three first-order simultaneous equations

$$\frac{df}{d\eta} = p, \qquad \frac{dp}{d\eta} = q, \qquad \frac{dq}{d\eta} = -\tfrac{1}{2}fq \qquad (3.3.15)$$

Then the Runge-Kutta formulas (1.1.9) are applied to each of them to get

$$\Delta_1 f_i = hp_i$$
$$\Delta_1 p_i = hq_i$$
$$\Delta_1 q_i = -\tfrac{1}{2}hf_i q_i$$
$$\Delta_2 f_i = h(p_i + \tfrac{1}{2}\Delta_1 p_i)$$
$$\Delta_2 p_i = h(q_i + \tfrac{1}{2}\Delta_1 q_i)$$
$$\Delta_2 q_i = -\tfrac{1}{2}h(f_i + \tfrac{1}{2}\Delta_1 f_i)(q_i + \tfrac{1}{2}\Delta_1 q_i)$$
$$\Delta_3 f_i = h(p_i + \tfrac{1}{2}\Delta_2 p_i)$$
$$\Delta_3 p_i = h(q_i + \tfrac{1}{2}\Delta_2 q_i)$$
$$\Delta_3 q_i = -\tfrac{1}{2}h(f_i + \tfrac{1}{2}\Delta_2 f_i)(q_i + \tfrac{1}{2}\Delta_2 q_i)$$
$$\Delta_4 f_i = h(p_i + \Delta_3 p_i)$$
$$\Delta_4 p_i = h(q_i + \Delta_3 q_i)$$
$$\Delta_4 q_i = -\tfrac{1}{2}h(f_i + \Delta_3 f_i)(q_i + \Delta_3 q_i) \qquad (3.3.16)$$

Self-Similar Laminar Boundary-Layer Flows

231

Finally, the values of f, f', and f'' are computed at η_{i+1} by analogy with (1.1.8).

$$f_{i+1} = f_i + \tfrac{1}{6}(\Delta_1 f_i + 2\,\Delta_2 f_i + 2\,\Delta_3 f_i + \Delta_4 f_i)$$
$$p_{i+1} = p_i + \tfrac{1}{6}(\Delta_1 p_i + 2\,\Delta_2 p_i + 2\,\Delta_3 p_i + \Delta_4 p_i)$$
$$q_{i+1} = q_i + \tfrac{1}{6}(\Delta_1 q_i + 2\,\Delta_2 q_i + 2\Delta_3 q_i + \Delta_4 q_i) \tag{3.3.17}$$

However, numerical integration of (3.3.15) cannot be started at $\eta = 0$ because q (i.e., f'') is not known there. Boundary conditions (3.3.13) and (3.3.14) provide only two of the three values, f and p (or f'), that are required at $\eta = 0$, but provide another value of p at infinity. If the problem is to be solved by the Runge-Kutta method, it may be combined with the half-interval method already introduced in Section 1.5, as follows.

First, the boundary condition (3.3.14) at infinity is not convenient for programming, since every value specified or computed in a program must be finite. Instead of extending to infinity, we limit our range of numerical integration up to a reasonably large height η_{max}, and approximate (3.3.14) by

$$1 - p \leq \epsilon \qquad \text{at} \qquad \eta = \eta_{max} \tag{3.3.18}$$

where ϵ represents a positive value much less than unity whose magnitude controls the accuracy of the solution. ϵ must be positive because p is the dimensionless parallel velocity component in the boundary layer that approaches the free-stream value of 1 from below.

At the beginning of our numerical computation a value q_0 is arbitrarily guessed and a positive increment $\Delta_1 q_0$ is picked. By letting $f = p = 0$ and $q = q_0$ at $\eta = 0$, (3.3.15) are integrated using Runge-Kutta formulas until η reaches η_{max}. The last value of p so computed is designated p_{max} and is represented by the leftmost point on the q_0 versus p_{max} plot of Fig. 3.3.2. If the ordinate of this point is below the value 1, as shown, we replace the starting value q_0 of q by $q_0 + \Delta_1 q_0$ and repeat the numerical integration from $\eta = 0$. The process would be repeated again if the same final situation were encountered at η_{max}. Suppose at the end of the third time a point above the dashed horizontal line of unit height is obtained; we know that the starting value of q is now too large. To reverse the direction, we let $\Delta_2 q_0 = -\Delta_1 q_0 / 2$ and then replace q_0 by $q_0 + \Delta_2 q_0$. Repeat with this negative value of $\Delta_2 q_0$ until the data point in Fig. 3.3.2 comes below the horizontal dashed line. Then we reverse the direction again by letting $\Delta_3 q_0 = -\Delta_2 q_0 / 2$ and replacing q_0 by $q_0 + \Delta_3 q_0$. In programming for such computations the subscripts to Δ are actually not needed. The conclusion of this discussion is that the value of Δq_0 is to be changed to $-\Delta q_0 / 2$ for the next iteration either when p_{max} is greater than 1 while Δq_0 is positive, or when p_{max} is less than 1 while Δq_0 is negative.

At the end of each iteration we define $(1 - p_{max})$ as the error as indicated in Fig. 3.3.2, and we check to see if error is positive and also if it is less than or

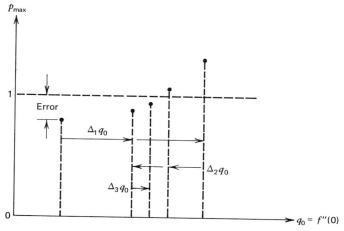

FIGURE 3.3.2 Half-interval method for the Blasius problem.

equal to the assumed ϵ. When both are true, the approximate boundary condition (3.3.18) is satisfied, and the iterative process is terminated.

Having obtained f and its derivatives as functions of η, we can now compute the distribution of velocity components in the boundary layer. (3.3.10) and (3.3.11) suggest dimensionless velocity components of the following form that are evaluated from

$$\frac{u}{U} = f' \tag{3.3.19}$$

$$v \bigg/ \left(\frac{\nu U}{x}\right)^{1/2} = \tfrac{1}{2}(\eta f' - f) \tag{3.3.20}$$

It is demonstrated in Program 3.2 that a boundary-value problem can be solved by using a numerical technique for solving an initial-value problem in combination with the half-interval method. This is not an efficient procedure for solving the present problem, however. An improved technique is presented later in Problem 3.2.

The tabulated numerical result agrees with that obtained by Howarth (1938). Solutions of higher accuracy can easily be computed by reducing both the value of ϵ and the size of h. The curve showing the variation of u/U in the boundary layer would coincide very well with Nikuradse's measured data points if they were also plotted. See Fig. 7.9 of Schlichting (1968).

Solution shows that the parallel velocity component u approaches the free-stream value following a smooth curve; thus the edge of a boundary

Self-Similar Laminar Boundary-Layer Flows

layer is actually an artificial terminology. If the boundary layer thickness δ is defined as the height where $u = 0.994U$, which occurs around $\eta = 5.2$, we have

$$\delta = 5.2 \left(\frac{\nu x}{U}\right)^{1/2} \qquad (3.3.21)$$

The edge of the boundary layer is therefore represented by a parabola, as sketched in Fig. 3.3.1.

Figure 3.3.3 displays the interesting phenomenon that the normal velocity component approaches a constant value when going away from the plate into the free stream. This may be explained by reasoning that the retardation of

List of Principal Variables in Program 3.2

Program Symbol	Definition
DQ0	Variable increment Δq_0 in q_0
D1F,D2F, etc.	The increments $\Delta_1 f_i$, $\Delta_2 f_i$, etc. in (3.3.16)
EPSLON	ϵ defined in (3.3.18)
ERROR	The difference $1 - p_{max}$ in each iteration
ETA	Similarity variable η
ETAMAX	The maximum value of η for numerical computation
F	Nondimensionalized stream function f defined in (3.3.9)
H	Increment in η
ITER	Iteration counter
P	Derivative $df/d\eta$
Q	Derivative $d^2 f/d\eta^2$
Q0	q_0, the starting value of q at $\eta = 0$
V	Dimensionless normal velocity defined in (3.3.20)

```
C           ***** PROGRAM 3.2 *****

C         SOLVING BLASIUS PROBLEM TO FIND VELOCITY DISTRIBUTION IN
C         A LAMINAR BOUNDARY LAYER ON A SEMI-INFINITE FLAT PLATE

          DIMENSION ETA(150),F(150),P(150),Q(150),V(150),LCHAR(3),NSCALE(4)

C         ..... SPECIFY INPUT DATA AND SET ITER EQUAL ONE .....
          DATA ETAMAX,H,EPSLON,Q0,DQ0/ 10.0, 0.1, 1.0E-6, 1.0, 0.2 /
          ITER = 1

C         ..... START ONE ITERATION AT ETA=0 .....
        1 I = 1
          ETA(I) = 0.0
          F(I)   = 0.0
          P(I)   = 0.0
                Q0 = Q0 + DQ0
          Q(I)   = Q0
```

```
C       ..... APPLY RUNGE-KUTTA FORMULAE (3.3.16) AND
C               (3.3.17) UNTIL ETAMAX IS REACHED .....
      2 D1F = H * P(I)
        D1P = H * Q(I)
        D1Q = -0.5 * H * F(I) * Q(I)

        D2F = H * (P(I)+D1P/2.)
        D2P = H * (Q(I)+D1Q/2.)
        D2Q = -0.5 * H * (F(I)+D1F/2.) * (Q(I)+D1Q/2.)

        D3F = H * (P(I)+D2P/2.)
        D3P = H * (Q(I)+D2Q/2.)
        D3Q = -0.5 * H * (F(I)+D2F/2.) * (Q(I)+D2Q/2.)

        D4F = H * (P(I)+D3P)
        D4P = H * (Q(I)+D3Q)
        D4Q = -0.5 * H * (F(I)+D3F) * (Q(I)+D3Q)

        F(I+1) = F(I) + (D1F + 2.*D2F + 2.*D3F + D4F)/6.0
        P(I+1) = P(I) + (D1P + 2.*D2P + 2.*D3P + D4P)/6.0
        Q(I+1) = Q(I) + (D1Q + 2.*D2Q + 2.*D3Q + D4Q)/6.0
        ETA(I+1) = ETA(I) + H
        I = I + 1
        IF( ETA(I)-ETAMAX ) 2, 3, 3

C       ..... KEEP ITERATING UNTIL THE APPROXIMATE BOUNDARY CONDITION
C               (3.3.18) IS SATISFIED. HALF-INTERVAL METHOD IS USED TO
C               MODIFY THE STARTING VALUE OF Q .....
      3 ERROR = 1.0 - P(I)
        IF( ERROR.GT.0. .AND. ERROR.LE.EPSLON ) GO TO 4
        IF( (P(I).GT.1. .AND. DQ0.GT.0.) .OR.
      A     (P(I).LT.1. .AND. DQ0.LT.0.) ) DQ0 = -DQ0/2.
        ITER = ITER + 1
        GO TO 1


C       ..... PRINT THE NUMBER OF ITERATIONS AND THEN PRINT F, F*, AND F**
C               AS FUNCTIONS OF ETA, WHERE AN * INDICATES DIFFERENTIATION
C               WITH RESPECT TO ETA. THUS F* IS P AND F** IS Q .....
      4 WRITE(6,5) ITER
      5 FORMAT( 1H1, 23X, 22HNUMBER OF ITERATIONS =, I3 // 15X,3HETA, 10X,
      A        1HF, 13X,2HF*, 11X,3HF**/ 15X,4H----, 3(5X,9H---------)/ )
        WRITE(6,6) ( (ETA(I),F(I),P(I),Q(I)), I=1,101,2 )
      6 FORMAT( 12X, F7.2, 3F14.7 )

C       ..... COMPUTE DIMENSIONLESS VERTICAL VELOCITY V, THEN PLOT BOTH
C               VELOCITY COMPONENTS ACROSS THE BOUNDARY LAYER .....
        DO 7 I = 1, 101
      7 V(I) = 0.5 * ( ETA(I)*P(I)-F(I) )

        DATA NSCALE/ 0, 1, 0, 1 /
        CALL PLOT1( NSCALE, 6, 8, 3, 27, 1H-, 1H1 )
        CALL PLOT2( 1.0, 0.0, 5.0, 0.0, .TRUE. )
        CALL PLOT3( 1HU, P, ETA, 1, 51, 1 )
        CALL PLOT3( 1HV, V, ETA, 1, 51, 1 )
        WRITE(6,8)
      8 FORMAT( 1H1 )
        LCHAR(1) = 3HETA
        CALL PLOT4( LCHAR, 3 )
        WRITE(6,9)
      9 FORMAT( /// 12X, 44HFIGURE 3.3.3   DISTRIBUTION OF DIMENSIONLESS,
      A         9H VELOCITY / 27X, 22HCOMPONENTS U AND V IN ,
      B         17HA BOUNDARY LAYER. )
        END
```

ETA	F	F*	F**
0.00	0.0000000	0.0000000	.3320572
.20	.0066410	.0664078	.3319837
.40	.0265599	.1327641	.3314697
.60	.0597346	.1989372	.3300790
.80	.1061082	.2647090	.3273891
1.00	.1655717	.3297799	.3230070
1.20	.2379487	.3937759	.3165891
1.40	.3229815	.4562616	.3078653
1.60	.4203206	.5167566	.2966634
1.80	.5295179	.5747579	.2829309
2.00	.6500242	.6297655	.2667515
2.20	.7811931	.6813101	.2483508
2.40	.9222898	.7289816	.2280917
2.60	1.0725056	.7724547	.2064546
2.80	1.2309769	.8115093	.1840066
3.00	1.3968077	.8460441	.1613603
3.20	1.5690944	.8760811	.1391281
3.40	1.7469495	.9017608	.1178763
3.60	1.9295245	.9233293	.0980863
3.80	2.1160290	.9411176	.0801260
4.00	2.3057456	.9555178	.0642342
4.20	2.4980388	.9669567	.0505198
4.40	2.6923600	.9758705	.0389727
4.60	2.8882470	.9826831	.0294839
4.80	3.0853196	.9877891	.0218713
5.00	3.2832725	.9915415	.0159069
5.20	3.4818664	.9942452	.0113419
5.40	3.6809178	.9961549	.0079278
5.60	3.8802893	.9974774	.0054321
5.80	4.0798805	.9983751	.0036485
6.00	4.2796195	.9989725	.0024021
6.20	4.4794558	.9993622	.0015502
6.40	4.6793550	.9996114	.0009807
6.60	4.8792942	.9997675	.0006081
6.80	5.0792581	.9998635	.0003696
7.00	5.2792370	.9999213	.0002202
7.20	5.4792250	.9999554	.0001286
7.40	5.6792182	.9999751	.0000736
7.60	5.8792145	.9999863	.0000413
7.80	6.0792124	.9999926	.0000227
8.00	6.2792113	.9999960	.0000122
8.20	6.4792107	.9999978	.0000065
8.40	6.6792104	.9999987	.0000034
8.60	6.8792102	.9999992	.0000017
8.80	7.0792101	.9999995	.0000008
9.00	7.2792100	.9999996	.0000004
9.20	7.4792099	.9999996	.0000002
9.40	7.6792098	.9999997	.0000001
9.60	7.8792098	.9999997	.0000000
9.80	8.0792097	.9999997	.0000000
10.00	8.2792096	.9999997	.0000000

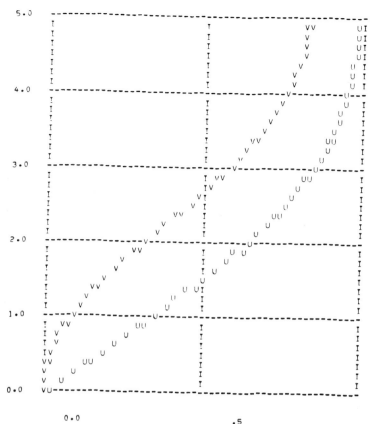

FIGURE 3.3.3 Distribution of dimensionless velocity components U and V in a boundary layer.

flow speed by viscosity causes a negative $\partial u / \partial x$ in the boundary layer, with which a positive $\partial v / \partial y$ is associated in virtue of the continuity equation. Integration of $\partial v / \partial y$ across the boundary layer results in a finite v in the region far above the plate. Thus the presence of a thin plate can be detected, theoretically, by a person right above it by measuring the upcoming velocity component. Notice that the dimensionless velocity components plotted in Fig. 3.3.3 are not based on the same reference velocities. If plotted in dimensional form v would be much smaller than u. That v is finite at infinity clarifies why only the parallel velocity component is specified in the boundary condition (3.3.6) away from the plate.

Let us consider a specific numerical example of an air flow traveling at 30 m/s. At sea level the kinematic viscosity of air is $\nu = 1.49 \times 10^{-5} \, \text{m}^2/\text{s}$. At a

Self-Similar Laminar Boundary-Layer Flows

station 1 m downstream from the leading edge of an aligned plate, the boundary-layer thickness is 0.37 cm according to (3.3.21), and the normal velocity far above the plate is 0.018 m/s according to (3.3.20), which is a negligibly small fraction of the free-stream speed. The Reynolds number at that station based on the distance from the leading edge is Ux/ν, or 2.01×10^6. A large Reynolds number is a requirement for the validation of the boundary-layer theory.

The shear stress in the fluid is

$$\tau = \mu \frac{\partial u}{\partial y} = \mu U \left(\frac{U}{\nu x}\right)^{1/2} f'' \tag{3.3.22}$$

Thus f'' describes the dimensionless shear stress distribution in the boundary layer, and the particular value of f'' at $\eta = 0$, which is the q_0 we looked for using half-interval method in Program 3.2, represents the dimensionless shear stress on the flat plate.

Problem 3.2   Program 3.2 shows that 31 iterations are needed to obtain a solution having the desired accuracy, and most of the computational effort has been wasted in the trial-and-error process. The reason for following such a time-consuming scheme is that f'' is not given at the beginning of the numerical integration.

A better method can be derived for finding f'' at $\eta = 0$. Let us introduce a linear transformation for the independent variable, $\eta = kz$, where k is a constant to be determined; in the meantime, we introduce a function g such that $f(\eta) = g(z)/k$. It can easily be verified that

$$\frac{d^n f}{d\eta^n} = \frac{1}{k^{n+1}} \frac{d^n g}{dz^n} \qquad \text{for } n \geqslant 1 \tag{3.3.23}$$

If a prime is used to denote differentiation with respect to the argument, the foregoing relations transform the system consisting of (3.3.12) to (3.3.14) into

$$g''' + \tfrac{1}{2} g g'' = 0 \tag{3.3.24}$$

$$g = g' = 0 \qquad \text{at} \quad z = 0 \tag{3.3.25}$$

$$g' \to k^2 \qquad \text{as} \quad z \to \infty \tag{3.3.26}$$

Since k is arbitrary and appears only in the condition at infinity, there is no restriction on the magnitude of g'' at the plate. Let us choose a convenient value of unity for $g''(0)$.

$$g'' = 1 \qquad \text{at} \quad z = 0 \tag{3.3.27}$$

This assumed condition transforms, according to (3.3.23), into

$$f'' = k^{-3} \quad \text{at} \quad \eta = 0 \qquad (3.3.28)$$

Write a computer program for integration of (3.3.24) starting at $z = 0$ with the prescribed initial conditions (3.3.25) and (3.3.27) up to a reasonably large value of z. The value of k is calculated by taking the square root of the last value of g' according to (3.3.26). The solution is obtained by using the transformations $f(\eta) = g(z)/k$ and $\eta = kz$.

However, if the solution for f is to be evaluated only at specific locations separated by $\Delta\eta$, an alternative procedure may be followed. Now f'' at $\eta = 0$ becomes known after substitution of k into (3.3.28). This value, together with $f = f' = 0$ at $\eta = 0$, enables us to integrate (3.3.12) immediately, and the solution will automatically satisfy the boundary condition (3.3.14) far away from the plate.

Problem 3.3   Injection or suction of fluid through the body surface is one of the techniques used to alter the structure of the boundary layer along a body. As a consequence, the drag of the body and the heat exchange rate between the body and the surrounding fluid are modified and, furthermore, the boundary-layer flow may be made to separate or attach to the body surface.

A boundary-layer flow in the presence of injection may become self-similar if the injection speed is properly distributed. As mentioned before, a flow is called self-similar if both the governing equation and the boundary conditions can be transformed into such a form that there is only one similarity parameter as the sole independent variable. Suppose a *uniform* flow is injected into the boundary layer from Blasius' semiinfinite plate. Using the transformations (3.3.8) and (3.3.9), we still obtain the same ordinary differential equation (3.3.12). However, the boundary condition that v be a constant on the plate transforms into one containing x explicitly. It implies mathematically that a solution cannot be found containing η alone and, physically, that the velocity profiles measured at different stations along the plate can no longer be collapsed into one profile no matter how the y-axis is stretched. In this case a similarity solution does not exist and the flow is not self-similar.

Now, if the injection speed v_0 is a function of the distance from the leading edge and is described by the law

$$v_0 = C \left(\frac{\nu U}{x} \right)^{1/2} \tag{3.3.29}$$

where C is a constant (Fig. 3.3.4), the transformed boundary conditions at the plate become, in virtue of (3.3.20),

$$f' = 0 \quad \text{and} \quad f = -2C \quad \text{at} \quad \eta = 0 \tag{3.3.30}$$

The governing equation and the boundary condition at infinity remain the same form as (3.3.12) and (3.3.14), respectively. In this case we have found another self-similar flow. The dimensionless injection speed C is used to control the strength and direction of the normal flow. Positive and negative C imply injection and suction, respectively.

To study the effects of injection and suction on velocity profile, boundary-layer thickness, and shear stress, write a program and then run it for C varying from -0.6 to $+0.6$ at intervals of size 0.1.

From the printed data estimate the boundary layer thickness δ. What is the conclusion after plotting δ and $f''(0)$ as functions of C, and what is the physical interpretation when $f''(0)$ vanishes? Explain why difficulties develop in the computation if a value of C greater than 0.6 is assigned. The result will show that for any positive C an inflection point exists in the velocity profile at which $\partial^2 u / \partial y^2 = 0$ or, equivalently, $f''' = 0$. Such a

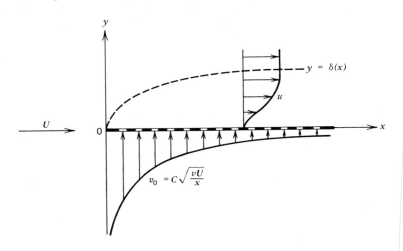

FIGURE 3.3.4 Flat plate with injection of speed v_0.

velocity profile would have a general appearance of the one sketched in Fig. 3.3.4. The presence of an inflection point in the profile is an indication of flow instability (Lin, 1955, Section 4.3), and turbulence is expected to develop in the boundary layer when fluid is injected from the plate.

The transformation described in Problem 3.2 does not work for the problem at hand, because the undetermined constant k appears in one of the boundary conditions at the plate as well as in the condition at infinity, and therefore $g''(0)$ can no longer be chosen arbitrarily.

When the half-interval method is used to find $f''(0)$, the algorithm must be modified from that used previously in Program 3.2. For the Blasius problem we assume in Fig. 3.3.2 (correctly by a mere coincidence) that p_{max} increases with increasing q_0. This is not always true, however. To insure that any arbitrary situation can be handled, the following technique is suggested. If at the end of a certain iteration the difference $(1 - p_{max})$ is computed and is named error, let the difference evaluated in the previous iteration be called error'. By examining every possible situation, you may conclude that the half-interval method is used (i.e. to replace Δq_0 by $-\Delta q_0/2$) when either the product (error)(error') or the difference $|error'| - |error|$ becomes negative.

Problem 3.4 A class of self-similar boundary-layer flows in the presence of a pressure gradient was found by Falkner and Skan (1930). If the flow parallel to a semiinfinite flat plate outside the boundary layer is given by

$$U(x) = U_0 x^m \qquad (3.3.31)$$

the pressure gradient along the plate can be computed from Euler's equation so that

$$\frac{\partial p}{\partial x} = -m\rho U_0^2 x^{2m-1} \qquad (3.3.32)$$

In a region of decreasing pressure along the flow direction, the net pressure force acting on a fluid element tends to accelerate it; the pressure gradient is called a favorable one. In a region in which pressure increases along the flow direction, it is called an adverse pressure gradient. (3.3.32) states that the pressure gradient on the flat plate can be made either favorable or adverse by assigning either positive or negative values to m.

With positive values of m, (3.3.31) represents a realistic two-dimensional potential flow at the surface of a wedge of wedge angle $2m\pi/(m + 1)$ placed symmetrically in an originally uniform stream. Negative values of m correspond to flow over a flat plate that is inclined away from the flow far upstream, although it is nonrealistic at $x = 0$ where flow speed becomes unbounded.

Adding the missing term $\rho^{-1}\partial p/\partial x$ on the right side of (3.3.3) and using the transformations that

$$\eta = y \left(\frac{m + 1}{2} \frac{U_0}{\nu} x^{m-1}\right)^{1/2}, \tag{3.3.33}$$

$$\psi = \left(\frac{2\nu U_0}{m + 1} x^{m+1}\right)^{1/2} f(\eta) \tag{3.3.34}$$

one can show that the resulting governing equation is

$$f''' + ff'' + \frac{2m}{m + 1}(1 - f'^2) = 0 \tag{3.3.35}$$

and the boundary conditions are exactly the same as those for Blasius problem, that is, (3.3.13) and (3.3.14).

Construct a computer program for solving the Falkner-Skan problem. Obtain solutions for $m = 0$, 0.5, and 1.0 and then for $m = -0.05$ and -0.09. Inflection point shows in the dimensionless velocity profiles $f'(=u/U)$ for negative values of m. Note that $f''(0)$ is nearly zero when $m = -0.0904$.

3.4 Flat-Plate Thermometer Problem—
Ordinary Boundary-Value Problems
Involving Derivative Boundary Conditions

Let us consider the energy aspect of the boundary-layer flow over a flat plate. Converted from the work done by frictional forces, heat is generated within the boundary layer. This heat is diffused away from a fluid element by thermal conductivity and, at the same time, is convected downstream by the fluid motion. If the flat plate is originally at the same temperature as that of the fluid far upstream, it will be heated up and its surface temperature will finally approach a constant value when a steady state is reached; under a steady state there is no heat transfer from the fluid to the wall, or $(\partial T/\partial y)_{y=0} = 0$. The steady-state temperature T_w of the equivalently insulated plate is called the *adiabatic* or *recovery temperature*. Because in a boundary layer fluid particles decelerate through an irreversible process, the temperature T_w at the wall where velocity is zero is generally not the stagnation temperature of the free stream. Furthermore, the variation of temperature from T_w at the plate to T_1 in the free stream takes place within a short distance, forming a thermal boundary layer in addition to the velocity boundary layer on the plate.

If the structure of the thermal boundary layer is known, the flat plate just described will serve the purpose of a thermometer when mounted on a moving body such as a flying aircraft, because then the ambient temperature T_1 can be calculated from T_w measured on the plate. For low speeds the density and physical properties of the fluid may be treated as constants. In addition, we postulate that the thermal boundary layer is also thin and its behavior is similar to that of the velocity boundary layer, so that $\partial T/\partial x \ll \partial T/\partial y$. In the absence of a pressure gradient, the governing boundary-layer equations at steady state are

$$\frac{\partial u}{\partial x} + \frac{\partial v}{\partial y} = 0 \tag{3.4.1}$$

$$u \frac{\partial u}{\partial x} + v \frac{\partial u}{\partial y} = \nu \frac{\partial^2 u}{\partial y^2} \tag{3.4.2}$$

$$\rho c_p \left(u \frac{\partial T}{\partial x} + v \frac{\partial T}{\partial y} \right) = k \frac{\partial^2 T}{\partial y^2} + \mu \left(\frac{\partial u}{\partial y} \right)^2 \tag{3.4.3}$$

The boundary conditions for velocity and temperature fields are

$$u = v = \frac{\partial T}{\partial y} = 0 \qquad \text{at} \quad y = 0 \tag{3.4.4}$$

$$u \to U, \qquad T \to T_1 \qquad \text{as} \quad y \to \infty \tag{3.4.5}$$

As already mentioned at the end of Section 3.1, the assumption of incompressibility decouples the energy equation from continuity and momentum equations. To solve the energy equation we first obtain expressions for velocity components from the Blasius solution, which is the solution to simultaneous equations (3.4.1) and (3.4.2), subject to their three boundary conditions stated in (3.4.4) and (3.4.5). Substituting u and v into (3.4.3) in terms of the similarity variables η and f, and introducing a dimensionless temperature difference $\theta(\eta)$ defined by

$$\theta = \frac{T - T_1}{U^2/2c_p} \tag{3.4.6}$$

we obtain

$$\frac{d^2\theta}{d\eta^2} + \frac{1}{2} Pr f \frac{d\theta}{d\eta} = -2Pr(f'')^2 \tag{3.4.7}$$

Flat-Plate Thermometer Problem

where $Pr = \mu c_p/k$ is the Prandtl number. The transformed boundary conditions are

$$\frac{d\theta}{d\eta} = 0 \qquad \text{at} \quad \eta = 0 \tag{3.4.8}$$

$$\theta \to 0 \qquad \text{as} \quad \eta \to \infty \tag{3.4.9}$$

The result shows that the temperature field is also self-similar.

We now have a boundary-value problem consisting of a second-order ordinary differential equation, whose coefficients contain tabulated functions f and f'', and two boundary conditions specified across the boundary layer. Solution could be obtained using the Runge-Kutta method again by guessing the starting value of θ at $\eta = 0$ and then modifying it following the interval-halving technique used in solving the Blasius problem. In so doing the boundary-value problem is treated as if it were an initial-value problem. Here we prefer to solve the problem using the numerical method developed in Section 2.2 for boundary-value problems, which is to be modified to handle the derivative boundary condition involved in this case.

Once more, a reasonably large $\eta_{\max}$ is chosen to approximate infinity so that the condition (3.4.9) becomes

$$\theta = 0 \qquad \text{at} \quad \eta = \eta_{\max} \tag{3.4.10}$$

We let the range between 0 and $\eta_{\max}$ of the independent variable be divided into n equally spaced intervals of size h, and name the end points consecutively $\eta_1, \eta_2, \ldots, \eta_{n+1}$, starting at the lower end. To represent the present problem on Fig. 2.2.1, x in that figure is to be replaced by η and f by θ. Recall that the original figure was used to describe a boundary-value problem with f known at the ends x_0 and x_{n+1} of the physical space, and that range of x was divided into $n + 1$ intervals of size h. When the same figure is used for the present case with a derivative boundary condition at $\eta = 0$, the physical space, divided into n intervals, is contained between η_1 and η_{n+1}, and η_0 becomes a fictitious point at a distance h to the left of η_1.

Comparing the differential equation (3.4.7) with the standard form (2.2.1), we obtain $A = \frac{1}{2} Pr f$, $B = 0$, and $D = -2Pr(f'')^2$. Their known values at η_i enable us to evaluate the coefficients C_{i1}, C_{i2}, C_{i3}, and C_{i4} from (2.2.11). When the finite-difference equation (2.2.10), used to replace the original differential equation, is applied at η_n, it becomes

$$C_{n1}\theta_{n-1} + C_{n2}\theta_n + C_{n3}\theta_{n+1} = C_{n4}$$

in which the third term is known from the boundary condition at $\eta_{\max}$. This boundary condition is incorporated in the numerical scheme, as already shown in Section 2.2 by the following assignments.

$$C_{n4} \leftarrow (C_{n4} - C_{n3}\theta_{n+1})$$
(3.4.11)

$$C_{n3} \leftarrow 0$$
(3.4.12)

Applying (2.2.10) at η_1 gives

$$C_{11}\theta_0 + C_{12}\theta_1 + C_{13}\theta_2 = C_{14}$$

where θ_0 at the fictitious point is not prescribed but, instead, its value can be related to those at some interior points through the boundary condition at $\eta = 0$. After replacing $d\theta/d\eta$ by the central-difference approximation (2.2.8), (3.4.8) states that $\theta_0 = \theta_2$, so that this derivative boundary condition is incorporated by the following assignments.

$$C_{13} \leftarrow (C_{11} + C_{13})$$
(3.4.13)

$$C_{11} \leftarrow 0$$
(3.4.14)

Having modified four of the coefficients C_{ij} according to (3.4.11) to (3.4.14), we can compute the solution $\theta_1, \theta_2, \ldots, \theta_n$ by solving n simultaneous equations arranged in the tridiagonal matrix form of (2.2.13) with the help of subroutine TRID, which was constructed in Section 2.3 for this particular purpose.

In Program 3.3 the functions f and f'' are first computed. Starting with the value $f''(0) = 0.3320572$ taken from the result of Program 3.2, these functions are obtained immediately after integrating (3.3.12) from $\eta = 0$ to $\eta = \eta_{max}$. Here $\eta_{max} = 10$ is still used in order to be consistent with Program 3.2.

Three cases are considered for three different fluid media, water, air, and mercury, whose Prandtl numbers are 6.750, 0.714, and 0.044, respectively. The variable names used in Program 3.3 in connection with the Blasius solution are exactly the same as those in Program 3.2. The coefficients defined in (2.2.11) are still named C(I,J), and PR is an array used to store Prandtl numbers of the three fluids just named. θ_i is represented by THETA(I) in the program. When the Kth case is finished, the values of THETA(I) are stored in a two-dimensional array T(K,I). In this way the temperature distributions for all three cases can be printed together in a single table for direct comparison at the end of numerical computations.

The numerical value of θ at $\eta = 0$ has a special physical meaning. Consider for the moment a steady isentropic flow. In the absence of viscosity and conductivity, by combining the energy equation (3.1.3) with an equation resulting from the dot product of velocity vector and the momentum equation

Flat-Plate Thermometer Problem

```fortran
C                 ***** PROGRAM 3.3 *****

C         FINDING SIMILARITY TEMPERATURE PROFILES FOR WATER, AIR, AND
C         MERCURY IN A BOUNDARY LAYER OVER A SEMI-INFINITE FLAT PLATE

      DIMENSION ETA(101),F(101),P(101),Q(101),PR(3),
     A           C(100,4),THETA(101),T(3,101)
C     ..... SPECIFY INPUT DATA.  PRANDTL NUMBERS FOR WATER, AIR,
C           AND MERCURY ARE STORED IN THE ARRAY PR .....
      DATA ETAMAX,N, PR/ 10.0,  100,    6.750, 0.714, 0.044 /
      H = ETAMAX / FLOAT(N)
      NP1 = N+1

C     ..... SOLVE BLASIUS PROBLEM USING RUNGE-KUTTA METHOD.  THE
C           STARTING VALUE OF Q IS TAKEN FROM PROGRAM 3.2 .....
      ETA(1) = 0.0
      F(1) = 0.0
      P(1) = 0.0
      Q(1) = 0.3320572
      DO 1  I = 2, NP1
         D1F = H * P(I-1)
         D1P = H * Q(I-1)
         D1Q = -0.5 * H * F(I-1) * Q(I-1)
      D2F = H * (P(I-1)+D1P/2.)
      D2P = H * (Q(I-1)+D1Q/2.)
      D2Q = -0.5 * H * (F(I-1)+D1F/2.) * (Q(I-1)+D1Q/2.)
         D3F = H * (P(I-1)+D2P/2.)
         D3P = H * (Q(I-1)+D2Q/2.)
         D3Q = -0.5 * H * (F(I-1)+D2F/2.) * (Q(I-1)+D2Q/2.)
      D4F = H * (P(I-1)+D3P)
      D4P = H * (Q(I-1)+D3Q)
      D4Q = -0.5 * H * (F(I-1)+D3F) * (Q(I-1)+D3Q)
         F(I) = F(I-1) + (D1F + 2.*D2F + 2.*D3F + D4F)/6.0
         P(I) = P(I-1) + (D1P + 2.*D2P + 2.*D3P + D4P)/6.0
         Q(I) = Q(I-1) + (D1Q + 2.*D2Q + 2.*D3Q + D4Q)/6.0
    1 ETA(I) = ETA(I-1) + H

C     ..... CONSIDER 3 CASES RESPECTIVELY FOR WATER, AIR, AND MERCURY...
      DO 4  K = 1, 3

C     ..... TO SOLVE THE ENERGY EQUATION (3.4.7), WE COMPUTE FIRST
C           THE COEFFICIENTS C(I,J) ACCORDING TO (2.2.11) .....
      DO 2  I = 1, N
      C(I,1) = 1.0 - H/4.*PR(K)*F(I)
      C(I,2) = -2.0
      C(I,3) = 1.0 + H/4.*PR(K)*F(I)
    2 C(I,4) = -2.0 * H**2 * PR(K) * Q(I)**2

C     ..... MODIFY COEFFICIENTS ACCORDING TO (3.4.11 - 14) TO
C           INCORPORATE BOUNDARY CONDITIONS .....
      THETA(NP1) = 0.0
      C(1,3) = C(1,1) + C(1,3)
      C(1,1) = 0.0
      C(N,4) = C(N,4) - C(N,3)*THETA(NP1)
      C(N,3) = 0.0

C     ..... COMPUTE THETA(I) FOR I=1,2,....,N.  THEN STORE THE DATA
C           IN ARRAY T(K,I) FOR PRINTING LATER .....
      CALL TRID( C, THETA, N )
      DO 3  I = 1, NP1
    3 T(K,I) = THETA(I)
    4 CONTINUE

C     ..... PRINT TEMPERATURE PROFILES FOR ALL THREE CASES .....
      WRITE(6,5)
```

246

```
   5 FORMAT( 1H1 // 15X,3HETA, 8X,5HTHETA, 9X,5HTHETA, 9X,5HTHETA /
   A          25X,7H(WATER), 8X,5H(AIR), 7X,9H(MERCURY) /
   B          15X,4H----, 3(5X,9H---------) / )
     DO 6  I = 1, NP1, 2
   6 WRITE(6,7)  ETA(I), (T(K,I),K=1,3)
   7 FORMAT( 12X, F7.2, 3F14.7 )
     END

     SUBROUTINE TRID( C, F, N )
C    ..... TAKEN FROM PROGRAM 2.1 .....
```

ETA	THETA (WATER)	THETA (AIR)	THETA (MERCURY)
0.00	2.4876693	.8442782	.1589981
.20	2.4579077	.8411293	.1588041
.40	2.3690066	.8316902	.1582222
.60	2.2230963	.8160033	.1572537
.80	2.0255915	.7941794	.1559023
1.00	1.7862500	.7664292	.1541745
1.20	1.5193073	.7330905	.1520811
1.40	1.2423639	.6946474	.1496377
1.60	.9740440	.6517402	.1468647
1.80	.7309548	.6051603	.1437875
2.00	.5248839	.5558318	.1404364
2.20	.3612128	.5047779	.1368448
2.40	.2390631	.4530742	.1330489
2.60	.1529575	.4017946	.1290856
2.80	.0951832	.3519542	.1249911
3.00	.0579261	.3044569	.1207998
3.20	.0346004	.2600518	.1165425
3.40	.0203054	.2193052	.1122464
3.60	.0116917	.1825882	.1079341
3.80	.0065873	.1500807	.1036244
4.00	.0036205	.1217889	.0993320
4.20	.0019360	.0975723	.0950683
4.40	.0010050	.0771767	.0908421
4.60	.0005056	.0602683	.0866603
4.80	.0002463	.0464655	.0825283
5.00	.0001160	.0353673	.0784502
5.20	.0000528	.0265758	.0744297
5.40	.0000232	.0197135	.0704699
5.60	.0000099	.0144347	.0665734
5.80	.0000040	.0104327	.0627429
6.00	.0000016	.0074421	.0589804
6.20	.0000006	.0052393	.0552881
6.40	.0000002	.0036400	.0516678
6.60	.0000001	.0024954	.0481213
6.80	.0000000	.0016880	.0446501
7.00	.0000000	.0011265	.0412556
7.20	.0000000	.0007417	.0379391
7.40	.0000000	.0004817	.0347016
7.60	.0000000	.0003086	.0315439
7.80	.0000000	.0001949	.0284669
8.00	.0000000	.0001214	.0254711
8.20	.0000000	.0000745	.0225569
8.40	.0000000	.0000450	.0197247
8.60	.0000000	.0000267	.0169744
8.80	.0000000	.0000156	.0143062
9.00	.0000000	.0000088	.0117198
9.20	.0000000	.0000049	.0092149
9.40	.0000000	.0000025	.0067911
9.60	.0000000	.0000012	.0044478
9.80	.0000000	.0000004	.0021844
10.00	0.0000000	0.0000000	0.0000000

(3.1.2), we obtain a simple energy relation: the sum of enthalpy and kinetic energy per unit mass is constant following a fluid particle; that is,

$$c_p T + \tfrac{1}{2} V^2 = \text{constant along a streamline} \qquad (3.4.15)$$

If a fluid particle, originally in the free stream of temperature T_1 and speed U, were decelerated isentropically to zero speed at the surface of a plate where its temperature rose to $(T_w)_{\text{isen}}$, which is, in fact, the stagnation temperature of the flow, according to (3.4.15), the total energy per unit mass of the fluid at the wall would be the same as that in the free stream of magnitude

$$c_p (T_w)_{\text{isen}} = c_p T_1 + \tfrac{1}{2} U^2 \qquad (3.4.16)$$

It shows that the kinetic energy would be fully recovered at the wall after an isentropic deceleration. In a real fluid with nonvanishing k and μ, however, the fluid temperature at the surface is T_w and the total energy per unit mass there becomes, according to (3.4.6),

$$c_p T_w = c_p T_1 + \theta(0) \cdot \tfrac{1}{2} U^2 \qquad (3.4.17)$$

The interpretation is that when a free-stream fluid particle slows down through an irreversible process in the boundary layer and finally becomes stationary at the wall, $\theta(0)$ times its original kinetic energy is recovered there and is converted into thermal energy. $\theta(0)$ is therefore called the *recovery factor*.

Program 3.3 reveals that the recovery factor is less than unity for air and mercury, whose Prandtl numbers are below 1, and it is greater than unity for water, whose Prandtl number is above 1. A sketchy explanation of this phenomenon is that for a fluid with a smaller Prandtl number, the thermal conductivity is relatively large; more heat is conducted out of a fluid element than is produced within it by friction, so that its energy becomes lower. The opposite is true for a fluid of larger Prandtl number, and the energy of a fluid particle at the wall may become higher than that outside of the boundary layer.

What would happen to the wall temperature if the Prandtl number of the fluid were unity? When this value is used in Program 3.3, its output gives a rounded figure $\theta(0) = 1$. In this particular case the energy and temperature at the wall are equal to the energy and stagnation temperature of the free stream, respectively, even if the transition has gone through an irreversible process. Following the preceding reasoning, it may be said that for $Pr = 1$, the heat produced just balances the heat diffused away.

You may verify by using Program 3.3 that $\theta(0)$ approximately equals $\sqrt{Pr}$ for Prandtl numbers in the neighborhood of unity. The present result for air is a good example. This agrees with the conclusion obtained by Pohlhausen (1921).

The numerical result of Program 3.3 shows that the thermal boundary-layer thickness increases with decreasing Prandtl number due to the increased conductivity. The height $\eta_{max} = 10$ is barely enough in the computation for air, but is definitely not sufficient for mercury. The accuracy for these two cases can be improved by choosing a larger η_{max}.

Problem 3.5 We have seen that the total fluid energy at the wall may be lower or higher than that in the free stream, depending on whether the Prandtl number is less than or greater than unity. One may compute to find out how the total energy is distributed across the boundary layer when either of the two situations occurs.

Let H denote the total energy or the total enthalpy of the fluid per unit mass, which is the sum of specific enthalpy and kinetic energy per unit mass. H_1 in the free stream has the magnitude given by (3.4.16). In a boundary layer the enthalpy can be expressed in terms of free-stream conditions and θ through (3.4.6), and the kinetic energy is contributed mainly from the motion parallel to the plate. Thus, in the boundary layer on an aligned, semiinfinite flat plate, the total energy is

$$H = (c_p T_1 + \tfrac{1}{2} U^2 \theta) + \tfrac{1}{2} u^2$$
$$= c_p T_1 + \tfrac{1}{2} U^2 (\theta + f'^2)$$

A dimensionless energy difference may be defined as

$$\frac{H - H_1}{U^2/2} = \theta + f'^2 - 1$$

Plot this quantity as a function of η for water, air, and mercury, and then give a physical interpretation of the result.

Problem 3.6 Consider a layer of fluid around an infinitely long circular cylinder of radius r_b. Radial gravitational forces hold the fluid in a state of static equilibrium, forming a free surface of radius r_a, as shown in Fig. 3.4.1. The fluid has a constant density ρ and viscosity coefficient μ. When the cylinder is given a rotational motion about its axis, a fluid motion v_θ is induced in the azimuthal direction. The θ component of the Navier-Stokes

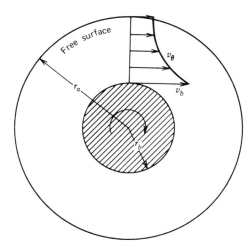

FIGURE 3.4.1 Fluid motion around a rotating cylinder.

equation in cylindrical coordinates is (from Hughes and Gaylord, 1964, p. 25)

$$\rho \left(\frac{\partial v_\theta}{\partial t} + v_r \frac{\partial v_\theta}{\partial r} + \frac{v_\theta}{r} \frac{\partial v_\theta}{\partial \theta} + v_z \frac{\partial v_\theta}{\partial z} + \frac{v_r v_\theta}{r} \right)$$

$$= -\frac{1}{r} \frac{\partial p}{\partial \theta} + \mu \left(\frac{\partial^2 v_\theta}{\partial r^2} + \frac{1}{r} \frac{\partial v_\theta}{\partial r} + \frac{1}{r^2} \frac{\partial^2 v_\theta}{\partial \theta^2} + \frac{\partial^2 v_\theta}{\partial z^2} + \frac{2}{r^2} \frac{\partial v_r}{\partial \theta} - \frac{v_\theta}{r^2} \right)$$

$$(3.4.18)$$

Because of the axisymmetry of the flow on the r-θ plane, derivatives with respect to θ and z are all zero, and there are no motions in radial and axial directions. When a steady state is reached, all terms in (3.4.18) vanish, except the first two and the last one in the parentheses on the right-hand side.

If the constant tangential speed at the surface of the cylinder is v_b, we may introduce a dimensionless radial distance $R = r/r_b$ and a dimensionless speed $V = v_\theta/v_b$, so that the resultant equation becomes

$$\frac{d^2 V}{dR^2} + \frac{1}{R} \frac{dV}{dR} - \frac{V}{R^2} = 0 \qquad (3.4.19)$$

The boundary conditions are, with R_a representing r_a/r_b,

$$V = 1 \qquad \text{at} \quad R = 1 \qquad (3.4.20)$$

$$\frac{dV}{dR} = 0 \qquad \text{at} \quad R = R_a \qquad (3.4.21)$$

The second condition simply states that shear stress must vanish at the free surface. If such a stress did exist, no tangential force could be generated above the free surface to balance it, and the equilibrium condition could never be attained.

Divide the fluid region into 50 equally spaced radial intervals and find the velocity distribution for $R_a = 2$ by use of the subroutine TRID. Compute local percent errors from the exact solution that

$$V = \frac{R_a}{1 + R_a^2} \left(\frac{R}{R_a} + \frac{R_a}{R} \right) \qquad (3.4.22)$$

Notice that if a diagram similar to Fig. 2.2.1 is constructed, the fluid region should be contained between R_0 and R_n, leaving R_{n+1} as a fictitious point to handle the derivative boundary condition at the free surface.

It is interesting to see that, from (3.4.22) or the numerical result, the fluid layer will not rotate with the cylinder like a solid body under the steady state. This result analogously explains why, even in the absence of differential solar heating and many other factors, our atmosphere can never achieve a solid-body rotation with the earth.

3.5 Pipe and Open-Channel Flows

With properly chosen geometries, the equations governing some viscous flows may become linearized. One example is given by Problem 3.6 in the preceding section; it concerns the angular motion of fluid between two concentric circles. In this section a class of flows is considered whose governing equation is linear but involves two independent variables.

We consider the general problem of steady incompressible flow through a straight pipe of uniform but arbitrary cross section. The flow may be caused by either an applied pressure gradient, the gravitational force, the motion of a part of the pipe wall relative to the rest, or any combination of these factors. The flow may be enclosed by a rigid wall or it may have a free surface. Moreover, the flow may contain multiply connected regions formed by axial inner tubes.

Let the infinitely long pipe be parallel to the x-axis along which $\partial v / \partial x = 0$. Because of this particular geometry, the flow has only one nonvanishing velocity component u in the axial direction, so that the continuity equation (3.1.6) is satisfied automatically and the nonlinear terms on the left side of momentum equation (3.1.7) disappear. There is only one equation governing u of the form

$$\frac{\partial^2 u}{\partial y^2} + \frac{\partial^2 u}{\partial z^2} = \frac{1}{\mu}\left(\frac{dp}{dx} - f_x\right) \tag{3.5.1}$$

where the x component of the gravitational force per unit volume, f_x, has been added. It turns out that we have a Poisson equation to which the iteration methods developed in Section 2.10 apply.

If $f_x = 0$ and the pipe cross section is circular, (3.5.1) describes the classical problem of the Poiseuille flow, whose analytical solution can be obtained immediately. Although (3.5.1) is a linear equation, to find its analytical solution satisfying a set of arbitrary boundary conditions is still a difficult task. Numerical solution of (3.5.1) is generally required except for a few cases in which the cross-sectional shape can be described by some simple geometries.

As an illustrative example for solving this class of problems, we consider a tilted open channel of square cross section making an angle θ with the horizontal (Fig. 3.5.1). The x-axis is still chosen to be parallel to the channel, and the angle θ is assumed to be so small that the free surface is everywhere parallel to the channel base. Since there is no applied pressure gradient along the channel, the only forcing function on the right side of (3.5.1) is that due to $f_x = \rho g \sin \theta$, g being the gravitational acceleration. With the

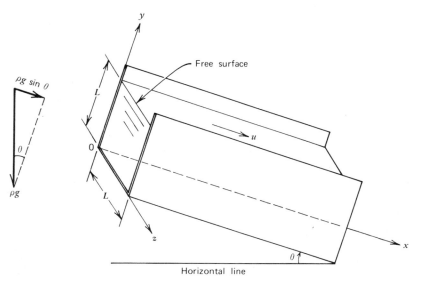

FIGURE 3.5.1 An open-channel flow.

introduction of dimensionless variables

$$Y = \frac{y}{L}, \qquad Z = \frac{z}{L}, \qquad U = \frac{u}{L^2 \rho g \, \sin \theta / \mu} \qquad (3.5.2)$$

where L is the width or height of the square cross section, (3.5.1) becomes

$$\frac{\partial^2 U}{\partial Y^2} + \frac{\partial^2 U}{\partial Z^2} = -1 \qquad (3.5.3)$$

Boundary conditions require that velocity vanishes at solid walls and shear stress is zero at the free surface, or, mathematically,

$$U = 0 \qquad \text{at} \quad Y = 0, \qquad \text{and at} \quad Z = 0 \text{ and } 1 \qquad (3.5.4)$$

$$\frac{\partial U}{\partial Y} = 0 \qquad \text{at} \quad Y = 1 \qquad (3.5.5)$$

For a numerical solution a square grid system is set up to cover the region occupied by the fluid in the Y-Z plane. With the grid size $h = 0.05$, we obtain 21 ($=m$) vertical grid lines and 21 ($=n$) horizontal grid lines. Because of the derivative boundary condition (3.5.5), a fictitious horizontal line is needed at a distance h above the free surface. Comparing (3.5.3) with the generalized form (2.10.1) of the Poisson equation, we can write down its finite-difference computational scheme based on the successive overrelaxation formula (2.10.13).

$$U_{i,j} = (1 - \omega) U_{i,j} + \frac{\omega}{4} (U_{i-1,j} + U_{i+1,j} + U_{i,j-1} + U_{i,j+1} + h^2) \qquad (3.5.6)$$

Here the superscripts used in (2.10.13) to indicate the number of iterations are omitted; they are not needed if in each iteration (3.5.6) is applied successively at interior points starting from the lower left corner of the grid. The optimum value for the relaxation parameter ω is determined according to (2.10.14).

In index notation the boundary conditions (3.5.4) become

$$U_{i,1} = 0, \qquad\qquad i = 1, 2, \ldots, m \qquad (3.5.7)$$

$$U_{1,j} = U_{m,j} = 0, \qquad j = 1, 2, \ldots, n \qquad (3.5.8)$$

After replacing the derivative in (3.5.5) by its central-difference approximation, the derivative boundary condition is reduced to

$$U_{i,n+1} = U_{i,n-1}, \qquad i = 2, 3, \ldots, m - 1 \qquad (3.5.9)$$

In the program to follow the conditions (3.5.7) and (3.5.8) are assigned at the beginning and are kept the same for all iterations, while (3.5.9) are used in the computation of (3.5.6) in every iteration whenever j reaches the value n.

Pipe and Open-Channel Flows

Knowing the velocity distribution across a section, we can calculate the volume of fluid passing through the channel per unit time. Let the solid lines in Fig. 3.5.2 represent schematically the grid system covering the channel cross section. Dashed vertical and horizontal lines are then drawn to bisect the solid-line square meshes.

At each of the grid points marked at the intersections of solid lines, the dimensionless velocity U is either given or computed. These points may be grouped into three categories according to their locations, the interior points, the boundary points, and the corner points, as indicated in Fig. 3.5.2. For small mesh sizes we may assume that the velocity in the shaded area containing a grid point is approximately the same as that at the point itself. As shown in the figure, the shaded small area for an interior point is of magnitude h^2, that for a boundary point is $h^2/2$, and that for a corner point is $h^2/4$. The volume flow rate is calculated approximately by summing the products of local velocity and area at all grid points.

In the output of Program 3.4 some of the numerical values of U are tabulated and, in addition, the contours of velocity bands are plotted in Fig. 3.5.3 using the technique described in Section 2.11. Variables used in this program are named according to their original forms; the name VOLRAT is used for the

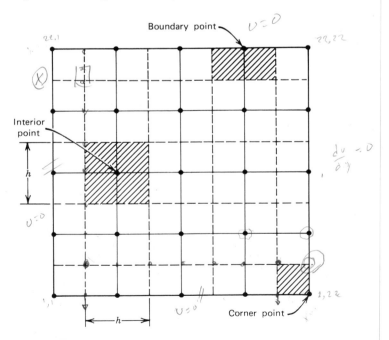

FIGURE 3.5.2 Evaluation of volume flow rate.

```fortran
C                      ***** PROGRAM 3.4 *****

C         FINDING VELOCITY DISTRIBUTION AND VOLUME FLOW RATE OF
C         FLUID IN AN OPEN-CHANNEL OF RECTANGULAR CROSS SECTION

      REAL LX,LY
      DIMENSION U(21,22),SYMBOL(7),GRAPH(70)

C         ..... ASSIGN INPUT DATA .....
      H = 0.05
      M = 1.0/H + 1.001
      N = 1.0/H + 1.001
      MM1 = M - 1
      NM1 = N - 1

C         ..... CALCULATE ALPHA AND THEN THE OPTIMUM
C               VALUE OF OMEGA FROM (2.10.14) .....
      PI = 4.0 * ATAN(1.0)
      ALPHA = COS(PI/M) + COS(PI/N)
      OMEGA = (8.-4.*SQRT(4.-ALPHA**2)) / ALPHA**2

C         ..... APPLY BOUNDARY CONDITIONS (3.5.7) AND (3.5.8) .....
      DO 1  I = 1, M
    1 U(I,1) = 0.0
      DO 2  J = 1, N
      U(1,J) = 0.0
    2 U(M,J) = 0.0

C         ..... ASSIGN INITIALLY GUESSED VALUES TO U .....
      DO 3  I = 2, MM1
      DO 3  J = 2, N
    3 U(I,J) = 0.04

C         ..... SUCCESSIVE OVERRELAXATION METHOD IS USED BY APPLYING (3.5.6)
C               AT ALL INTERIOR GRID POINTS.   THE DERIVATIVE BOUNDARY
C               CONDITION (3.5.9) IS APPLIED WHEN J=N.   ITERATION STOPS WHEN
C               THE SUM OF ABSOLUTE ERRORS IS LESS THAN OR EQUAL TO 0.001...
      ITER = 0
    4 ITER = ITER + 1
      ERROR = 0.0
      DO 6  I = 2, MM1
      DO 6  J = 2, N
      UOLD = U(I,J)
      IF( J.NE.N )  GO TO 5
      U(I,J+1) = U(I,J-1)
    5 U(I,J) = (1.0-OMEGA)*U(I,J) + OMEGA/4.0*( H**2
     A         + U(I-1,J) + U(I+1,J) + U(I,J-1) + U(I,J+1) )
    6 ERROR = ERROR + ABS( U(I,J)-UOLD )
      IF( ERROR.GT.0.001 )  GO TO 4

C         ..... COMPUTE VOLUME FLOW RATE THROUGH THE CHANNEL .....
      VOLRAT = 0.0
      DO 7  I = 2, MM1
      DO 7  J = 2, NM1
    7 VOLRAT = VOLRAT + U(I,J)*H**2
```

255

```
         DO 8  I = 2, MM1
       8 VOLRAT = VOLRAT + (U(I,1)+U(I,N))*H**2/2.0
         DO 9  J = 2, NM1
       9 VOLRAT = VOLRAT + (U(1,J)+U(M,J))*H**2/2.0
         VOLRAT = VOLRAT + (U(1,1)+U(1,N)+U(M,1)+U(M,N))*H**2/4.0

C     ..... PRINT VELOCITY DISTRIBUTION .....
         WRITE(6,10)  ITER,VOLRAT
      10 FORMAT( 1H1 /// 8X, 41HVELOCITY DISTRIBUTION ACROSS THE CHANNEL ,
        A          9HBASED ON , I2, 11H ITERATIONS //
        B          28X, 18HVOLUME FLOW RATE =, F6.4 // )
         DO 11  K = 1, N, 2
         J = N - K + 1
      11 WRITE(6,12)  (U(I,J), I=1,M,2)
      12 FORMAT( /// 11F7.3 )

C     ..... PLOT VELOCITY DISTRIBUTION USING THE METHOD DESCRIBED IN
C          SECTION 2.11. SEE PROGRAM 2.8 FOR DEFINITION OF NAMES .....
         DATA XRANGE,YRANGE,NX/ 1.0, 1.0, 70 /
         DATA SYMBOL/ 1H0, 1H , 1H1, 1H , 1H2, 1H , 1H3 /
         WRITE(6,13)
      13 FORMAT( 1H1 /// 23X, 32H0  REPRESENTS  0.000 ≤ U < 0.018 /
        A          23X, 32H1  REPRESENTS  0.036 ≤ U < 0.054 /
        B          23X, 32H2  REPRESENTS  0.072 ≤ U < 0.090 /
        C          23X, 32H3  REPRESENTS  0.108 ≤ U < 0.126 /// )
         NY = 0.6 * NX * YRANGE / XRANGE
         DELX = XRANGE / NX
         DELY = YRANGE / NY
         DELTA = 0.018
         DO 15  JSYMBL = 1, NY
         Y = YRANGE - (JSYMBL-0.5)*DELY
         J = 1 + Y/H
             LY = Y - (J-1)*H
             HMLY = H - LY
         DO 14  ISYMBL = 1, NX
         X = (ISYMBL-0.5) * DELX
         I = 1 + X/H
             LX = X - (I-1)*H
             HMLX = H - LX
             A1 = HMLX * HMLY
             A2 = HMLX * LY
             A3 = LX * LY
             A4 = LX * HMLY
         UC = ( A1*U(I,J) + A2*U(I,J+1) + A3*U(I+1,J+1)
        A        + A4*U(I+1,J) ) / H**2
         NRANGE = 1 + UC/DELTA
      14 GRAPH(ISYMBL) = SYMBOL(NRANGE)
      15 WRITE(6,16)  GRAPH
      16 FORMAT( 4X, 70A1 )
         WRITE(6,17)
      17 FORMAT( /// 11X, 44HFIGURE 3.5.3   VELOCITY DISTRIBUTION ACROSS
        A          7HAN OPEN / 26X, 32HCHANNEL OF SQUARE CROSS SECTION. )
         END
```

0.000	.042	.073	.096	.109	.114	.109	.096	.073	.042	0.000
0.000	.041	.073	.095	.109	.113	.109	.095	.073	.041	0.000
0.000	.041	.072	.094	.107	.112	.107	.094	.072	.041	0.000
0.000	.040	.070	.092	.104	.108	.104	.092	.070	.040	0.000
0.000	.038	.067	.088	.100	.104	.100	.088	.067	.038	0.000
0.000	.036	.063	.082	.093	.097	.093	.082	.063	.036	0.000
0.000	.033	.058	.075	.084	.087	.084	.075	.058	.033	0.000
0.000	.029	.050	.064	.072	.074	.072	.064	.050	.029	0.000
0.000	.023	.039	.049	.055	.056	.055	.049	.039	.023	0.000
0.000	.014	.023	.028	.031	.032	.031	.028	.023	.014	0.000
0.000	0.000	0.000	0.000	0.000	0.000	0.000	0.000	0.000	0.000	0.000

volume flow rate. In the last part of the program for contour plotting, the variable names are exactly the same as those used in Program 2.8, and their definitions can be found there.

The result shows that the maximum velocity occurs at the middle of the free surface having the magnitude $0.114L^2\rho g \sin \theta/\mu$, and the volume flow rate is $0.0569L^4\rho g \sin \theta/\mu$. It is interesting to compare this result with the analytical result obtained for the same free-surface flow with the two side walls removed. The velocity profile and volume flow rate in such a channel of

Pipe and Open-Channel Flows

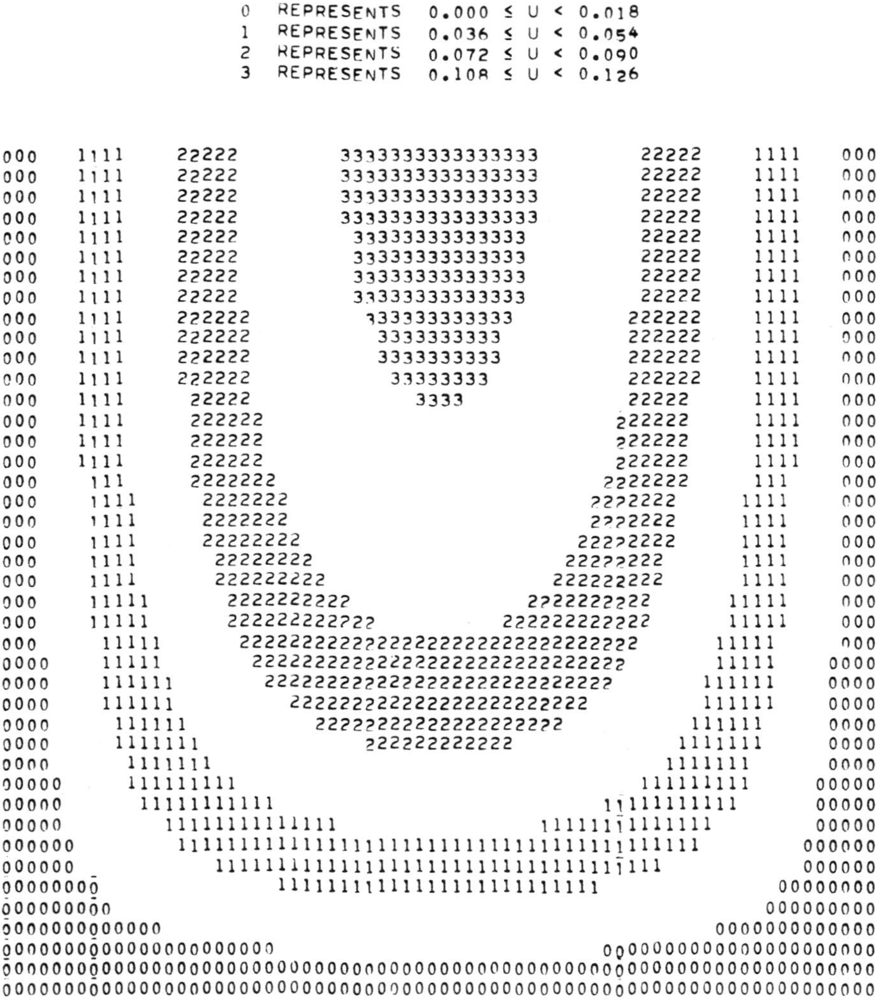

```
        0   REPRESENTS   0.000 ≤ U < 0.018
        1   REPRESENTS   0.036 ≤ U < 0.054
        2   REPRESENTS   0.072 ≤ U < 0.090
        3   REPRESENTS   0.108 ≤ U < 0.126


000   1111   22222    3333333333333333         22222   1111   000
000   1111   22222    3333333333333333         22222   1111   000
000   1111   22222    3333333333333333         22222   1111   000
000   1111   22222    3333333333333333         22222   1111   000
000   1111   22222     33333333333333          22222   1111   000
000   1111   22222     33333333333333          22222   1111   000
000   1111   22222     3333333333333           22222   1111   000
000   1111   22222     33333333333333          22222   1111   000
000   1111   222222     333333333333           222222  1111   000
000   1111   222222      3333333333            222222  1111   000
000   1111   222222      3333333333            222222  1111   000
000   1111   222222       33333333             222222  1111   000
000   1111   22222          3333               22222   1111   000
000   1111   222222                            222222  1111   000
000   1111   222222                            222222  1111   000
000   1111   222222                            222222  111    000
000    111   2222222                           2222222 1111   000
000   1111   2222222                           2222222 1111   000
000   1111   22222222                          2222222 1111   000
000   1111   22222222                          22222222 1111  000
000   1111   222222222                         22222222 1111  000
000   11111  2222222222       2222222222       11111   000
000   11111  222222222222     22222222222       11111   000
000   11111  22222222222222222222222222222222    11111   000
0000  11111  22222222222222222222222222222222    11111   0000
0000  111111 22222222222222222222222222222222    111111  0000
0000  111111 2222222222222222222222222          111111   0000
0000  111111 222222222222222222222             111111    0000
0000  1111111 222222222222                    1111111    0000
00000  1111111                               1111111    0000
00000  111111111                            111111111   00000
00000  1111111111                          111111111    00000
00000  11111111111111        11111111111111111111       00000
000000 11111111111111111111111111111111111111111        000000
000000 11111111111111111111111111111111111111111111     000000
00000000 1111111111111111111111111111111111111           00000000
000000000                                                 000000000
000000000                                                 000000000
000000000000000                                    00000000000000000
00000000000000000000000                    0000000000000000000000000
000000000000000000000000000000000000000000000000000000000000000000000
000000000000000000000000000000000000000000000000000000000000000000000
```

FIGURE 3.5.3 Velocity distribution across an open channel of square cross section.

width L are, from p. 183, Batchelor (1967), expressed in our notation,

$$u = \frac{\rho g \sin \theta}{2\mu} \, y(2L - y) \tag{3.5.10}$$

$$Q = \frac{\rho g \sin \theta}{3\mu} \, L^4 \tag{3.5.11}$$

The maximum velocity at the free surface is then $0.5L^2\rho g \sin\theta/\mu$, which is 4.4 times the value in the presence of two side walls separated by a distance equal to the fluid depth, and the flow rate becomes 5 times bigger. The comparison reveals that the side walls have a tremendous retardative effect on channel flows.

Problem 3.7 Consider a channel with a square cross section of area L^2 having a movable upper wall. Find the velocity distribution and the volume flow rate of the steady flow caused by moving the upper wall at a constant velocity u_0 in the direction parallel to the channel length.

Assume that the gaps between the moving and the stationary side walls are smaller than the grid width h. The boundary condition on the uppermost grid line is that the velocity is u_0 at all grid points except those at the corners, where the velocity is zero.

Problem 3.8 Solve the problem in which the upper movable wall described in Problem 3.7 is replaced by one of width $L/2$, leaving the uncovered portion of the leveled upper fluid surface free. Study the variation of flow rate with the horizontal position of the upper plate whose axial velocity is kept at the same value u_0.

Problem 3.9 A steady flow is established in a long pipe after a constant pressure gradient dp/dx has been applied along the axis for a long time. Compute the velocity distribution and the volume flow rate of an incompressible fluid in pipes of the cross-sectional shapes described in Fig. 3.5.4.

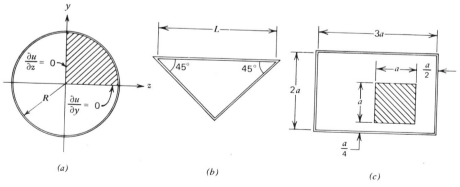

FIGURE 3.5.4 Various tube cross sections.

1. A circular tube of radius R. From symmetry only the first quadrant is needed for the numerical computation. The boundary conditions at the two straight edges of the fan-shaped domain are that the variations of velocity normal to the edges are zero. To handle the curved boundary, the method of Program 2.9 may be used.

Compare the numerical result with the analytical solution for Poiseuille flow (Batchelor, 1967, p. 180) that

$$u = -\frac{1}{4\mu}\frac{dp}{dx}(R^2 - r^2)$$

$$Q = -\frac{\pi R^4}{8\mu}\frac{dp}{dx}$$

2. A triangular tube whose two slant walls make 45° angles with the third.

3. A rectangular tube containing a square inner tube.

3.6 Explicit Methods for Solving Parabolic Partial Differential Equations— Generalized Rayleigh Problem

In studying the development of a boundary layer on a body moving through an incompressible fluid, Rayleigh (1911) considered the unsteady motion of an infinitely extended fluid in response to an infinite flat plate suddenly set in motion along its own plane. If the plate is normal to the y-axis and the motion is in the x direction, the continuity equation (3.1.6) is satisfied automatically and the incompressible Navier-Stokes equation (3.1.7) is simplified to

$$\frac{\partial u}{\partial t} = \nu \frac{\partial^2 u}{\partial y^2} \tag{3.6.1}$$

Sometimes this equation, governing arbitary unsteady planar fluid motions, is expressed in terms of vorticity $\zeta (= -\partial u/\partial y)$ in the form

$$\frac{\partial \zeta}{\partial t} = \nu \frac{\partial^2 \zeta}{\partial y^2} \tag{3.6.2}$$

which describes the diffusion of vorticity through a one-dimensional space. According to the discussions of Section 2.9, both (3.6.1) and (3.6.2) are classified as parabolic partial differential equations. Here we will construct a numerical scheme for solving (3.6.1), examine its computational stability, and then apply it to a particular physical problem.

In numerical computations the space coordinates must be finite. Let us assume that the fluid above the plate at $y = 0$ is bounded below a finite depth that is divided into $m - 1$ equally spaced intervals of size h. If the time axis is divided into steps of size τ, a grid system is formed, as shown in Fig. 3.6.1. To approximate (3.6.1) by a finite difference equation at the grid point (i, j), the second-order spatial derivative is replaced by the central-difference formula (2.2.9) and the time derivative is replaced by the forward-difference formula (2.2.6). After rearrangement the equation has the final form

$$u_{i,j+1} = u_{i,j} + R(u_{i-1,j} - 2u_{i,j} + u_{i+1,j}) \tag{3.6.3}$$

in which $R = \nu\tau/h^2$ is a dimensionless parameter. The equation states that the solution at a certain height at time interval τ later can be computed based on the present informations at the local and two neighboring stations. For given boundary conditions expressed as known functions of time, the solution at time level t_2 is computed explicitly from the initial condition at t_1 by using (3.6.3). Repeating the procedure for the successive time steps, the solution at any desired time level can be obtained. For this property the method in which (3.6.3) is applied is called an *explicit method* for solving the parabolic equation (3.6.1).

Playing a similar role as the Courant number C in (2.12.4), the parameter R in (3.6.3) cannot be arbitrarily chosen, and the limitation imposed on its magnitude is to be determined from a stability analysis of the numerical

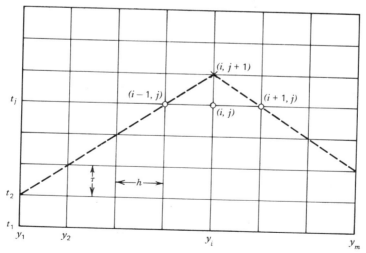

FIGURE 3.6.1 An explicit method for solving parabolic equations.

Explicit Methods for Solving Parabolic Partial Differential Equations

scheme. Following the technique illustrated in Section 2.12, we assume

$$u_{i,j} = U_j e^{Iikh} \tag{3.6.4}$$

and obtain, after substituting into (3.6.3),

$$U_{j+1} = [1 - 2R(1 - \cos kh)]U_j \tag{3.6.5}$$

The quantity contained within the brackets is the amplification factor λ. If $|\lambda| > 1$, $|U_{j+1}| > |U_j|$ and the amplitude of the solution becomes unbounded as $j \to \infty$. This is called an unstable situation. Thus, for stability we require $\lambda^2 \leq 1$ or, consequently, after expanding the left-hand side,

$$R \leq \frac{1}{1 - \cos kh}$$

Since the lowest value of the expression on the right-hand side is 1/2 when $\cos kh = -1$, the stability criterion derived for (3.6.3) is

$$\frac{\nu\tau}{h^2} \leq \frac{1}{2} \tag{3.6.6}$$

When the upper limiting value is used for this parameter, (3.6.3) has a particularly simple form.

$$u_{i,j+1} = \tfrac{1}{2}(u_{i-1,j} + u_{i+1,j}) \tag{3.6.7}$$

This is called the *Bender-Schmidt recurrence equation*, which determines the solution at (y_i, t_{j+1}) as the average of the values right and left of y_i at a time t_j. However, more accurate results are obtained by using (3.6.3) for $R < 1/2$.

The differential equation (3.6.1) and its finite-difference approximation (3.6.3) apply to any unsteady planar flows bounded by two parallel infinite plates performing arbitrary parallel motions along their own planes. One of the plates may be replaced by a free surface. Furthermore, with modifications to suit cylindrical coordinates, the resulting equations apply to flows between concentric cylinders. Solving for the velocity and the related fields of these flows may be classified as the generalized Rayleigh problem.

For illustrative purposes we consider water contained between two originally stationary flat plates separated by a distance of 1 m. At an initial instant $t = 0$, the upper plate has suddenly acquired a constant speed $u_0 (= 1 \text{ m/s})$ while the lower plate is kept stationary all the time. The sudden motion of the upper plate creates a sharp velocity change there, forming a concentrated vortex sheet right below the plate. The vorticity is diffused downward, according to (3.6.2), into a region practically free of vorticity, and the velocity is redistributed accordingly. We like to find numerically the velocity distribution across the channel at different times.

Viscous Fluid Flows

In terms of the notations of Fig. 3.6.1, the initial velocity distribution is

$$u_{i,1} = 0 \quad \text{for} \quad i = 1, 2, \ldots, m-1$$

$$u_{m,1} = u_0$$

and the boundary conditions are

$$u_{i,j} = 0 \quad \text{for} \quad j > 1$$

$$u_{m,j} = u_0 \quad \text{for} \quad j > 1$$

For water $\nu = 1 \times 10^{-6} \, \text{m}^2/\text{s}$, approximately. If the space between plates is divided into 20 equal intervals, $m = 21$ and $h = 0.05$ m. Let us choose $R = 1/4$, which determines the time interval $\tau = Rh^2/\nu$ or 625 s. This time step size seems to be rather large. But it is a reasonable size for a laminar shear flow in which vorticity or velocity gradient is diffused purely by intermolecular activities characterized by a small kinematic viscosity.

In the form shown in (3.6.3), the velocity field is a two-dimensional array. This is not necessary, however, in programming for the computation. In Program 3.5 we use one-dimensional arrays UOLD(I) and UNEW(I) to denote $u_{i,j}$ and $u_{i,j+1}$, respectively. Immediately after completing the computation for one time level, we store the values of UNEW under the name UOLD, print them only at certain selected time levels, and then repeat the same computation to find UNEW at the next time level. This is an often used technique when the size of the field length of a program becomes too large to be handled by the register of a computer.

Because of the large amount of output data, the solution for u is printed every 20 time steps and plotted every 40 time steps, with only the first five curves shown in Fig. 3.6.2. The selective printing and plotting are achieved by assigning these two time-step values to variable names NPRINT and NPLOT, respectively, and doing what we did in Program 2.10—using the special property of integer division.

The output shows that a velocity discontinuity cannot exist in a viscous fluid and is smoothed out immediately by viscous diffusion. As time progresses the velocity profile approaches a linear distribution that varies from 0 at the lower plate to 1 m/s at the upper and corresponds to the solution for the Couette flow between two parallel plates in a steady shear motion.

```
C                      ***** PROGRAM 3.5 *****

C          EXPLICIT METHOD IS USED TO FIND THE UNSTEADY VELOCITY
C          DISTRIBUTION IN WATER CONTAINED BETWEEN A STATIONARY
C          LOWER PLATE AND AN UPPER PLATE IN AN IMPULSIVE MOTION

       DIMENSION UOLD(21),UNEW(21),Y(21),NSCALE(4),
      A           LABEL(3),PCHAR(5),TPLOT(5)
       REAL NU
```

Explicit Methods for Solving Parabolic Partial Differential Equations

```
C       ..... SPECIFY INPUT DATA .....
        DATA H,M,UO,NU,R/ 0.05, 21, 1.0, 1.0E-6, 0.25 /
        DATA TMAX,NPRINT,NPLOT/ 2.5E+5, 20, 40 /
        TAU = R * H**2 / NU
        MM1 = M - 1

C       ..... DEFINE Y(I) AND SPECIFY INITIAL VELOCITY DISTRIBUTION .....
        J = 1
        T = 0.0
        Y(1) = 0.0
        DO 1  I = 2, M
      1 Y(I) = Y(I-1) + H
        DO 2  I = 1, MM1
      2 UOLD(I) = 0.0
        UOLD(M) = 1.0

C       ..... PRINT HEADING AND THE INITIAL CONDITION .....
        WRITE(6,3)
      3 FORMAT( 1H1 /// 5X4HTIME, 41X31HVELOCITIES AT DIFFERENT HEIGHTS,
       A          / 3X, 9H(SECONDS), 46X, 15H(METERS/SECOND) // )
        WRITE(6,4)  T, (UOLD(I), I=1,M,2)
      4 FORMAT( F10.0, 10X, 11F8.4 / )

C       ..... SET UP FRAME FOR PLOTTING .....
        DATA NSCALE/ 0, 1, 0, 1 /
        DATA PCHAR/ 1H1, 1H2, 1H3, 1H4, 1H5 /
        LABEL(1) = 10HHEIGHT IN
        LABEL(2) = 6HMETERS
        CALL PLOT1( NSCALE, 6, 9, 6, 12, 1H., 1H. )
        CALL PLOT2( 1.0, 0.0, 1.0, 0.0, .TRUE. )

C       ..... COMPUTE SOLUTION AT THE NEXT TIME LEVEL.
C             BOUNDARY CONDITIONS REMAIN UNCHANGED .....
      5 J = J + 1
        T = T + TAU
        DO 6  I = 2, MM1
      6 UNEW(I) = UOLD(I) + R*( UOLD(I-1) - 2.*UOLD(I) + UOLD(I+1))

C       ..... STORE THE VALUES OF UNEW IN UOLD.  FOR T≤TMAX,
C             PRINT THE SOLUTION EVERY NPRINT TIME STEPS AND
C             PLOT EVERY NPLOT TIME STEPS .....
        DO 7  I = 2, MM1
      7 UOLD(I) = UNEW(I)
        IF( (J-1)/NPRINT*NPRINT .NE. J-1 )  GO TO 8
        WRITE(6,4)  T, (UOLD(I), I=1,M,2)
      8 K = (J-1) / NPLOT
        IF( K*NPLOT.NE.J-1) .OR. (K.GT.5) )  GO TO 9
        TPLOT(K) = T
        CALL PLOT3( PCHAR(K), UOLD, Y, 1, M, 1 )
      9 IF( T.LT.TMAX )  GO TO 5

C       ..... PRINT THE FIGURE AND THE CAPTION .....
        WRITE(6,10)   ((PCHAR(K),TPLOT(K)), K=1,5)
     10 FORMAT( 1H1 / (37X, A1, 8H FOR T =, F7.0, 8H SECONDS) )
        CALL PLOT4( LABEL, 16 )
        WRITE(6,11)
     11 FORMAT( / 42X, 14HVELOCITY (M/S) /// 21X, 15HFIGURE 3.6.2    ,
       A        41HVELOCITY DISTRIBUTION AT DIFFERENT TIMES. )
        END
```

TIME (SECONDS)											
0	0.0000	0.0000	0.0000	0.0000	0.0000	0.0000	0.0000	0.0000	0.0000	0.0000	1.0000
12500	0.0000	.0000	.0000	.0001	.0015	.0115	.0596	.2110	.5327		1.0000
25000	0.0000	.0000	.0003	.0017	.0073	.0257	.0748	.1821	.3742	.6570	1.0000
37500	0.0000	.0009	.0034	.0106	.0287	.0686	.1455	.2753	.4672	.7163	1.0000
50000	0.0000	.0039	.0113	.0270	.0582	.1147	.2072	.3443	.5285	.7527	1.0000
62500	0.0000	.0091	.0231	.0478	.0902	.1581	.2590	.3974	.5727	.7779	1.0000
75000	0.0000	.0157	.0371	.0703	.1216	.1974	.3027	.4396	.6064	.7967	1.0000
87500	0.0000	.0230	.0520	.0928	.1513	.2324	.3397	.4741	.6333	.8114	1.0000
100000	0.0000	.0304	.0666	.1143	.1786	.2635	.3715	.5029	.6552	.8234	1.0000
112500	0.0000	.0375	.0806	.1343	.2033	.2909	.3990	.5274	.6736	.8332	1.0000
125000	0.0000	.0442	.0935	.1526	.2256	.3152	.4230	.5484	.6893	.8416	1.0000
137500	0.0000	.0503	.1053	.1692	.2455	.3367	.4439	.5666	.7027	.8487	1.0000
150000	0.0000	.0559	.1159	.1840	.2632	.3557	.4622	.5824	.7144	.8549	1.0000
162500	0.0000	.0609	.1255	.1973	.2790	.3724	.4784	.5963	.7245	.8603	1.0000
175000	0.0000	.0653	.1340	.2091	.2930	.3873	.4926	.6085	.7334	.8650	1.0000
187500	0.0000	.0693	.1416	.2196	.3053	.4004	.5051	.6192	.7413	.8691	1.0000
200000	0.0000	.0729	.1483	.2289	.3163	.4119	.5162	.6286	.7481	.8727	1.0000
212500	0.0000	.0760	.1543	.2371	.3260	.4222	.5259	.6370	.7542	.8759	1.0000
225000	0.0000	.0788	.1596	.2444	.3346	.4312	.5345	.6443	.7595	.8787	1.0000
237500	0.0000	.0812	.1643	.2508	.3422	.4392	.5422	.6508	.7642	.8812	1.0000
250000	0.0000	.0834	.1684	.2565	.3489	.4463	.5499	.6565	.7684	.8834	1.0000

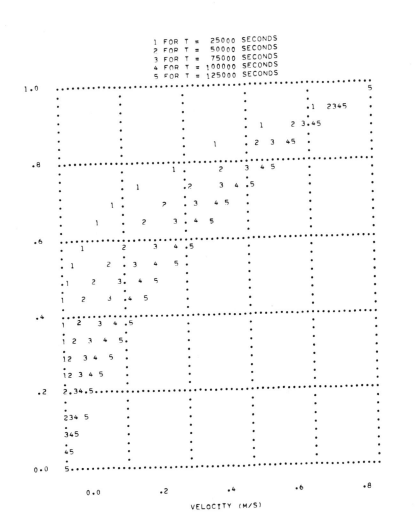

```
            1  FOR  T  =    25000  SECONDS
            2  FOR  T  =    50000  SECONDS
            3  FOR  T  =    75000  SECONDS
            4  FOR  T  =   100000  SECONDS
            5  FOR  T  =   125000  SECONDS
```

FIGURE 3.6.2 Velocity distribution at different times.

Problem 3.10 Assign a value to R that is greater that 0.5, and then run Program 3.5 to watch the growth of the solution to some unrealistic magnitudes. The result proves the validity of the stability criterion (3.6.6).

Problem 3.11 Find the velocity distribution at increasing times in the originally stationary fluid around a circular cylinder with a free surface (Fig. 3.4.1) after the cylinder is suddenly given a

Viscous Fluid Flows

rotation of tangential speed v_b at the surface. As time approaches a very large value, the solution should approach that for Problem 3.6.

Problem 3.12 Find the velocity distribution in the channel described in Program 3.5 with the upper plate replaced by one oscillating at the speed $u_0 \sin \omega t$, where $u_0 = 1$ m/s and $\omega = 1/1000$ s^{-1}.

Problem 3.13 In approximating the differential equation (3.6.1) by the finite-difference equation (3.6.3), we used forward difference in time and central difference in space, so that the truncation error of the approximation is $O(\tau, h^2)$. An improved explicit scheme with truncation error of $O(\tau^2, h^2)$ can be constructed by also using central difference in time and by replacing the second $u_{i,j}$ on the right side of (3.6.3) by the time-average $(u_{i,j-1} + u_{i,j+1})/2$. Thus

$$\frac{u_{i,j+1} - u_{i,j-1}}{2\tau} = \nu \frac{u_{i-1,j} - u_{i,j-1} - u_{i,j+1} + u_{i+1,j}}{h^2}$$

$$(1 + 2R)u_{i,j+1} = 2R(u_{i-1,j} + u_{i+1,j}) + (1 - 2R)u_{i,j-1} \quad (3.6.8)$$

This is called the *DuFort–Frankel formula*, which involves three time levels, as does the formula (2.12.13) derived for hyperbolic differential equations.

Show that (3.6.8) is an unconditionally stable numerical scheme.

Let us take a close look at the explicit formula (3.6.3). The solution at the grid pont $(i, j + 1)$ is computed, using this formula, from the solutions evaluated at three grid points $(i - 1, j)$, (i, j), and $(i + 1, j)$. These values, in turn, are computed from solutions in their neighborhood at the previous time step. In this way we can trace out the region of dependence of the point $(i, j + 1)$, which is confined between the two dashed lines shown in Fig. 3.6.1. It means that the disturbance created at any other height in the fluid reaches the height y_i with a finite speed h/τ. This contradicts the real situation in an incompressible fluid, in which a disturbance at any point is felt immediately by all parts throughout the fluid. Thus, to improve the accuracy of (3.6.3), we may reduce the size of τ or the value of R. In so doing the dashed lines will approach the horizontal grid line passing through $(i, j + 1)$ and, in the meantime, more time steps will be needed in the computation to reach the same time level. The improved accuracy in the explicit method is therefore obtained at the expense of an increased amount of computer time.

Explicit Methods for Solving Parabolic Partial Differential Equations

3.7 Implicit Methods for Solving Parabolic Partial Differential Equations—Starting Flow in a Channel

The deficiency associated with the explicit methods that the solution computed at one point is not affected immediately by the conditions at all other points in the fluid can be avoided by devising an alternative numerical scheme for solving the same diffusion equation (3.6.1). If we still use centered difference in space but use backward, instead of forward, difference in time, a finite-difference equation is obtained at (i, j) of the form

$$\frac{1}{\tau}(u_{i,j} - u_{i,j-1}) = \frac{\nu}{h^2}(u_{i-1,j} - 2u_{i,j} + u_{i+1,j}) \tag{3.7.1}$$

It becomes, after regrouping,

$$Ru_{i-1,j} - (1 + 2R)u_{i,j} + Ru_{i+1,j} = -u_{i,j-1} \tag{3.7.2}$$

where $R = \nu\tau/h^2$ is the same dimensionless parameter as that defined in Section 3.6.

The slight modification in approximating the time derivative causes a radical change in the procedure for obtaining a solution. Suppose the solution at $t = t_j$ is to be computed based on the solution known at the previous time step $t = t_{j-1}$; (3.7.2) shows that every three neighboring unknown values are interrelated through this linear algebraic equation. Applying (3.7.2) at all grid points interior to the boundaries at one time level gives a system of simultaneous equations that can be solved for all the unknowns at that time instant. In this way the velocities at different heights are not independent of one another; a change at one point will be felt immediately by all other points. Thus this numerical scheme is more sound than the explicit scheme on physical grounds. By using the numerical scheme the solution can no longer be computed explicitly as before, so (3.7.2) is called a formula for the *implicit method*.

The computational stability of (3.7.2) can be examined again with von Neumann's stability analysis by assuming the form already shown in (3.6.4) for the numerical solution. It can easily be verified that the resulting relationship from that analysis is

$$U_j = \frac{1}{1 + 2R(1 - \cos kh)} U_{j-1} = \lambda U_{j-1} \tag{3.7.3}$$

As $\cos kh$ varies from -1 to $+1$, the value of the amplification factor λ changes from $1/(1 + 4R)$ to 1 and can never exceed 1. Therefore this numerical scheme is stable for all positive values of R.

Although for computational stability there is no restriction on the magnitude of R as long as it is positive, a smaller value of R results in a more accurate numerical solution. The reason for this is that after multiplying (3.7.1) through by τ, the truncated higher-order terms on the right-hand side are all multiplied by R.

We now apply the implicit method to solve a problem concerning the development of a channel flow caused by the application of a constant pressure gradient. The initially stationary incompressible fluid contained between two parallel infinite plates is set in motion by a suddenly imposed pressure gradient dp/dx along the channel. Simplified for the present geometry, the equation of motion (3.1.7) becomes

$$\frac{\partial u}{\partial t} = -\frac{1}{\rho}\frac{dp}{dx} + \nu \frac{\partial^2 u}{\partial y^2} \tag{3.7.4}$$

If the distance between plates is $2L$ and the origin of the coordinate system is placed at the middle of the channel, the boundary and initial conditions are

$$u = 0 \quad\text{at}\quad y = \pm L \quad\text{for all } t \tag{3.7.5}$$

$$u = 0 \quad\text{at}\quad t = 0 \quad\text{for } -L \leqslant y \leqslant L \tag{3.7.6}$$

We know that as time increases, the velocity profile will approach its steady-state parabolic distribution

$$u_s = -\frac{1}{2\mu}\frac{dp}{dx}(L^2 - y^2) \tag{3.7.7}$$

which is a particular solution to (3.7.4) satisfying the boundary conditions (3.7.5). By introducing the dimensionless variables

$$T = \frac{t}{L^2/\nu}, \quad Y = \frac{y}{L}, \quad U = u \Big/ \left(-\frac{L^2}{2\mu}\frac{dp}{dx}\right) \tag{3.7.8}$$

and a dimensionless velocity difference

$$W = (u_s - u)\Big/\left(-\frac{L^2}{2\mu}\frac{dp}{dx}\right) = (1 - Y^2) - U \tag{3.7.9}$$

the governing equation (3.7.4) is simplified to

$$\frac{\partial W}{\partial T} = \frac{\partial^2 W}{\partial Y^2} \tag{3.7.10}$$

with boundary and initial conditions

$$W = 0 \quad\text{at}\quad Y = \pm 1 \quad\text{for all } T \tag{3.7.11}$$

$$W = 1 - Y^2 \quad\text{at}\quad T = 0 \quad\text{for } -1 \leqslant Y \leqslant 1 \tag{3.7.12}$$

The implicit numerical scheme for solving (3.7.10) is, according to (3.7.2),

$$R\, W_{i-1,j} - (1 + 2R)W_{i,j} + R\, W_{i+1,j} = -W_{i,j-1} \tag{3.7.13}$$

Implicit Methods for Solving Parabolic Partial Differential Equations **269** where $R = \tau/h^2$, τ and h being the interval sizes in the nondimensionalized time and space coordinates, respectively. (3.7.13) is, in fact, a special case of the finite-difference equation (2.2.10) with constant coefficients C_{ij} on the left-hand side, so that, after it is applied to all interior grid points, the resulting tridiagonal system of simultaneous equations can be solved by calling the subroutine TRID constructed in Program 2.1. Comparing (3.7.13) with the standard form (2.2.10) reveals that the coefficients are

$$C_{i1} = R, \qquad C_{i2} = -1 - 2R, \qquad C_{i3} = R, \qquad C_{i4} = -W_{i,j-1}$$

The subscripts j and $j - 1$ in (3.7.13) can be dropped in the computer program if we compute the coefficients C_{i4} using the values of W_i at hand, call TRID to get the solution at the next time step that will be stored under the same name W_i, and then advance in time by repeating the same process. Moreover, because the flow is symmetric about the x-axis, it suffices to find a solution only in the upper half of the channel. Accordingly, the conditions (3.7.11) and (3.7.12) are modified to

$$W = 0 \qquad \text{at} \quad Y = 1 \quad \text{for all } T \qquad\qquad (3.7.14)$$

$$\frac{\partial W}{\partial Y} = 0 \qquad \text{at} \quad Y = 0 \quad \text{for all } T \qquad\qquad (3.7.15)$$

$$W = 1 - Y^2 \qquad \text{at} \quad T = 0 \quad \text{for } 0 \leqslant Y \leqslant 1 \qquad\qquad (3.7.16)$$

The derivative boundary condition (3.7.15) indicates that a fictitious grid point is needed below $Y = 0$. Thus, in order to use subroutine TRID, which solves n simultaneous equations, we divide the range $0 \leqslant Y \leqslant 1$ into n segments of equal width h, denote the grid points by $Y_1, Y_2, \ldots, Y_{n+1}$, respectively, and finally add a fictitious point Y_0 at $Y = -h$. (3.7.15) then becomes $W_0 = W_2$, and the boundary conditions are exactly analogous to those for the temperature profile on the flat-plate thermometer in Section 3.4. Following the discussion in that section, before TRID is called, the coefficient matrix must be modified using the following assignment statements.

$$C_{13} \leftarrow (C_{11} + C_{13}) \qquad\qquad (3.7.17)$$

$$C_{11} \leftarrow 0 \qquad\qquad (3.7.18)$$

$$C_{n4} \leftarrow (C_{n4} - C_{n3} W_{n+1}) \qquad\qquad (3.7.19)$$

$$C_{n3} \leftarrow 0 \qquad\qquad (3.7.20)$$

The dimensionless velocity U at any time level is calculated from (3.7.9) once W becomes known for that time.

Velocity profiles are printed and plotted selectively in Program 3.6, computed for $n = 20$ and $R = 0.4$. In the plot the nondimensionalized steady-state velocity distribution U_s is added to compare with the profiles at other time instants.

One thousand time steps have been marched in the computation of Program 3.6 without encountering any instability, as expected of the implicit method. Fig. 3.7.1 displays the development of the flow from a stationary state to the steady-state parabolic velocity profile as time approaches infinity. Right after the pressure gradient is applied, the flow acceleration is large, but the acceleration diminishes with increasing time. At $T = 1$ the velocity at the center of the channel is 91% the steady-state value. For water ($\nu = 1 \times 10^{-6}$ m^2/s) in a channel 2 cm high, this dimensionless time corresponds to 100 s in real time. The slow readjustment of the vorticity distribution is a characteristic of laminar flows with small kinematic viscosity.

The governing equation (3.7.10) and the boundary and initial conditions (3.7.11) and (3.7.12) for W show that $W(Y, T)$ describes the deceleration of a channel flow with a parabolic velocity profile when the pressure gradient used to sustain the flow is suddenly removed.

For the problem considered here, an analytical solution to (3.7.4) is possible by assuming

$$u = -\frac{1}{2\mu} \frac{dp}{dx} (L^2 - y^2) + \sum_{m=1}^{\infty} A_m \exp\left(-\lambda_m^2 \frac{\nu t}{L}\right) \cos\left(\lambda_m \frac{y}{L}\right)$$

```
C                    ***** PROGRAM 3.6 *****
C
C          IMPLICIT METHOD IS USED TO FIND THE SOLUTION FOR A CHANNEL
C          FLOW CAUSED BY A SUDDENLY IMPOSED CONSTANT PRESSURE GRADIENT

      DIMENSION W(21),U(21),US(21),Y(21),C(20,4),
     A          NSCALE(4),LABEL(3),PCHAR(5),TPLOT(5)

C      ..... SPECIFY INPUT DATA .....
      DATA N,R,TMAX,NPRINT,NPLOT/ 20, 0.4, 1.0, 50, 200 /
      H = 1.0 / FLOAT(N)
      TAU = R * H**2
      NP1 = N + 1

C      ..... DEFINE Y(I) AND SPECIFY INITIAL VELOCITY DISTRIBUTION .....
      J = 1
      T = 0.0
      Y(1) = 0.0
      DO 1  I = 2, NP1
    1 Y(I) = Y(I-1) + H
      DO 2  I = 1, NP1
      US(I) = 1.0 - Y(I)**2
      W(I) = US(I)
    2 U(I) = US(I) - W(I)

C      ..... PRINT HEADING AND THE INITIAL CONDITION .....
      WRITE(6,3)
    3 FORMAT( 1H1///6X4HTIME, 40X31HVELOCITIES AT DIFFERENT HEIGHTS // )
      WRITE(6,4)  T, (U(I), I=1,NP1,2)
    4 FORMAT( F10.2, 10X, 11F8.4 / )
```

Implicit Methods for Solving Parabolic Partial Differential Equations

```
C     ..... SET UP FRAME FOR PLOTTING
C           AND PLOT THE INITIAL PROFILE .....
      DATA NSCALE/ 0, 1, 0, 1 /
      DATA PCHAR/ 1H0, 1H1, 1H2, 1H3, 1H4 /
      LABEL(1) = 10HDIMENSIONL
      LABEL(2) = 10HESS HEIGHT
      CALL PLOT1( NSCALE, 6, 9, 6, 12, 1H., 1H. )
      CALL PLOT2( 1.0, 0.0, 1.0, 0.0, .TRUE. )
      CALL PLOT3( PCHAR(1), U, Y, 1, NP1, 1 )
      TPLOT(1) = T

C     ..... EVALUATE COEFFICIENTS C(I,J) FOR J = 1, 2, 3, AND 4 .....
    5 DO 6  I = 1, N
      C(I,1) =  R
      C(I,2) = -1.0 - 2.0*R
      C(I,3) =  R
    6 C(I,4) = -W(I)

C     ..... MODIFY COEFFICIENTS ACCORDING TO (3.7.17 - 20)
C           TO INCORPORATE BOUNDARY CONDITIONS .....
      C(1,3) = C(1,1) + C(1,3)
      C(1,1) = 0.0
      C(N,4) = C(N,4) - C(N,3)*W(NP1)
      C(N,3) = 0.0

C     ..... SOLVE SIMULTANEOUS EQUATIONS FOR W AT NEXT TIME LEVEL .....
      J = J + 1
      T = T + TAU
      CALL TRID( C, W, N )

C     ..... FOR T≤TMAX PRINT SOLUTION EVERY NPRINT TIME STEPS
C           AND PLOT EVERY NPLOT TIME STEPS .....
      DO 7  I = 1, N
    7 U(I) = US(I) - W(I)
      JM1 = J - 1
      IF( JM1/NPRINT*NPRINT .NE. JM1 )  GO TO 8
      WRITE(6,4)  T, (U(I), I=1,NP1,2)
    8 K = JM1 / NPLOT
      IF( (K*NPLOT.NE.JM1) .OR. (K.GT.4) )  GO TO 9
      TPLOT(K+1) = T
      CALL PLOT3( PCHAR(K+1), U, Y, 1, NP1, 1 )
    9 IF( T.LT.TMAX )  GO TO 5

C     ..... ADD A CURVE FOR STEADY-STATE PROFILE, PRINT CURVE
C           DESIGNATIONS, SHOW THE FIGURE, AND THEN PRINT CAPTION .....
      CALL PLOT3( 1H*, US, Y, 1, NP1, 1 )
      WRITE(6,10)  ((PCHAR(K),TPLOT(K)), K=1,5)
   10 FORMAT( 1H1 / (41X, A1, 8H FOR T =, F5.2) )
      WRITE(6,11)
   11 FORMAT( 39X, 18H* FOR STEADY STATE )
      CALL PLOT4( LABEL, 20 )
      WRITE(6,12)
   12 FORMAT( / 38X, 22HDIMENSIONLESS VELOCITY // 33X,
     A          33HFIGURE 3.7.1  VELOCITY PROFILES. )
      END




      SUBROUTINE TRID( C, F, N )
C     ..... TAKEN FROM PROGRAM 2.1 .....
```

VELOCITIES AT DIFFERENT HEIGHTS

TIME											
0.00	0.0000	0.0000	0.0000	0.0000	0.0000	0.0000	0.0000	0.0000	0.0000	0.0000	0.0000
.05	.0999	.0999	.0997	.0993	.0983	.0960	.0915	.0828	.0672	.0411	0.0000
.10	.1975	.1970	.1953	.1919	.1860	.1765	.1617	.1395	.1072	.0618	0.0000
.15	.2882	.2868	.2825	.2747	.2625	.2446	.2194	.1848	.1384	.0778	0.0000
.20	.3699	.3676	.3605	.3482	.3297	.3038	.2689	.2232	.1647	.0911	0.0000
.25	.4426	.4395	.4298	.4133	.3889	.3556	.3121	.2567	.1876	.1027	0.0000
.30	.5071	.5032	.4912	.4708	.4412	.4014	.3502	.2862	.2076	.1129	0.0000
.35	.5642	.5596	.5455	.5217	.4874	.4418	.3838	.3121	.2253	.1218	0.0000
.40	.6147	.6094	.5935	.5667	.5283	.4775	.4135	.3351	.2409	.1297	0.0000
.45	.6593	.6535	.6360	.6065	.5644	.5091	.4398	.3553	.2547	.1367	0.0000
.50	.6988	.6925	.6735	.6416	.5963	.5370	.4630	.3733	.2669	.1429	0.0000
.55	.7337	.7270	.7067	.6727	.6246	.5617	.4835	.3891	.2777	.1483	0.0000
.60	.7646	.7575	.7361	.7002	.6495	.5835	.5016	.4031	.2872	.1532	0.0000
.65	.7918	.7844	.7620	.7245	.6716	.6028	.5176	.4155	.2957	.1574	0.0000
.70	.8160	.8082	.7850	.7460	.6911	.6199	.5318	.4264	.3031	.1612	0.0000
.75	.8373	.8293	.8053	.7650	.7084	.6349	.5444	.4361	.3097	.1645	0.0000
.80	.8561	.8479	.8232	.7818	.7236	.6483	.5554	.4447	.3155	.1675	0.0000
.85	.8728	.8644	.8390	.7967	.7371	.6601	.5652	.4523	.3207	.1701	0.0000
.90	.8875	.8789	.8531	.8098	.7490	.6705	.5739	.4589	.3253	.1701	0.0000
.95	.9006	.8918	.8654	.8214	.7596	.6797	.5816	.4649	.3293	.1724	0.0000
1.00	.9121	.9032	.8764	.8317	.7689	.6878	.5883	.4701	.3328	.1762	0.0000

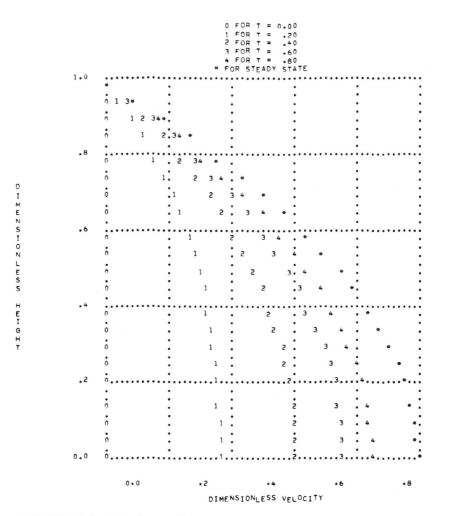

FIGURE 3.7.1 Velocity profiles.

The eigenvalues λ_m are to be determined from the boundary conditions (3.7.5), and the coefficients A_m are to be evaluated after substituting this expression into the initial condition (3.7.6) and using the orthogonal relationships among cosine functions.

The truncation error resulting from approximating the differential equation (3.6.1) by the implicit formula (3.7.2) is $O(\tau, h^2)$. Following a procedure very similar to that used to obtain the DuFort-Frankel formula (3.6.8) for improvement of the explicit method, we may derive an improved implicit scheme with a truncation error of $O(\tau^2, h^2)$. Let us replace the differential equation by a finite-difference equation at the midpoint $(i, j - \frac{1}{2})$ between grid points (i, j) and $(i, j - 1)$, as shown in Fig. 3.7.2. The time derivative in (3.6.1) is approximated by the central-difference formula (2.2.8) to give

$$\left(\frac{\partial u}{\partial t}\right)_{i,j-(1/2)} = \frac{u_{i,j} - u_{i,j-1}}{2(\tau/2)} + O\left[\left(\frac{\tau}{2}\right)^2\right]$$

The second-order spatial derivative at $(i, j - \frac{1}{2})$ is approximated by the average of those at (i, j) and $(i, j - 1)$, each of which is expressed in the central-difference form. Thus we have

$$\left(\frac{\partial^2 u}{\partial y^2}\right)_{i,j-(1/2)} = \frac{1}{2}\left[\frac{u_{i-1,j} - 2u_{i,j} + u_{i+1,j}}{h^2} + \frac{u_{i-1,j-1} - 2u_{i,j-1} + u_{i+1,j-1}}{h^2}\right] + O(h^2)$$

Substituting these two approximations into (3.6.1) yields, after rearranging and dropping terms $O(\tau^2)$ and $O(h^2)$,

$$Ru_{i-1,j} - 2(1 + R)u_{i,j} + Ru_{i+1,j} = -Ru_{i-1,j-1} - 2(1 - R)u_{i,j-1} - Ru_{i+1,j-1}$$

$$(3.7.21)$$

The form of this more accurate formula differs from the form of (3.7.2) in that it involves the evaluation of u at three grid points instead of at one on the right-hand side. The corresponding increase in computational work is merely in the evaluation of the coefficients C_{i4}, which is apparently an insignificant portion of the total computational effort. The numerical technique using (3.7.21) is called the *Crank-Nicolson method*. It can easily be shown that this is also an unconditionally stable scheme.

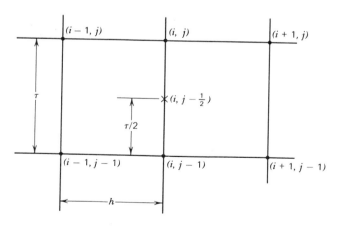

FIGURE 3.7.2 Crank-Nicolson method.

Implicit Methods for Solving Parabolic Partial Differential Equations

Problem 3.14 For the unsteady flow in an infinitely long tube of circular cross section along the x-axis, the governing equation is

$$\frac{\partial u}{\partial t} = -\frac{1}{\rho}\frac{\partial p}{\partial x} + \nu\left(\frac{\partial^2 u}{\partial r^2} + \frac{1}{r}\frac{\partial u}{\partial r}\right) \qquad (3.7.22)$$

Derive an implicit scheme for this differential equation in a form analogous to (3.7.2), examine its stability, and then use it to find the solution for the flow caused by the sudden application of a constant pressure gradient along a tube of radius A. The analytical solution of this problem can be found in Section 4.3 of Batchelor (1967), expressed in terms of a Fourier-Bessel series.

Problem 3.15 Find the solution for a flow in a tube of radius A caused by the application of an alternating pressure gradient

$$\frac{\partial p}{\partial x} = -K\cos\omega t$$

where K is the constant amplitude and ω is the circular frequency of the oscillation.

For this problem it is more convenient to introduce dimensionless variables

$$T = \omega t, \qquad R = \frac{r}{A}, \qquad U = \frac{u}{KA^2/4\mu}$$

so that the governing equation (3.7.22) becomes, with $\Omega = \omega\rho A^2/\mu$,

$$\Omega\frac{\partial U}{\partial T} = 4\cos T + \frac{\partial^2 U}{\partial R^2} + \frac{1}{R}\frac{\partial U}{\partial R}$$

Using the numerical scheme constructed in the previous problem, find $U(R, T)$ for $\Omega = 0.1, 7$, and 35 and compare them with the analytical solution plotted in Fig. 1 of Chow and Lai (1972).

3.8 Numerical Solution of Biharmonic Equations—Stokes' Flows

In the previous three sections, examples with special geometries were chosen so that the nonlinear terms contained in the substantial derivative on the left side of (3.1.7) disappeared. The same result can be achieved by requiring that the Reynolds number of the flow be small.

Assuming that the incompressible flow is characterized by a length L and a speed U, we may introduce the following dimensionless variables.

$$t' = \frac{t}{L/U}, \qquad (x', y', z') = \left(\frac{x}{L}, \frac{y}{L}, \frac{z}{L}\right), \qquad V' = \frac{V}{U}, \qquad p' = \frac{p}{\rho U^2}$$

In terms of the new variables, the Navier-Stokes equation (3.1.7) becomes

$$\frac{DV'}{Dt'} = -\nabla' p' + \frac{1}{Re} \nabla'^2 V \tag{3.8.1}$$

where the primed operators are those in dimensionless form and $Re = \rho UL/\mu$ is the Reynolds number.

As stated earlier in Section 2.1, the Reynolds number measures the relative importance of inertial force and viscous force in a flow. If Re is large, the viscous force terms, or the second group on the right side of (3.8.1), become small in comparison with the inertial force terms on the left. In this case the viscous force is important only in a thin region on the surface of a body in motion relative to the fluid. This thin region is the boundry layer considered in Sections 3.3 and 3.4. On the other hand, if Re is much smaller than unity, viscous force has a dominating influence on the fluid motion. By dropping the inertial force term and returning back to dimensional notation, we obtain for the steady state the linear equation

$$\mu \nabla^2 V = \nabla p \tag{3.8.2}$$

Flows for which $Re \ll 1$ are called the *Stokes' flows* or the *creeping flows*. The flow of molasses and the flow induced by the motion of a microorganism are examples. Taking the curl of (3.8.2) gives

$$\nabla^2 \zeta = 0 \tag{3.8.3}$$

in which $\zeta = \nabla \times V$ is the vorticity vector. Alternatively, divergence of (3.8.2) yields, remembering that $\nabla \cdot V = 0$ for incompressible fluids,

$$\nabla^2 p = 0 \tag{3.8.4}$$

It states that the pressure in an incompressible Stokes' flow satisfies the Laplace equation.

Here we consider the special case of two-dimensional Stokes' flows parallel to the x-y plane. If the stream function introduced in Section 2.1 is used so that

$$u = \frac{\partial \psi}{\partial y} \qquad \text{and} \qquad v = -\frac{\partial \psi}{\partial x} \tag{3.8.5}$$

the only nonvanishing vorticity component ζ in the z direction can be written, according to (2.10.16), as

$$\nabla^2 \psi = -\zeta \tag{3.8.6}$$

Numerical Solution of Bibarmonic Equations–Stokes' Flows

and the vector equation (3.8.3) reduces to a scalar equation

$$\nabla^2 \zeta = 0 \tag{3.8.7}$$

Combining (3.8.6) and (3.8.7) results in

$$\nabla^4 \psi = 0 \tag{3.8.8}$$

where the operator is defined as

$$\nabla^4 = \nabla^2 \nabla^2 = \frac{\partial^4}{\partial x^4} + 2\frac{\partial^4}{\partial x^2 \partial y^2} + \frac{\partial^4}{\partial y^4} \tag{3.8.9}$$

Because the Laplace equation is also called the harmonic equation, (3.8.8) is usually referred to as the *biharmonic equation*.

Numerical solution of the biharmonic equation (3.8.8) seems to be simple. It may be treated as two simultaneous second-order equations (3.8.6) and (3.8.7) to each of which the numerical methods developed in Section 2.10 for solving elliptic equations may be applied. However, difficulties arise in specifying the boundary conditions for the vorticity equation (3.8.7). On the boundaries of a flow region we can only prescribe the velocity components and/or some of their derivatives or, equivalently, the stream function and/or some of its derivatives. Vorticity is to be computed from the velocity field, but it cannot be specified at the boundaries before the problem is solved. Thus the numerical solution of the biharmonic equation is not as simple as it first appears to be, and special techniques are needed to utilize the numerical methods, devised originally for solving Poisson and Laplace equations, to solve the present boundary-value problem.

An example with an especially simple geometry is used here to demonstrate how such a problem is handled. A square cavity *DABC* is shown in Fig. 3.8.1 within which a steady fluid motion is generated by sliding an infinitely long plate lying on top of the cavity. Suppose that all variables are normalized so that the size of the cavity is 1×1 and the sliding velocity is 1 in the positive x direction, and that the Reynolds number is so low that the induced motion can be classified as a Stokes' flow.

If there is no fluid squeezed out of the cavity below the moving plate, the fluid motion forms closed paths within the cavity. The surfaces *DA*, *AB*, *BC*, and *CD* are then segments of the bounding streamline designated by $\psi = 0$. This is equivalent to specifying that the velocities normal to these four surfaces are all zero. To require that the tangential velocity be vanishing on all surfaces except on the top plate where it is 1, we obtain four additional boundary conditions in terms of derivatives of ψ, as indicated in Fig. 3.8.1, along the four sides. In summary, the boundary conditions are

$$\psi = 0 \quad \text{and} \quad \frac{\partial \psi}{\partial x} = 0 \quad \text{on } DA \tag{3.8.10}$$

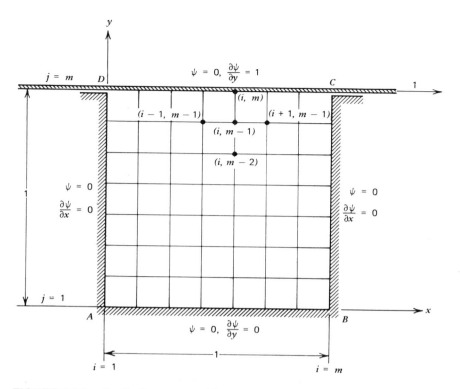

FIGURE 3.8.1 Cavity flow caused by a moving plate.

$$\psi = 0 \quad \text{and} \quad \frac{\partial \psi}{\partial y} = 0 \quad \text{on } AB \qquad (3.8.11)$$

$$\psi = 0 \quad \text{and} \quad \frac{\partial \psi}{\partial x} = 0 \quad \text{on } BC \qquad (3.8.12)$$

$$\psi = 0 \quad \text{and} \quad \frac{\partial \psi}{\partial y} = 1 \quad \text{on } CD \qquad (3.8.13)$$

This problem is similar to the one considered in Section 7.2 of Greenspan (1974) except in that analysis the inertial terms are retained. Furthermore, we will use a different procedure to compute the vorticity at the boundaries.

The fluid region is covered with a square mesh of size h having m horizontal grid lines and an equal number of vertical grid lines, as shown schematically in Fig. 3.8.1. Next, we derived expressions for boundary values of ζ in terms of ψ, which are needed for solving (3.8.7).

Let us first consider an arbitrary grid point (i, m) at the moving plate on top

Numerical Solution of Biharmonic Equations–Stokes' Flows

of the cavity. Our purpose is to calculate the vorticity at this point based on the local velocity and on the information of ψ at four neighboring grid points marked in Fig. 3.8.1. Similar to the expression (2.11.7), we now assume the following form for vorticity at (i, m).

$$\zeta_{i,m} = -\left(\frac{\partial^2 \psi}{\partial x^2} + \frac{\partial^2 \psi}{\partial y^2}\right)_{i,m}$$

$$= \alpha_1 \psi_{i-1,m-1} + \alpha_2 \psi_{i,m-1} + \alpha_3 \psi_{i+1,m-1} + \alpha_4 \psi_{i,m-2} + \alpha_5 \left(\frac{\partial \psi}{\partial y}\right)_{i,m} \qquad (3.8.14)$$

Substituting from the Taylor's series expansions,

$$\psi_{i\pm1,m-1} = \psi_{i,m} \pm h \left(\frac{\partial \psi}{\partial x}\right)_{i,m} + \frac{h^2}{2}\left(\frac{\partial^2 \psi}{\partial x^2}\right)_{i,m} - h \left(\frac{\partial \psi}{\partial y}\right)_{i,m} + \frac{h^2}{2}\left(\frac{\partial^2 \psi}{\partial y^2}\right)_{i,m} + O(h^3)$$

$$\psi_{i,m-n} = \psi_{i,m} - nh \left(\frac{\partial \psi}{\partial y}\right)_{i,m} + \frac{1}{2} n^2 h^2 \left(\frac{\partial^2 \psi}{\partial y^2}\right)_{i,m} + O(h^3)$$

and retaining only terms up to $O(h^2)$, (3.8.14) becomes

$$-\left(\frac{\partial^2 \psi}{\partial x^2} + \frac{\partial^2 \psi}{\partial y^2}\right)_{i,m} = (\alpha_1 + \alpha_2 + \alpha_3 + \alpha_4)\psi_{i,m} + (\alpha_3 - \alpha_1)h \left(\frac{\partial \psi}{\partial x}\right)_{i,m}$$

$$+ \left(\frac{\alpha_5}{h} - \alpha_1 - \alpha_2 - \alpha_3 - 2\alpha_4\right)h \left(\frac{\partial \psi}{\partial y}\right)_{i,m} + (\alpha_1 + \alpha_3)\frac{h^2}{2}\left(\frac{\partial^2 \psi}{\partial x^2}\right)_{i,m}$$

$$+ (\alpha_1 + \alpha_2 + \alpha_3 + 4\alpha_4)\frac{h^2}{2}\left(\frac{\partial^2 \psi}{\partial y^2}\right)_{i,m}$$

The constants α's are determined by equating the coefficients of like terms on the two sides of this equation. Substitution of these values into (3.8.14) gives

$$\zeta_{i,m} = \frac{1}{h^2}\left(-\psi_{i-1,m-1} + \frac{8}{3}\psi_{i,m-1} - \psi_{i+1,m-1} - \frac{2}{3}\psi_{i,m-2}\right) - \frac{2}{3h}\left(\frac{\partial \psi}{\partial y}\right)_{i,m} \qquad (3.8.15)$$

where $(\partial \psi/\partial y)_{i,m}$ has the constant value of 1 according to the second boundary condition in (3.8.13). By analogy we can write down the expressions for vorticity at the three stationary boundaries as

$$\zeta_{1,j} = \frac{1}{h^2}\left(-\psi_{2,j-1} + \frac{8}{3}\psi_{2,j} - \psi_{2,j+1} - \frac{2}{3}\psi_{3,j}\right) \qquad (3.8.16)$$

$$\zeta_{m,j} = \frac{1}{h^2}\left(-\psi_{m-1,j-1} + \frac{8}{3}\psi_{m-1,j} - \psi_{m-1,j+1} - \frac{2}{3}\psi_{m-2,j}\right) \qquad (3.8.17)$$

$$\zeta_{i,1} = \frac{1}{h^2}\left(-\psi_{i-1,2} + \frac{8}{3}\psi_{i,2} - \psi_{i+1,2} - \frac{2}{3}\psi_{i,3}\right) \qquad (3.8.18)$$

in which the derivative boundary conditions in (3.8.10) to (3.8.12) have been employed.

The vorticity boundary conditions just derived enable us to solve (3.8.7) provided that ψ is known at some interior points. However, the determination of ψ from (3.8.6) depends on the distribution of vorticity within the bounded domain. Thus ψ and ζ are coupled, and an iterative scheme will be constructed to find the solution.

A stationary state is first assumed so that $\psi = 0$ everywhere in the fluid region. Based on this initial assumption, the boundary values computed from (3.8.15) to (3.8.18) show that vorticity is initially generated at the moving plate. This concentrated vorticity starts to diffuse into the cavity, resulting in a temporary vorticity distribution that is the solution of (3.8.7) that satisfies the present boundary conditions. This computed vorticity distribution causes a modification to the assumed ψ after solving (3.8.6) subject to the restriction that $\psi = 0$ on the boundary. In this way we have completed the first iteration. To start the next iteration, the boundary values of vorticity are recomputed based on the modified stream function, and the same procedure is repeated to obtain a new solution for ψ and ζ. During each iteration the difference between the newly computed ζ and the previous value at every grid point is recorded as the local error, and the sum of absolute errors at all grid points is called ERZETA. The same is done for ψ, and the corresponding sum is called ERPSI. Iteration is terminated when both ERZETA and ERPSI are smaller than a specified small positive value EPSLON; the solution at this stage is then considered to be satisfactory, since it has the desired accuracy.

In Program 3.7 an iteration counter ITER is introduced whose value is printed only when the final result has been obtained. Solving the Poisson equation (3.8.6) is achieved by applying Liebmann's iterative formula (2.10.6) repeatedly at all interior points. The iteraction method is programmed in a subroutine named LIEBMN for a square domain. The same subroutine is called to solve the Laplace equation (3.8.7), which is, in fact, a Poisson equation with a vanishing right-hand side.

When Liebmann's method is used, the values of vorticity at the corners A, B, C, and D of the cavity (Fig. 3.8.1) are not involved in the numerical computation for solving equations (3.8.6) and (3.8.7). These values are computed from the final vorticity distribution by taking the average of the values at two neighboring grid points.

After printing the numerical values of ψ and ζ for the final solution, the flow pattern is plotted in Fig. 3.8.2 using the technique described in Section 2.11. It turns out that the stream function is either negative or zero in the present problem, so that the absolute value of ψ is used instead in determination of the plotting symbols for various ranges of the stream function.

The output of Program 3.7 shows that with EPSLON = 0.001, 109 iterations are needed to have the numerical result converge to the desired solution. During the last iteration, ERPSI, or the sum of absolute errors in the

Numerical Solution of Bibarmonic Equations–Stokes' Flows

```
C                   ***** PROGRAM 3.7 *****

C           SOLVING BIHARMONIC EQUATION FOR STOKES FLOW IN
C           A SQUARE CAVITY CAUSED BY A MOVING UPPER PLATE

      DIMENSION PSI(11,11),PSI1(11,11), ZETA(11,11),ZETA1(11,11),
     A          RHS(11,11),SYMBOL(7),GRAPH(70)
      COMMON H,MM1
      REAL LX,LY

C     ..... SPECIFY INPUT DATA .....
      M = 11
      MM1 = M - 1
      H = 1.0 / MM1
      HSQ = H**2

C     ..... ASSUME AN INITIAL STATIONARY STATE FOR THE FLOW .....
      ITER = 0
      DO 1 I = 1, M
      DO 1 J = 1, M
      PSI(I,J) = 0.0
    1 ZETA(I,J) = 0.0

C     ..... COMPUTE BOUNDARY VALUES FOR VORTICITY .....
    2 ITER = ITER + 1
      DO 3 J = 2, MM1
      ZETA(1,J) = ( -PSI(2,J+1) + 8./3.*PSI(2,J) - PSI(2,J-1)
     A              -2./3.*PSI(3,J) ) / HSQ
    3 ZETA(M,J) = ( -PSI(MM1,J+1) + 8./3.*PSI(MM1,J) - PSI(MM1,J-1)
     A              -2./3.*PSI(M-2,J) ) / HSQ
      DO 4 I = 2, MM1
      ZETA(I,1) = ( -PSI(I+1,2) + 8./3.*PSI(I,2) - PSI(I-1,2)
     A              -2./3.*PSI(I,3) ) / HSQ
    4 ZETA(I,M) = ( -PSI(I+1,MM1) + 8./3.*PSI(I,MM1) - PSI(I-1,MM1)
     A              -2./3.*PSI(I,M-2) ) / HSQ  - 2./(3.*H)

C     ..... FIND SOLUTION TO (3.8.7) FOR VORTICITY BY CALLING
C           SUBROUTINE LIEBMN FOR SOLVING POISSON EQUATION,
C           THEN COMPUTE SUM OF ABSOLUTE ERRORS FOR VORTICITY .....
      DO 5 I = 1, M
      DO 5 J = 1, M
      RHS(I,J) = 0.0
    5 ZETA1(I,J) = ZETA(I,J)
      CALL LIEBMN( ZETA, RHS, M, 0.001 )
      ERZETA = 0.0
      DO 6 I = 1, M
      DO 6 J = 1, M
    6 ERZETA = ERZETA + ABS( ZETA(I,J)-ZETA1(I,J) )

C     ..... FIND SOLUTION TO (3.8.6) FOR STREAM FUNCTION BY CALLING
C           LIEBMN, THEN COMPUTE THE SUM OF ERRORS FOR PSI .....
      DO 7 I = 2, MM1
      DO 7 J = 2, MM1
      RHS(I,J) = -ZETA(I,J)
    7 PSI1(I,J) = PSI(I,J)
      CALL LIEBMN( PSI, RHS, M, 0.001 )
```

```
      ERPSI = 0.0
      DO 8  I = 2, MM1
      DO 8  J = 2, MM1
    8 ERPSI = ERPSI + ABS( PSI(I,J)-PSI1(I,J) )

C     ..... RESULT IS CONSIDERED SATISFACTORY WHEN BOTH ERZETA
C           AND ERPSI ARE LESS THAN OR EQUAL TO EPSLON .....
      EPSLON = 0.001
      IF( ERZETA.GT.EPSLON .OR. ERPSI.GT.EPSLON )  GO TO 2

C     ..... VORTICITY AT EACH OF THE FOUR CORNERS IS RECOMPUTED
C           AS THE AVERAGE OF THE NEIGHBORING VALUES .....
      ZETA(1,1) = ( ZETA(1,2) + ZETA(2,1) ) / 2.0
      ZETA(M,1) = ( ZETA(M,2) + ZETA(MM1,1) ) / 2.0
      ZETA(1,M) = ( ZETA(2,M) + ZETA(1,MM1) ) / 2.0
      ZETA(M,M) = ( ZETA(M,MM1) + ZETA(MM1,M) ) / 2.0

C     ..... PRINT ITER, ERPSI, AND THE STREAM FUNCTION .....
      WRITE(6,9)  ITER, ERPSI
    9 FORMAT( 1H1 /// 20X, 25HSTREAM FUNCTION BASED ON , I3,
     A          11H ITERATIONS / 32X, 7HERPSI =, F7.5 )
      DO 10  K = 1, M
      J = M - K + 1
   10 WRITE(6,11)  ( PSI(I,J), I=1,M )
   11 FORMAT( /// 1X, 11F7.3 )

C     ..... PRINT ERZETA AND THE VORTICITY DISTRIBUTION .....
      WRITE(6,12)  ERZETA
   12 FORMAT( 1H1 /// 28X, 22HVORTICITY DISTRIBUTION /
     A          32X, 8HERZETA =, F7.5 )
      DO 13  K = 1, M
      J = M - K + 1
   13 WRITE(6,11)  ( ZETA(I,J), I=1,M )

C     ..... PLOT STREAM FUNCTION USING THE METHOD DESCRIBED IN
C           SECTION 2.11.  SEE PROGRAM 2.8 FOR DEFINITION OF NAMES.....
      DATA XRANGE,YRANGE,NX/ 1.0, 1.0, 70/
      DATA SYMBOL/ 1H0, 1H , 1H3, 1H , 1H6, 1H , 1H9 /
      WRITE(6,14)
   14 FORMAT( 1H1 /// 22X, 34H9  REPRESENTS   -.105 < PSI ≤ -.090 /
     A          22X, 34H6  REPRESENTS   -.075 < PSI ≤ -.060 /
     B          22X, 34H3  REPRESENTS   -.045 < PSI ≤ -.030 /
     C          22X, 34H0  REPRESENTS   -.015 < PSI ≤ 0.000 /// )
      NY = 0.6 * NX * YRANGE / XRANGE
      DELX = XRANGE / NX
      DELY = YRANGE / NY
      DELTA = 0.015
      DO 16  JSYMBL = 1, NY
      Y = YRANGE - (JSYMBL-0.5)*DELY
      J = 1 + Y/H
      LY = Y - (J-1)*H
      HMLY = H - LY
      DO 15  ISYMBL = 1, NX
      X = (ISYMBL-0.5) * DELX
      I = 1 + X/H
      LX = X - (I-1)*H
      HMLX = H - LX
```

```
            A1 = HMLX * HMLY
            A2 = HMLX * LY
            A3 = LX * LY
            A4 = LX * HMLY
      PSIC = ( A1*PSI(I,J) + A2*PSI(I,J+1) + A3*PSI(I+1,J+1)
     A            + A4*PSI(I+1,J) ) / HSQ
      NRANGE = 1 + ABS(PSIC)/DELTA
   15 GRAPH(ISYMBL) = SYMBOL(NRANGE)
   16 WRITE(6,17)  ( GRAPH(I), I=1,NX )
   17 FORMAT( 4X, 70A1 )
      WRITE(6,18)
   18 FORMAT( ///// 18X, 36HFIGURE 3.8.2    STREAM FUNCTION PLOT. )

      END


      SUBROUTINE LIEBMN( F, Q, M, ERRMAX )
C     ..... LIEBMANN,S ITERATIVE METHOD IS APPLIED TO SOLVE THE
C           EQUATION (2.10.1),   LAP(F) = Q ,   WHERE LAP REPRESENTS
C           THE LAPLACIAN OPERATOR. M IS THE TOTAL NUMBER OF THE
C           HORIZONTAL AS WELL AS THAT OF THE VERTICAL GRID LINES IN
C           A SQUARE DOMAIN. ITERATION IS STOPPED WHEN THE SUM OF
C           ABSOLUTE ERRORS IS LESS THAN OR EQUAL TO ERRMAX.   THE
C           BOUNDARY VALUES OF F ARE DEFINED IN THE CALLING PROGRAM .....

      DIMENSION F(M,M),Q(M,M)
      COMMON H,MM1

C     ..... ASSIGN GUESSED VALUES TO F AT INTERIOR POINTS .....
      DO 1  I = 2, MM1
      DO 1  J = 2, MM1
    1 F(I,J) = 0.0

C     ..... LIEBMANN,S ITERATIVE FORMULA (2.10.6) IS APPLIED AT ALL
C           INTERIOR POINTS UNTIL DESIRED ACCURACY IS REACHED .....
    2 ERROR = 0.0
      DO 3  I = 2, MM1
      DO 3  J = 2, MM1
      FOLD = F(I,J)
      F(I,J) = 0.25*( F(I-1,J) + F(I+1,J) + F(I,J-1)
     A            + F(I,J+1) - H**2*Q(I,J) )
    3 ERROR = ERROR + ABS( F(I,J)-FOLD )
      IF( ERROR.GT.ERRMAX )  GO TO 2
      RETURN
      END
```

           STREAM FUNCTION BASED ON 109 ITERATIONS
                       ERPSI = .00004

0.000 0.000 0.000 0.000 0.000 0.000 0.000 0.000 0.000 0.000 0.000

0.000 -.029 -.047 -.058 -.064 -.065 -.064 -.058 -.047 -.029 0.000

0.000 -.032 -.059 -.077 -.088 -.092 -.088 -.077 -.059 -.032 0.000

0.000 -.027 -.054 -.076 -.089 -.093 -.089 -.076 -.054 -.027 0.000

284

```
0.000   -.020   -.044   -.064   -.077   -.081   -.077   -.064   -.044   -.020   0.000

0.000   -.014   -.032   -.049   -.060   -.064   -.060   -.049   -.032   -.014   0.000

0.000   -.009   -.022   -.034   -.042   -.045   -.043   -.034   -.022   -.009   0.000

0.000   -.005   -.013   -.021   -.027   -.029   -.027   -.021   -.013   -.005   0.000

0.000   -.002   -.006   -.011   -.014   -.015   -.014   -.011   -.006   -.002   0.000

0.000   -.000   -.002   -.004   -.005   -.005   -.005   -.004   -.002   -.000   0.000

0.000   0.000   0.000   0.000   0.000   0.000   0.000   0.000   0.000   0.000   0.000
```

VORTICITY DISTRIBUTION
ERZETA = .00094

```
-4.415  -7.497  -6.612  -5.875  -5.420  -5.268  -5.420  -5.875  -6.611  -7.497  -4.415

-1.334  -3.586  -4.236  -4.331  -4.286  -4.257  -4.286  -4.331  -4.236  -3.586  -1.334

 .995   -1.279  -2.414  -2.928  -3.135  -3.189  -3.135  -2.928  -2.414  -1.279   .996

1.658   -.112   -1.213  -1.832  -2.137  -2.227  -2.137  -1.832  -1.213  -.111   1.658

1.622    .388   -.493   -1.052  -1.353  -1.448  -1.353  -1.052  -.493    .388   1.623

1.328    .533   -.095   -.528   -.777   -.857   -.776   -.528   -.095    .533   1.328

 .973    .510    .108   -.190   -.368   -.427   -.368   -.189    .108    .510    .973

 .649    .426    .205    .030   -.078   -.115   -.078    .030    .205    .427    .649

 .400    .342    .256    .184    .139    .124    .139    .184    .256    .342    .400

 .257    .286    .294    .309    .326    .333    .326    .309    .294    .286    .258

 .254    .251    .325    .433    .523    .557    .523    .434    .325    .251    .254
```

```
9   REPRESENTS   -.105 < PSI ≤ -.090
6   REPRESENTS   -.075 < PSI ≤ -.060
3   REPRESENTS   -.045 < PSI ≤ -.030
0   REPRESENTS   -.015 < PSI ≤ 0.000
```

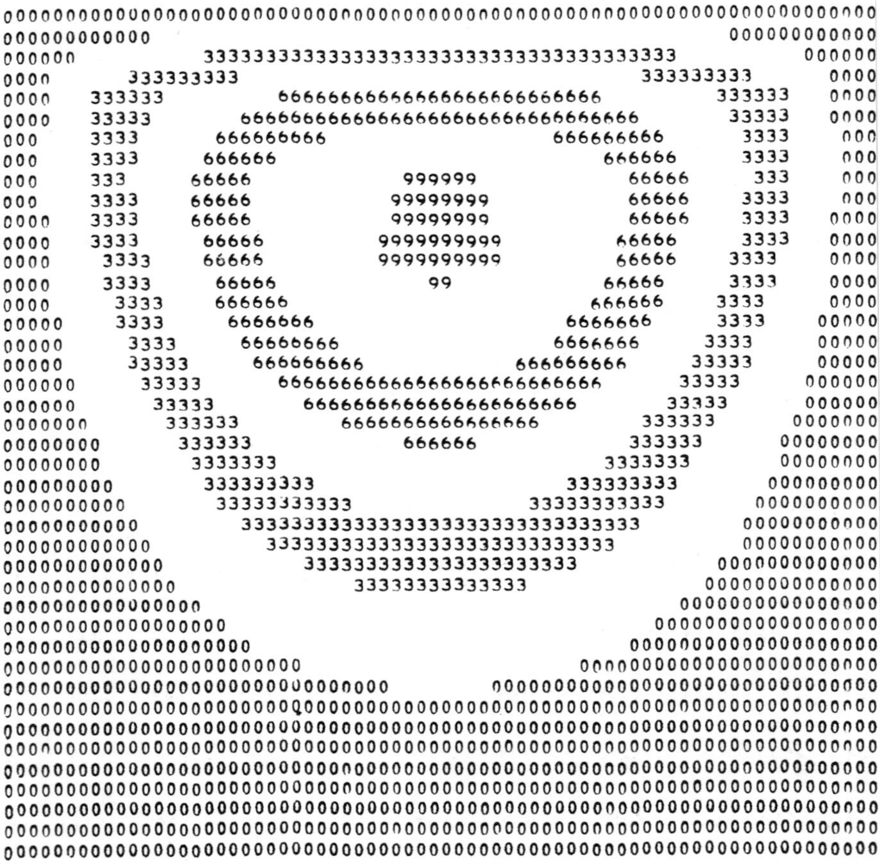

```
000000000000000000000000000000000000000000000000000000000000000000000000000000
000000000000                                                    000000000000
000000          333333333333333333333333333333333333333          000000
0000      333333333                                  333333333      0000
0000    333333        666666666666666666666666666          333333      0000
0000    33333      66666666666666666666666666666666666       33333      0000
000     3333    666666666              666666666    3333      000
000     3333    666666                    666666    3333      000
000     333    66666          999999          66666    333      000
000     3333    66666         99999999          66666    3333      000
0000    3333    66666         99999999          66666    3333      0000
0000    3333    66666        9999999999         66666    3333      0000
0000    3333    66666        9999999999         66666    3333      0000
0000    3333    66666           99             66666    3333      0000
0000    3333    666666                        666666    3333      0000
00000    3333    6666666                      6666666    3333      00000
00000    3333    66666666                    6666666    3333      00000
00000    33333    666666666              666666666    33333      00000
000000    33333    66666666666666666666666666666    33333      000000
000000     33333    6666666666666666666666666    33333      000000
0000000     333333    66666666666666666    333333      0000000
00000000     333333        666666        333333      00000000
00000000     3333333                    3333333      00000000
000000000     333333333              333333333      000000000
0000000000     33333333333        33333333333      0000000000
00000000000     3333333333333333333333333333333      00000000000
000000000000     333333333333333333333333333      000000000000
0000000000000     333333333333333333333333      0000000000000
00000000000000     33333333333333      00000000000000
000000000000000000                    0000000000000000
00000000000000000000000              000000000000000000000
0000000000000000000000000          0000000000000000000000000
000000000000000000000000000000000000              00000000000000000000000000000
00000000000000000000000000000000000000000000000000000000000000000000000000000
00000000000000000000000000000000000000000000000000000000000000000000000000000
00000000000000000000000000000000000000000000000000000000000000000000000000000
00000000000000000000000000000000000000000000000000000000000000000000000000000
00000000000000000000000000000000000000000000000000000000000000000000000000000
00000000000000000000000000000000000000000000000000000000000000000000000000000
00000000000000000000000000000000000000000000000000000000000000000000000000000
00000000000000000000000000000000000000000000000000000000000000000000000000000
```

FIGURE 3.8.2 Steam function plot.

computation for ψ, is 0.00004 whereas **ERZETA**, the sum of absolute errors in the computation for ζ, is 0.00094. After being distributed among 121 grid points, the average error at each grid point is really a very small quantity. The time spent on this program is 90 s on a CDC 6400 computer.

The printed values of ψ show that stream function is symmetric about the vertical line passing through the center of the cavity, but a slight asymmetry in vorticity is detected at some locations. The errors, which appear only at the third decimal place, can be reduced by assigning a number smaller than 0.001 to ERRMAX when the subroutine LIEBMN is called. But to obtain a solution of higher accuracy requires a longer computer time. Program 3.7 is constructed for the purpose of demonstration, and the accuracy of the result is not a main consideration here. On the other hand, results of higher resolutions can be obtained by increasing the total number of grid points.

Problem 3.16   Solve for the Stokes' flow in a square cavity shown in Fig. 3.8.1 with a free surface CD and a bottom surface AB moving at a constant unit velocity in the positive x direction.

chapter four
SECONDARY FLOWS AND FLOW INSTABILITIES

The nonlinear behavior of fluid flow is emphasized in this chapter. Here we examine the phenomena of secondary flow and flow instability through illustrative examples and assigned problems.

In the numerical methods that we have used so far, a solution is computed only at a set of selected grid points in the flow region. The Galerkin method introduced in Section 4.1 takes a different approach, in which the solution is approximated by a continuous function expressed in the form of a polynomial with arbitrary coefficients. After distributing the resultant error throughout the entire fluid domain, the problem of solving nonlinear partial differential equations is simplified to that of solving a set of nonlinear algebraic equations. The solution of nonlinear algebraic equations can be found iteratively by using the Newton-Raphson method. The solution procedure is illustrated by considering the flow of a viscous incompressible fluid past a sphere. Secondary flow is found to appear in the wake when the Reynolds number exceeds a critical value.

When a continuous vortex sheet is approximated by a number of discrete vortex lines, its motion resulting from the interaction among various parts of the sheet is governed by the Biot-Savart law. The displacement in time can be computed step by step using the fourth-order Runge-Kutta method described in Chapter 1. This method is applied in Section 4.2 to study the roll-up of the vortex sheet trailing behind a finite wing and in Section 4.3 to examine the Helmholtz instability of a velocity discontinuity.

To construct an effective numerical scheme for handling the nonlinear substantial derivative, which will be used repeatedly in later sections, a simplified one-dimensional model equation is considered in Section 4.4. Analyses show that only the upwind differencing scheme can give computational stability.

An upwind differencing formula suitable for approximating the substantial derivative in axisymmetric flow is introduced in Section 4.5. The problem solved using this numerical scheme is concerned with the secondary flow caused by a rotational electromagnetic force field. The same method works as well in other coordinate systems. It is applied in Section 4.6 to find the planar convective motion of a fluid layer under an unstable arrangement of heating from below. The method can be extended to study the behavior of rotating flows.

4.1 Flow Around a Sphere at Finite Reynolds Numbers—Galerkin Method

The Stokes' flows considered in Section 3.8 rarely occur in practical flow problems. For example, Fig. 1.2.5 shows that the Reynolds number of a free-falling water droplet of 0.5 mm diameter is 59, which is already far beyond the Stokes' flow range. If the geometry of a body is not as simple as those described in Sections 3.5 to 3.7, the inertial terms in the Navier-Stokes equation do not drop out, and the equation remains nonlinear for finite Reynolds numbers. Similar to the linearized equations, the full Navier-Stokes equation can also be solved using finite difference techniques that will be developed later in this chapter. However, in this section a completely different approach is taken in which the solution is expressed in the form of a continuous function instead of the form of numerical values at discrete points. The essence of this method, often called the *Galerkin method*, is revealed here by considering a special problem of incompressible flow past a sphere.

Let (r, θ, φ) be the spherical coordinates with the origin at the center of the sphere of radius a (Fig. 4.1.1). Far away from the sphere the flow is of a constant speed U along the x-axis. Because of the axisymmetric configuration, the flow field is independent of the azimuthal coordinate φ. For such a flow a *Stokes' stream function* ψ can be introduced, from which the r- and θ-components of the velocity are derived from the relationships

$$u_r = \frac{1}{r^2 \sin \theta} \frac{\partial \psi}{\partial \theta}, \qquad u_\theta = -\frac{1}{r \sin \theta} \frac{\partial \psi}{\partial r} \qquad (4.1.1)$$

To be generalized for curvilinear coordinates, the incompressible Navier-Stokes equation (3.1.7) is first rewritten in the form (Hughes and Gaylord, 1964, p. 24)

$$\rho \left[\frac{\partial \mathbf{V}}{\partial t} - \mathbf{V} \times (\nabla \times \mathbf{V}) \right] = -\nabla(p + \tfrac{1}{2} \rho V^2) - \mu \nabla \times \nabla \times \mathbf{V} \qquad (4.1.2)$$

After taking the curl of this equation, the leading terms on both sides

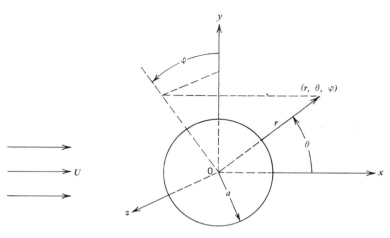

FIGURE 4.1.1 Uniform flow past a sphere.

disappear under the assumption of a steady flow. Upon substitution from (4.1.1), the resultant equation becomes

$$\frac{1}{r^2 \sin \theta} \left(\frac{\partial \psi}{\partial \theta} \frac{\partial}{\partial r} - \frac{\partial \psi}{\partial r} \frac{\partial}{\partial \theta} + 2 \cot \theta \frac{\partial \psi}{\partial r} - \frac{2}{r} \frac{\partial \psi}{\partial \theta} \right) D^2 \psi = \frac{2}{Re} D^4 \psi \qquad (4.1.3)$$

in which $Re = 2\rho U a / \mu$ is the Reynolds number based on the diameter of the sphere, ψ and r are the stream function and radial coordinate made non-dimensionalized by reference to $U a^2$ and a, respectively, and the operator D^2 is defined as

$$D^2 \equiv \frac{\partial}{\partial r^2} + \frac{\sin \theta}{r^2} \frac{\partial}{\partial \theta} \left(\frac{1}{\sin \theta} \frac{\partial}{\partial \theta} \right) \qquad (4.1.4)$$

The boundary conditions are that the fluid velocity components u_r and u_θ are both zero on the sphere, or

$$\psi(1, \theta) = 0 \qquad (4.1.5)$$

$$\frac{\partial \psi}{\partial r} (1, \theta) = 0 \qquad (4.1.6)$$

and the flow far away from the body is uniform, or

$$\psi(r, \theta) = \tfrac{1}{2} r^2 \sin^2 \theta \qquad \text{as } r \to \infty \qquad (4.1.7)$$

In order to facilitate mathematical manipulations, it is more convenient to introduce a new variable $\zeta \equiv \cos \theta$ instead of θ, so that (4.1.3) becomes

$$\frac{2}{Re} D^4 \psi + \frac{1}{r^2} \left(\frac{\partial \psi}{\partial \zeta} \frac{\partial}{\partial r} - \frac{\partial \psi}{\partial r} \frac{\partial}{\partial \zeta} - \frac{2\zeta}{1-\zeta^2} \frac{\partial \psi}{\partial r} - \frac{2}{r} \frac{\partial \psi}{\partial \zeta} \right) D^2 \psi = 0 \qquad (4.1.8)$$

Flow Around a Sphere at Finite Reynolds Numbers

where

$$D^2 \equiv \frac{\partial^2}{\partial r^2} + \frac{1-\zeta^2}{r^2} \frac{\partial^2}{\partial \zeta^2} \tag{4.1.9}$$

and the boundary condition (4.1.7) becomes

$$\psi(r, \zeta) = \tfrac{1}{2}r^2(1 - \zeta^2) \qquad \text{as } r \to \infty \tag{4.1.10}$$

Conditions (4.1.5) and (4.1.6) remain the same form, except θ is replaced by ζ.

According to the Galerkin method, the solution to (4.1.8) is approximated by a polynomial formed by a linear combination of a basic set of functions Φ_i.

$$\psi = \sum_{i=1}^{n} \Phi_i(r, \zeta) \tag{4.1.11}$$

The functions Φ_i must be linearly independent in the flow domain and must be differentiable to the extent that all terms in the differential equation and boundary conditions can be obtained. These functions are so chosen that all boundary conditions are satisfied exactly, although the solution (4.1.11) itself is only approximate. If Φ_i are the first n members of a complete set of orthogonal functions, the approximate solution is expected to approach the exact solution as $n \to \infty$.

For the problem at hand, let us choose for Φ_i the following form:

$$\Phi_i(r, \zeta) = f_i(r)[P_{i-1}(\zeta) - P_{i+1}(\zeta)], \qquad i = 1, 2, \dots \tag{4.1.12}$$

where $P_m(\zeta)$ are the Legendre polynomials defined by Rodrigues' formula

$$P_m(\zeta) = \frac{1}{2^m m!} \frac{d^m}{d\zeta^m} (\zeta^2 - 1)^m \tag{4.1.13}$$

of which the first four are

$$P_0(\zeta) = 1$$
$$P_1(\zeta) = \zeta$$
$$P_2(\zeta) = \tfrac{1}{2}(3\zeta^2 - 1)$$
$$P_3(\zeta) = \tfrac{1}{2}\zeta(5\zeta^2 - 3)$$

Φ_i defined in (4.1.12) form a complete set of independent functions in the sense that the integrals

$$\int_{-1}^{1} \Phi_i(r, \zeta)\Phi_j(r, \zeta)\, d\zeta$$

vanish for all i and j except when $i = j$, which follows from the orthogonal property of the Legendre polynomials.

Any appropriately chosen trial functions $f_i(r)$ may be used. As an example,

Secondary Flows and Flow Instabilities

we follow Kawaguti (1955) by choosing $n = 2$ and letting

$$\psi = \left(\frac{1}{2}r^2 + \frac{A_1}{r} + \frac{A_2}{r^2} + \frac{A_3}{r^3} + \frac{A_4}{r^4}\right)(1 - \zeta^2)$$

$$+ \left(\frac{B_1}{r} + \frac{B_2}{r^2} + \frac{B_3}{r^3} + \frac{B_4}{r^4}\right)\zeta(1 - \zeta^2) \tag{4.1.14}$$

which automatically satisfies the boundary condition (4.1.10) of uniform flow at infinity. The coefficients A's and B's are constants to be determined. The no-slip conditions (4.1.5) and (4.1.6) at the surface require

$$A_1 + A_2 + A_3 + A_4 = -\tfrac{1}{2} \tag{4.1.15}$$

$$A_1 + 2A_2 + 3A_3 + 4A_4 = 1 \tag{4.1.16}$$

$$B_1 + B_2 + B_3 + B_4 = 0 \tag{4.1.17}$$

$$B_1 + 2B_2 + 3B_3 + 4B_4 = 0 \tag{4.1.18}$$

When the stream function is approximated by the trial function (4.1.14), an error or residual (NS) results on the left-hand side of the Navier-Stokes equation (4.1.8), which reads

$$\begin{aligned}(NS) &= g_1(r)(1 - \zeta^2) + g_2(r)\zeta(1 - \zeta^2) + g_3(r)\zeta^2(1 - \zeta^2)\\ &+ g_4(r)(1 - \zeta^2)(1 - 3\zeta^2) + g_5(r)\zeta(1 - \zeta^2)(1 - 3\zeta^2)\\ &+ g_6(r)\zeta^3(1 - \zeta^2)\end{aligned} \tag{4.1.19}$$

With the notations that

$$\alpha_m \equiv \frac{A_m}{r^m} \quad \text{and} \quad \beta_m \equiv \frac{B_m}{r^m}$$

the functions in (4.1.19) are defined as

$$g_1(r) = \frac{4}{r^5}\left[\frac{2}{Re}r(9\alpha_2 + 35\alpha_3 + 90\alpha_4) - (2\alpha_2 + 5\alpha_3 + 9\alpha_4)(\beta_1 + \beta_2 + \beta_3 + \beta_4)\right]$$

$$\begin{aligned}g_2(r) = \frac{4}{r^5}&\left[-\frac{6}{Re}r(\beta_1 - 6\beta_3 - 21\beta_4) + (\tfrac{1}{2}r^2 + \alpha_1 + \alpha_2 + \alpha_3 + \alpha_4)(12\alpha_2 + 35\alpha_3 + 73\alpha_4)\right.\\ &\left.+ (\beta_1 + \beta_2 + \beta_3 + \beta_4)(2\beta_1 - 3\beta_3 - 7\beta_4)\right]\end{aligned}$$

$$\begin{aligned}g_3(r) = \frac{4}{r^5}&[(\tfrac{1}{2}r^2 + \alpha_1 + \alpha_2 + \alpha_3 + \alpha_4)(-6\beta_1 + 15\beta_3 + 42\beta_4) + (3\alpha_1 + 4\alpha_2 + 5\alpha_3 + 6\alpha_4)\\ &\cdot(-2\beta_1 + 3\beta_3 + 7\beta_4) + 3(2\alpha_2 + 5\alpha_3 + 9\alpha_4)(\beta_1 + \beta_2 + \beta_3 + \beta_4)]\end{aligned}$$

$$\begin{aligned}g_4(r) = \frac{2}{r^5}&[(r^2 - \alpha_1 - 2\alpha_2 - 3\alpha_3 - 4\alpha_4)(2\beta_1 - 3\beta_3 - 7\beta_4)\\ &- (8\alpha_2 + 25\alpha_3 + 54\alpha_4)(\beta_1 + \beta_2 + \beta_3 + \beta_4)]\end{aligned}$$

Flow Around a Sphere at Finite Reynolds Numbers

$$g_5(r) = \frac{2}{r^5}[(\beta_1 + \beta_2 + \beta_3 + \beta_4)(6\beta_1 - 15\beta_3 - 42\beta_4)$$

$$- (\beta_1 + 2\beta_2 + 3\beta_3 + 4\beta_4)(2\beta_1 - 3\beta_3 - 7\beta_4)]$$

$$g_6(r) = -\frac{4}{r^5}(4\beta_1 + 5\beta_2 + 6\beta_3 + 7\beta_4)(2\beta_1 - 3\beta_3 - 7\beta_4)$$

We require that the Navier-Stokes equation be satisfied exactly on the sphere by the trial function; that is, (NS) = 0 at $r = 1$ which leads to

$$9A_2 + 35A_3 + 90A_4 = 0 \qquad (4.1.20)$$

$$B_1 - 6B_3 - 21B_4 = 0 \qquad (4.1.21)$$

Away from the sphere the error is generally not zero. It is distributed throughout the flow field in the following manner:

$$\int_1^\infty \int_{-1}^1 \frac{1}{r} (\text{NS})[P_0(\zeta) - P_2(\zeta)] \, d\zeta \, dr = 0 \qquad (4.1.22)$$

$$\int_1^\infty \int_{-1}^1 \frac{1}{r^2} (\text{NS})[P_1(\zeta) - P_3(\zeta)] \, d\zeta \, dr = 0 \qquad (4.1.23)$$

from which we obtain

$$\frac{32}{Re}\left(\frac{6}{5}A_2 + 4A_3 + 9A_4\right) + \frac{8}{35}(B_1 + B_3 + 4B_4)$$

$$- \frac{512}{245}A_1B_1 - \frac{184}{35}A_2B_1 - \frac{448}{45}A_3B_1 - \frac{2848}{175}A_4B_1 - \frac{256}{105}A_2B_2$$

$$- \frac{32}{5}A_3B_2 - \frac{4608}{385}A_4B_2 + \frac{64}{21}A_1B_3 + \frac{48}{35}A_2B_3 - \frac{704}{385}A_3B_3$$

$$- \frac{232}{35}A_4B_3 + \frac{176}{25}A_1B_4 + \frac{2368}{385}A_2B_4 + \frac{392}{105}A_3B_4 - \frac{128}{455}A_4B_4 = 0 \qquad (4.1.24)$$

$$\frac{32}{21Re}(-2B_1 + 9B_3 + 28B_4) + \frac{16}{21}(2A_2 + 5A_3 + 9A_4)$$

$$+ \frac{128}{63}A_1A_2 + \frac{16}{3}A_1A_3 + \frac{768}{77}A_1A_4 + \frac{64}{35}A_2^2 + \frac{1504}{231}A_2A_3$$

$$+ \frac{32}{3}A_2A_4 + \frac{40}{9}A_3^2 + \frac{3424}{273}A_3A_4 + \frac{384}{49}A_4^2 - \frac{8}{63}B_1^2$$

$$- \frac{128}{567}B_1B_2 - \frac{16}{105}B_1B_3 - \frac{32}{693}B_1B_4 + \frac{64}{231}B_2B_3 + \frac{16}{27}B_2B_4$$

$$+ \frac{8}{21}B_3^2 + \frac{352}{273}B_3B_4 + \frac{64}{63}B_4^2 = 0 \qquad (4.1.25)$$

Secondary Flows and Flow Instabilities

Thus we have eight algebraic equations, (4.1.15) to (4.1.18), (4.1.20), (4.1.21), (4.1.24), and (4.1.25), which can be solved simultaneously for the eight unknown coefficients that appeared in the trial stream function described by (4.1.14). The first six equations enable us to express six of the unknowns in terms of the remaining two.

$$A_2 = -\frac{15}{29}(8 + 5A_1), \qquad B_2 = -\frac{23}{9}B_1$$

$$A_3 = \frac{9}{29}(17 + 7A_1), \qquad B_3 = \frac{19}{9}B_1$$

$$A_4 = -\frac{1}{58}(95 + 34A_1), \qquad B_4 = -\frac{5}{9}B_1 \qquad (4.1.26)$$

Substituting these expressions into (4.1.24) and (4.1.25) yields

$$\frac{9.931035}{Re}A_1 + 0.011713B_1 - 0.002546A_1B_1 + \frac{44.689656}{Re} = 0 \qquad (4.1.27)$$

$$0.050148A_1 + \frac{2.201058}{Re}B_1 + 0.002421A_1^2 - 0.000376B_1^2 + 0.379018 = 0 \qquad (4.1.28)$$

The solution of simultaneous nonlinear algebraic equations will be discussed later in this section. After we manage to obtain the values of A_1 and B_1, the values of the other coefficients are immediately computed from (4.1.26), so that the approximated ψ in (4.1.14) becomes a known function of r and θ.

Once the stream function becomes known, the flow pattern can be plotted and any other properties of the flow can be deduced. For example, substituting ψ into (4.1.1), we obtain the following expressions for velocity components:

$$\frac{u_r}{U} = \left(1 + 2\frac{A_1}{r^3} + 2\frac{A_2}{r^4} + 2\frac{A_3}{r^5} + 2\frac{A_4}{r^6}\right)\cos\theta$$

$$+ \left(\frac{B_1}{r^3} + \frac{B_2}{r^4} + \frac{B_3}{r^5} + \frac{B_4}{r^6}\right)(3\cos^2\theta - 1) \qquad (4.1.29)$$

$$\frac{u_\theta}{U} = \left(-1 + \frac{A_1}{r^3} + 2\frac{A_2}{r^4} + 3\frac{A_3}{r^5} + 4\frac{A_4}{r^6}\right)\sin\theta$$

$$+ \left(\frac{B_1}{r^3} + 2\frac{B_2}{r^4} + 3\frac{B_3}{r^5} + 4\frac{B_4}{4^6}\right)\sin\theta\cos\theta \qquad (4.1.30)$$

where u_r and u_θ are dimensional velocities and r is in dimensionless form.

One of the most important tasks is to compute the drag of the sphere. Let $(\tau_{rr})_{r=a}$ and $(\tau_{r\theta})_{r=a}$ be the normal and shear stresses, respectively, acting on an elementary surface area that is axisymmetric about the x-axis, as shown in Fig. 4.1.2. They are components of the *stress tensor* in spherical coordinates;

Flow Around a Sphere at Finite Reynolds Numbers

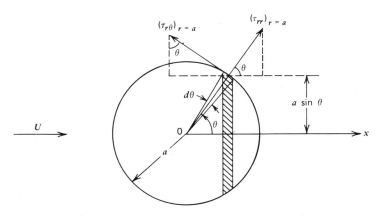

FIGURE 4.1.2 Stresses on a sphere.

the first index denotes the outward normal direction of the surface on which the stress is acting, and the second index denotes the direction of the stress. The total drag of the sphere consists of two parts.

$$\text{Drag} = D_p + D_s \tag{4.1.31}$$

where D_p is the pressure or form drag caused by normal stresses, and D_s is the skin friction caused by shear stresses on the body surface. They have the following expressions, derived by observing Fig. 4.1.2.

$$D_p = \int_0^\pi (\tau_{rr})_{r=a} \cos\theta \cdot 2\pi a \sin\theta \cdot a \, d\theta \tag{4.1.32}$$

$$D_s = -\int_0^\pi (\tau_{r\theta})_{r=a} \sin\theta \cdot 2\pi a \sin\theta \cdot a \, d\theta \tag{4.1.33}$$

The *dimensional* expressions for these two stress tensor components are (Hughes and Gaylord, 1964, p. 8)

$$\tau_{rr} = -p + 2\mu \frac{\partial u_r}{\partial r} \tag{4.1.34}$$

$$\tau_{r\theta} = \mu \left[r \frac{\partial}{\partial r} \left(\frac{u_\theta}{r} \right) + \frac{1}{r} \frac{\partial u_r}{\partial \theta} \right] \tag{4.1.35}$$

It can easily be shown by substituting from the *dimensionless* equations (4.1.29) and (4.1.30) that

$$\left(\frac{\partial u_r}{\partial r} \right)_{r=a} = 0 \tag{4.1.36}$$

$$(\tau_{r\theta})_{r=a} = \frac{2\mu U}{a} [(2A_2 + 5A_3 + 9A_4) \sin\theta + (2B_2 + 3B_3 + 3B_4) \sin\theta \cos\theta] \tag{4.1.37}$$

We now need to find an expression for the pressure. The lengthy derivation is briefly stated as follows. We start from the θ component of the vector equation of motion (4.1.2). It can be integrated immediately after substituting velocity components from (4.1.29) and (4.1.30). When evaluated at the surface, many terms drop out, and the expression is simplified to

$$(p)_{r=a} = P + \frac{\mu U}{a} [-2(8A_2 + 25A_3 + 54A_4) \cos \theta$$
$$+ 3(2B_2 + 7B_3 + 16B_4) \sin^2 \theta] \tag{4.1.38}$$

where P is the constant of integration. Since the constant term and the terms containing $\sin^2 \theta$ or $\sin \theta \cos \theta$ do not contribute to the integrals in (4.1.32) and (4.1.33), we obtain

$$D_p = \tfrac{8}{3}\pi\mu a U (8A_2 + 25A_3 + 54A_4) \tag{4.1.39}$$
$$D_s = \tfrac{16}{3}\pi\mu a U (2A_2 + 5A_3 + 9A_4) \tag{4.1.40}$$

The drag coefficient is defined as the drag force divided by $\tfrac{1}{2}\rho U^2 \cdot \pi a^2$, the product of the free-stream dynamic pressure and the projected area of the sphere. Based on this definition, we can compute the pressure drag coefficient, the drag coefficient from skin friction, and the total drag coefficient to arrive at the following expressions.

$$c_{dp} = \frac{32}{3Re} (8A_2 + 25A_3 + 54A_4) \tag{4.1.41}$$

$$c_{ds} = \frac{64}{3Re} (2A_2 + 5A_3 + 9A_4) \tag{4.1.42}$$

$$c_d = c_{dp} + c_{ds} \tag{4.1.43}$$

As the Reynolds number increases, the flow may become separated from the rear part of the sphere, and a secondary flow may appear in the form of a standing eddy. If the Reynolds number is not too large, the recirculating wake remains laminar and is confined within the streamline $\psi = 0$, as sketched in Fig. 4.1.3. The dimensionless radial distance r_s of the tip of the eddy, where the velocity vanishes, is a root of the equation

$$\frac{1}{2} + \frac{A_1 + B_1}{r^3} + \frac{A_2 + B_2}{r^4} + \frac{A_3 + B_3}{r^5} + \frac{A_4 + B_4}{r^6} = 0 \tag{4.1.44}$$

which is obtained from (4.1.29) by letting $u_r = 0$ at $\theta = 0$.

To find the root of this nonlinear algebraic equation, we will use the *Newton-Raphson method*. This method is first applied to find the root of a single equation and will be extended later to find the roots of a system of simultaneous equations.

Flow Around a Sphere at Finite Reynolds Numbers

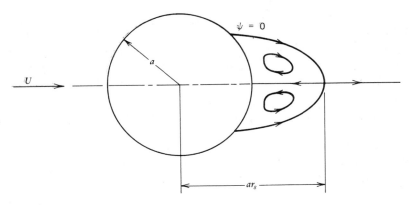

FIGURE 4.1.3 Secondary flow behind sphere.

Consider a function $f(x)$ intersecting the x-axis at a point x, which is a root of the equation

$$f(x) = 0 \qquad (4.1.45)$$

According to the Newton-Raphson method, an arbitrary value x_1 is first guessed for the solution of (4.1.45). The function evaluated at x_1 is f_1, which is generally not zero, as shown in Fig. 4.1.4. To improve the solution, the tangent to $f(x)$ is drawn at $x = x_1$ intersecting the x-axis at x_2. Since $\tan \theta_1 = f'(x_1) = f_1/(x_1 - x_2)$, where f' represents df/dx, we have

$$x_2 = x_1 - \frac{f(x_1)}{f'(x_1)}$$

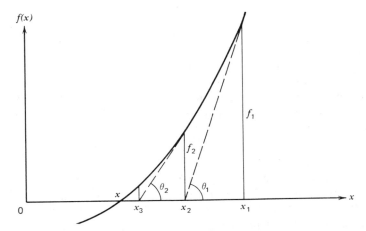

FIGURE 4.1.4 Newton-Raphson method.

which is closer than x_1 to the exact solution. By repeating the same procedure, successive improvements $x_3, x_4, \ldots$ of the root are computed from the generalized iterative formula

$$x_{k+1} = x_k - \frac{f(x_k)}{f'(x_k)} \qquad (4.1.46)$$

The iteration is terminated when the absolute value of the difference between two consecutive approximations is less than a designated small positive quantity ϵ; that is, when

$$|x_{k+1} - x_k| < \epsilon \qquad (4.1.47)$$

The Newton-Raphson method for finding the root of a single algebraic equation is programmed in the subroutine NEWTN1 that is attached to Program 4.1; by calling this subroutine, the roots of (4.1.44) are found for various Reynolds numbers.

At the final stage of the Galerkin method, the values of A_1 and B_1 must be calculated by solving simultaneously the nonlinear algebraic equations (4.1.27) and (4.1.28), which are represented in the following general form.

$$g(x, y) = 0 \qquad \text{and} \qquad h(x, y) = 0 \qquad (4.1.48)$$

The solution to the system of two equations may again be found using the Newton-Raphson method by first guessing arbitrary initial values x_1, y_1 and then applying successively the modified iterative formulas

$$x_{k+1} = x_k - \frac{g(x_k, y_k)}{(\partial g/\partial x)(x_k, y_k)} \qquad (4.1.49)$$

$$y_{k+1} = y_k - \frac{h(x_{k+1}, y_k)}{(\partial h/\partial y)(x_{k+1}, y_k)} \qquad (4.1.50)$$

Notice that the value of x obtained from the first formula is used on the right-hand side of the second formula in computing the improved value for y. The approximate solution is considered satisfactory when both

$$|x_{k+1} - x_k| < \epsilon \qquad \text{and} \qquad |y_{k+1} - y_k| < \epsilon \qquad (4.1.51)$$

The subroutine NEWTN2 is constructed for such a purpose.

Now we are ready to write a computer program for computing flow properties at various Reynolds numbers based on the result deduced from the Galerkin method. The procedure is outlined as follows.

Several FUNCTION subprograms are constructed. F(R) represents the function on the left side of (4.1.44) and DF(R) represents its first derivative. Similarly, the left-hand sides of (4.1.27) and (4.1.28) are represented by G(A1, B1) and H(A1, B1), respectively, whereas the partial derivatives $\partial G/\partial A_1$ and $\partial H/\partial B_1$ are respectively represented by DG(A1, B1) and

Flow Around a Sphere at Finite Reynolds Numbers

299

DH(A1, B1). The function PSIF(X, Y) is used to describe the trial stream function (4.1.14). Its arguments are changed from (r, ζ) to suit the subroutine SEARCH borrowed from Program 2.2 in Section 2.4. The purpose of the other two subroutines, NEWTN1 and NEWTN2, has already been stated.

Twenty different Reynolds numbers ranging from 5 to 1000 are considered. For each Reynolds number, the solution A_1 and B_1 of the system of simultaneous nonlinear algebraic equations (4.1.27) and (4.1.28) is obtained by calling NEWTN2. In the Newton-Raphson method an initial guess of the solution is needed. For fast convergence the starting values of A_1 and B_1 to be used in the present problem are those corresponding to $Re = 0$. The starting values may be changed when the same subroutine is used in solving other problems. After the remaining coefficients A's and B's have been computed from (4.1.26), drag coefficients are calculated according to (4.1.41) to (4.1.43), and the size of the standing eddy, r_s, which is a root of (4.1.44), is obtained by calling NEWTN1.

In addition to the data computed for all Reynolds numbers, we like to show in the output three flow patterns plotted respectively for $Re = 10$, 100, and 300. This is achieved by storing the values of the coefficients A's and B's in these three particular cases in two arrays AA and BB, and restoring them back to their original names after all data have been printed in the form of a table. Points along streamlines are obtained by calling SEARCH; their images are placed on a film by calling computer library subroutines in the way described in Section 2.7. The result is shown in Figs. 4.1.5 to 4.1.7.

Variable names to appear in Program 4.1 are consistent with those used in this section, so a list of names is not necessary.

The computed drag coefficient c_d agrees well with the measured curve within the same Reynolds number range, as plotted in Fig. 1.2.2. The printed data reveal the important fact that at low Reynolds numbers, the drag of a sphere is attributed mainly to the skin friction, whereas the pressure drag becomes more important for $Re > 90$. This suggests an efficient method for drag reduction at high Reynolds numbers by reducing the size of the separated flow region behind a body, because the pressure drag is caused primarily by the low pressure associated with the secondary flow in the wake.

According to the result of the Galerkin method, the boundary layer on a sphere separates at a Reynolds number slightly lower than 40 when r_s first becomes greater than unity. This is higher than the value of 24 measured by Taneda (1956). The lengths r_s at all the considered Reynolds numbers are shorter than those observed in the same experiment. Streamline plots show an almost fore-aft symmetric flow pattern at $Re = 10$. The expansion in the separated flow region as the Reynolds number increases from 100 to 300 can also be seen.

Kawaguti (1955) claimed that the trial stream function shown in (4.1.14) is

```
C                      ***** PROGRAM 4.1 *****
C
C           THE PROBLEM OF FLOW PAST A SPHERE IS SOLVED BY APPLYING
C           GALERKIN METHOD FOR REYNOLDS NUMBERS UP TO ONE THOUSAND

            DIMENSION AA(4,3),BB(4,3),X(81),Y(41),PSI(81,41),CONST(7),
         A      XL(8),YL(8),TL(8),XX(1000),YY(1000)
            COMMON /ONE/A1,A2,A3,A4,B1,B2,B3,B4 /TWO/RE /THREE/M,N,DY
            EXTERNAL F,DF, G,DG, H,DH
            DATA CONST/ 3.125, 2.0, 1.125, 0.5, 0.125, 0.0, -0.006 /

C           ..... PRINT HEADING FOR A TABLE ON TOP OF A PAGE .....
            WRITE(6,1)
          1 FORMAT( 1H1, 10/, 17X, 35HRESULT OF GALERKIN METHOD FOR FLOW ,
         A         13HPAST A SPHERE /// 9X,2HRE, 7X,2HA1, 8X,2HB1, 7X,3HCDP,
         B         7X,3HCDS, 8X,2HCD, 8X,2HRS / 8X,4H----, 6(3X,7H------)/ )

C           ..... ASSIGN VARIOUS VALUES TO REYNOLDS NUMBER .....
            RE = 0.0
            DRE = 5.0
            L = 0
            DO 3  I = 1, 20
            IF( I.GE.3 .AND. I.LE.11 )  DRE = 10.0
            IF( I.GT.11 )  DRE = 100.0
            RE = RE + DRE

C           ..... FOR EACH REYNOLDS NUMBER, FIND A1 AND B1 BY SOLVING
C               (4.1.27) AND (4.1.28) SIMULTANEOUSLY .....
            CALL NEWTN2( G, DG, H, DH, A1, B1, 1.E-6 )

C           ..... COMPUTE VALUES FOR OTHER A,S AND B,S FROM (4.1.26) .....
            A2 = -15./29. * ( 8.+ 5.*A1)
            A3 =   9./29. * (17.+ 7.*A1)
            A4 =  -1./58. * (95.+34.*A1)
            B2 = -23./9. * B1
            B3 =  19./9. * B1
            B4 =  -5./9. * B1

C           ..... COMPUTE DRAG COEFFICIENTS FROM (4.1.41 - 43) .....
            CDP = 32./(3.*RE) * ( 8.*A2 + 25.*A3 + 54.*A4 )
            CDS = 64./(3.*RE) * ( 2.*A2 +  5.*A3 +  9.*A4 )
            CD = CDP + CDS

C           ..... COMPUTE RADIAL DISTANCE RS OF THE STAGNATION POINT
C               IN THE WAKE BY FINDING THE ROOT OF (4.1.44) .....
            CALL NEWTN1( F, DF, RS, 1.E-6 )

C           ..... PRINT COMPUTED DATA .....
            WRITE(6,2)  RE, A1, B1, CDP, CDS, CD, RS
          2 FORMAT( 8X, F4.0, 6F10.4 )

C           ..... THE COEFFICIENTS A,S AND B,S FOR RE=10, 100, AND 300 ARE
C               STORED TO BE USED FOR PLOTTING THE FLOW PATTERN .....
            IF( I.NE.2 .AND. I.NE.11 .AND. I.NE.13 )  GO TO 3
            L = L + 1
            AA(1,L) = A1
            AA(2,L) = A2
            AA(3,L) = A3
            AA(4,L) = A4
            BB(1,L) = B1
            BB(2,L) = B2
            BB(3,L) = B3
            BB(4,L) = B4
          3 CONTINUE
```

```
C      ..... PREPARE FOR PLOTTING THREE FLOW PATTERNS.  SEE SECTION 2.7
C            FOR USE OF THE SUBROUTINES FOR FILM PLOT .....
       XL(1) = 10H
       YL(1) = 10H
       TL(2) = 10H    FIGURE
       TL(3) = 10H 4.1.5  FL
       TL(4) = 10HOW ABOUT A
       TL(5) = 10H SPHERE AT
       TL(6) = 10H RE = 10.
       M = 81
       X(1) = -3.0
       DX = 0.075
       DO 4  I = 2, M
   4   X(I) = X(I-1) + DX
       N = 41
       Y(1) = 0.0
       DY = 0.075
       DO 5  J = 2, N
   5   Y(J) = Y(J-1) + DY
       DO 13  L = 1, 3
       CALL EZLABL( 5HF10.0, 0, 5HF10.0, 6, 0, 6, 0 )
       GO TO ( 8, 6, 7 ),  L
   6   TL(3) = 10H 4.1.6  FL
       TL(6) = 10H RE = 100.
       GO TO 8
   7   TL(3) = 10H 4.1.7  FL
       TL(6) = 10H RE = 300.
   8   CONTINUE

C      ..... COMPUTE PSI AT ALL GRID POINTS .....
       A1 = AA(1,L)
       A2 = AA(2,L)
       A3 = AA(3,L)
       A4 = AA(4,L)
       B1 = BB(1,L)
       B2 = BB(2,L)
       B3 = BB(3,L)
       B4 = BB(4,L)
       DO 9  I = 1, M
       DO 9  J = 1, N
   9   PSI(I,J) = PSIF( X(I), Y(J) )

C      ..... SEARCH FOR POINTS ON SEVERAL STREAMLINES ABOVE THE X-AXIS
C            AND SHOW THEM ON THE FILM.  THEY ARE ROTATED ABOUT THE
C            X-AXIS TO SHOW THE STREAMLINES BELOW.  THE NUMBER OF
C            CURVES, NC, TO BE PLOTTED ON A FIGURE IS 2*KMAX .....
       CALL SEARCH( X, Y, PSI, CONST(1), XX, YY, JMAX )
       KMAX = 7
       IF( L.EQ.1 )  KMAX = 6
       NC = 2 * KMAX
       CALL EZPLOT( XX,YY,JMAX,1H*,-3.0,3.0,-3.0,3.0,XL,YL,TL,1,NC )
       DO 10  J = 1, JMAX
  10   YY(J) = -YY(J)
       CALL NXCURV( XX, YY, JMAX, 1H* )
       DO 12 K = 2, KMAX
       IF( L.EQ.3 )  CONST(7) = -0.08
       CALL SEARCH( X, Y, PSI, CONST(K), XX, YY, JMAX )
       CALL NXCURV( XX, YY, JMAX, 1H* )
       DO 11  J = 1, JMAX
  11   YY(J) = -YY(J)
  12   CALL NXCURV( XX, YY, JMAX, 1H* )
  13   CONTINUE
       END
```

```
      SUBROUTINE NEWTN1( F, DF, X, EPSLON )
C     ..... FINDING THE ROOT X OF THE ALGEBRAIC EQUATION  F(X) = 0 BY
C             USING NEWTON-RAPHSON ITERATIVE FORMULA (4.1.46) UNTIL THE
C             DIFFERENCE BETWEEN TWO CONSECUTIVE APPROXIMATIONS BECOMES
C             LESS THAN EPSLON.  DF(X) REPRESENTS DF/DX .....
      X = 2.0
    1 DX = - F(X) / DF(X)
      X = X + DX
      IF( ABS(DX).GE.EPSLON )  GO TO 1
      RETURN
      END



      SUBROUTINE NEWTN2( G, DG, H, DH, X, Y, EPSLON )
C     ..... FINDING ROOTS X AND Y OF SIMULTANEOUS ALGEBRAIC EQUATIONS
C             G(X,Y) = 0  AND  H(X,Y) = 0  USING NEWTON-RAPHSON ITERATIVE
C             FORMULAE (4.1.49) AND (4.1.50) UNTIL THE DIFFERENCE BETWEEN
C             TWO CONSECUTIVE APPROXIMATIONS FOR X AND THAT FOR Y BOTH
C             BECOME LESS THAN EPSLON.  DG(X,Y) AND DH(X,Y) REPRESENT THE
C             PARTIAL DERIVATIVES DG/DX AND DH/DY, RESPECTIVELY.  THE
C             STARTING VALUES OF X AND Y FOR THE PRESENT PROBLEM ARE THOSE
C             OF A1 AND B1 FOR VANISHING REYNOLDS NUMBER .....
      X = - 44.689656 / 9.931035
      Y = 0.0
    1 DX = - G(X,Y) / DG(X,Y)
      X = X + DX
      DY = - H(X,Y) / DH(X,Y)
      Y = Y + DY
      IF( ABS(DX).GE.EPSLON .OR. ABS(DY).GE.EPSLON )  GO TO 1
      RETURN
      END


      FUNCTION F( R )
C     ..... IT REPRESENTS THE LEFT SIDE OF (4.1.44) .....
      COMMON /ONE/A1,A2,A3,A4,B1,B2,B3,B4
      F = .5 + (A1+B1)/R**3 + (A2+B2)/R**4 + (A3+B3)/R**5 + (A4+B4)/R**6
      RETURN
      END


      FUNCTION DF( R )
C     ..... IT REPRESENTS DF(R)/DR .....
      COMMON /ONE/A1,A2,A3,A4,B1,B2,B3,B4
      DF = - 3.*(A1+B1)/R**4 - 4.*(A2+B2)/R**5 - 5.*(A3+B3)/R**6
     A     - 6.*(A4+B4)/R**7
      RETURN
      END


      FUNCTION G( A1, B1 )
C     ..... IT REPRESENTS THE LEFT SIDE OF (4.1.27) .....
      COMMON /TWO/RE
      G = 9.931035/RE*A1 + 0.011713*B1 - 0.002546*A1*B1 + 44.689656/RE
      RETURN
      END


      FUNCTION DG( A1, B1 )
C     ..... IT REPRESENTS PARTIAL DERIVATIVE DG(A1,B1)/DA1 .....
      COMMON /TWO/RE
      DG = 9.931035/RE - 0.002546*B1
      RETURN
      END
```

```
      FUNCTION H( A1, B1 )
C     ..... IT REPRESENTS THE LEFT SIDE OF (4.1.28) .....
      COMMON /TWO/RE
      H = 0.050148*A1 + 2.201058/RE*B1 + 0.002421*A1**2
     A    - 0.000376*B1**2 + 0.379018
      RETURN
      END


      FUNCTION DH( A1, B1 )
C     ..... IT REPRESENTS PARTIAL DERIVATIVE DH(A1,B1)/DB1 .....
      COMMON /TWO/RE
      DH = 2.201058/RE - 0.000752*B1
      RETURN
      END


      FUNCTION PSIF( X, Y )
C     ..... IT REPRESENTS THE STREAM FUNCTION SHOWN ON THE RIGHT SIDE
C           OF (4.1.14).  IT IS SET TO BE ZERO INSIDE THE SPHERE .....
      COMMON /ONE/A1,A2,A3,A4,B1,B2,B3,B4
      R = SQRT( X**2 + Y**2 )
      IF( R.LE.1. )  GO TO 1
      ZETA = X / R
      ZETA2 = ZETA**2
      PSIF = (R**2/2. + A1/R + A2/R**2 + A3/R**3 + A4/R**4) * (1.-ZETA2)
     A     + (B1/R + B2/R**2 + B3/R**3 + B4/R**4) * ZETA * (1.-ZETA2)
      GO TO 2
    1 PSIF = 0.0
    2 RETURN
      END


      SUBROUTINE SEARCH( X, Y, PSI, PSIA, XX, YY, KMAX )
C     ..... TAKEN FROM PROGRAM 2.2 WITH CHANGES IN THE
C           FOLLOWING STATEMENTS .....

      DIMENSION X(81),Y(41),PSI(81,41),XX(1000),YY(1000)
      COMMON /THREE/M,N,DY
```

RESULT OF GALERKIN METHOD FOR FLOW PAST A SPHERE

RE	A1	B1	CDP	CDS	CD	RS
5	-4.4946	-.4599	3.2225	6.4095	9.6319	1.0000
10	-4.4786	-.9208	1.6449	3.2189	4.8639	1.0000
20	-4.4145	-1.8494	.8896	1.6377	2.5273	1.0000
30	-4.3086	-2.7936	.6671	1.1230	1.7901	1.0000
40	-4.1621	-3.7604	.5771	.8746	1.4517	1.0144
50	-3.9771	-4.7560	.5393	.7323	1.2716	1.1574
60	-3.7563	-5.7851	.5265	.6427	1.1693	1.2759
70	-3.5037	-6.8500	.5270	.5828	1.1097	1.3813
80	-3.2242	-7.9502	.5344	.5408	1.0751	1.4779
90	-2.9235	-9.0813	.5450	.5102	1.0552	1.5681
100	-2.6083	-10.2357	.5566	.4870	1.0436	1.6529
200	.1831	-20.6758	.5709	.3667	.9376	2.2526
300	1.6289	-26.8160	.4817	.2870	.7687	2.5321
400	2.3858	-30.3180	.4009	.2320	.6329	2.6770
500	2.8401	-32.5260	.3398	.1936	.5334	2.7640
600	3.1410	-34.0355	.2937	.1658	.4594	2.8216
700	3.3546	-35.1301	.2581	.1448	.4029	2.8626
800	3.5139	-35.9592	.2300	.1284	.3584	2.8932
900	3.6372	-36.6087	.2073	.1154	.3227	2.9168
1000	3.7354	-37.1310	.1887	.1047	.2934	2.9357

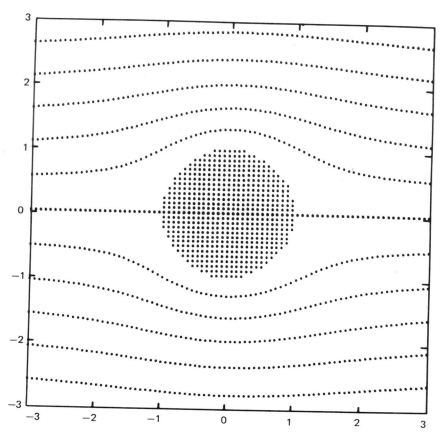

FIGURE 4.1.5 Flow about a sphere at $Re = 10$.

appropriate for the regime $10 < Re < 80$ as far as drag is concerned. For Re less than 2, he used another trial function:

$$\psi = \left(\frac{1}{2}r^2 + A_1 r + A_2 + \frac{A_3}{r} + \frac{A_4}{r^2}\right)(1 - \zeta^2)$$

$$+ \left(B_1 r + B_2 + \frac{B_3}{r} + \frac{B_4}{r^2}\right)\zeta(1 - \zeta^2) \qquad (4.1.52)$$

and the drag coefficient so computed agrees very well with that measured in that Reynolds number regime. It can be shown that (4.1.52) reduces to the Stokes approximation as the Reynolds number tends to zero.

In the present example we choose a trial stream function that satisfies boundary conditions exactly but the governing differential equation only

Flow Around a Sphere at Finite Reynolds Numbers

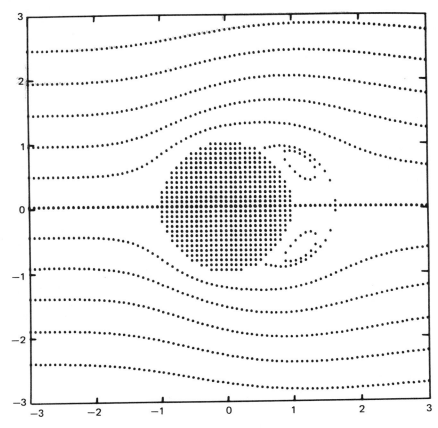

FIGURE 4.1.6 Flow about a sphere at $Re = 100$.

approximately. The resultant error is distributed throughout the flow region and is set to zero on the body surface. On the other hand, the trial function may be so chosen that the governing equation is satisfied exactly but the boundary conditions are fulfilled approximately. The errors resulting from the boundary conditions are then distributed over some appropriate regions. However, the latter method can hardly be used in problems whose governing differential equations are nonlinear, such as the one considered presently.

Many other flow problems involving spherical geometry have been solved using the Galerkin method. For example, Hamielec, Storey, and Whitehead (1963) computed the drag of an undeformed fluid sphere moving through another fluid medium at various Reynolds numbers. A rigid sphere with a radial mass efflux was examined by Hamielec, Hoffman, and Ross (1967),

Secondary Flows and Flow Instabilities

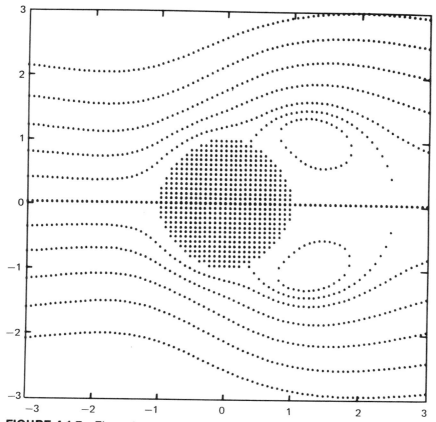

FIGURE 4.1.7 Flow about a sphere at $Re = 300$.

and a sphere made of an electrically conducting material moving through a fluid carrying an electric current was studied by Chow and Halat (1969).

Let us now apply the Galerkin method to solve a problem that is not related to flows about a spherical body. Consider the steady two-dimensional flow of an incompressible fluid between two inclined plane walls meeting at an angle 2α, as shown in Fig. 4.1.8. The flow is assumed to be radial, so that $u_r(r, \theta)$ is the only component of the velocity vector. The simplified Navier-Stokes equation may be written in the following component form (Hughes and Gaylord, 1964, p. 25).

$$\rho u_r \frac{\partial u_r}{\partial r} + \frac{\partial p}{\partial r} - \mu \left\{ \frac{\partial}{\partial r} \left[\frac{1}{r} \frac{\partial}{\partial r} (r u_r) \right] + \frac{1}{r^2} \frac{\partial^2 u_r}{\partial \theta^2} \right\} = 0 \qquad (4.1.53)$$

$$-\frac{1}{r} \frac{\partial p}{\partial \theta} + \frac{2\mu}{r^2} \frac{\partial u_r}{\partial \theta} = 0 \qquad (4.1.54)$$

Flow Around a Sphere at Finite Reynolds Numbers

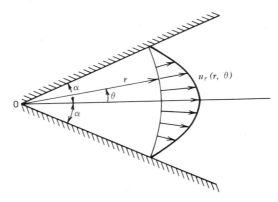

FIGURE 4.1.8 Radial flow between inclined walls.

If Q represents the specified mass flow rate per unit width of the channel, the equation of continuity can be written as

$$2\rho \int_0^\alpha u_r r \, d\theta = Q \tag{4.1.55}$$

Boundary conditions are that on the walls the velocity must be zero, or

$$u_r = 0 \quad \text{at } \theta = \pm\alpha \tag{4.1.56}$$

Snyder, Spriggs, and Stewart (1964) considered a number of trial functions and found that the set

$$u_r = \sum_{j=1}^n \frac{Q}{\rho r} C_j (\alpha^2 - \theta^2)^j \tag{4.1.57}$$

$$p - p_0 = \frac{Q\mu}{\rho r^2} \left[C_0 + 2 \sum_{j=1}^n C_j (\alpha^2 - \theta^2)^j \right] \tag{4.1.58}$$

yielded the best approximate solution when only the three leading terms were used in solving the problem. These trial functions already satisfy the θ component of the equation of motion, (4.1.54), and the boundary conditions (4.1.56). Upon substitution from (4.1.57), the continuity equation (4.1.55) becomes

$$2 \sum_{j=1}^n \int_0^\alpha C_j (\alpha^2 - \theta^2)^j \, d\theta = 1 \tag{4.1.59}$$

When the trial functions are substituted into the r component of the equation of motion, (4.1.53), an error $E(\theta)$ instead of zero is obtained on the left-hand side.

Secondary Flows and Flow Instabilities

$$E(\theta) = \sum_{i=1}^{n} \sum_{j=1}^{n} Re\ C_i C_j (\alpha^2 - \theta^2)^{i+j} + 2C_0$$

$$+ 4 \sum_{j=1}^{n} C_j (\alpha^2 - \theta^2)^j + \sum_{j=1}^{n} C_j \frac{d^2}{d\theta^2} (\alpha^2 - \theta^2)^j \qquad (4.1.60)$$

where Re is the Reynolds number defined as Q/μ. The error is then distributed throughout the flow region by making it orthogonal to $(\alpha^2 - \theta^2)^k$ in the following manner.

$$\int_0^\alpha E(\theta)(\alpha^2 - \theta^2)^k\ d\theta = 0 \qquad \text{for } k = 1, \ldots, n \qquad (4.1.61)$$

The linear equation (4.1.59) and n quadratic equations (4.1.61) are solved simultaneously for the $n + 1$ unknowns C_0 and C_j.

Problem 4.1 Evaluate analytically the integrals in (4.1.59) and (4.1.61) for $n = 2$, and write a computer program for $Re = 14.164$ and $\alpha = 0.36$ rad. After the coefficients C_0, C_1, and C_2 have been obtained, the dimensionless velocity $u_r \rho r/Q$ and the dimensionless pressure difference $(p - p_0)\rho r^2/Q\mu$ are computed as functions of θ. The velocity distribution for this case, as plotted by Snyder, Spriggs, and Stewart (1964), is found to be in excellent agreement with the exact solution found by Rosenhead (1940). The same problem was also studied analytically by Landau and Lifshitz (1959, p. 81) with the exact solution expressed in terms of elliptic integrals.

4.2 Rolling Up of the Trailing Vortex Sheet Behind a Finite Wing

In computing the lift for a wing of infinite span, the wing can be replaced by a vortex sheet coinciding with the mean camberline of the airfoil. If the total circulation of the vortex sheet is Γ and the wing flies at a constant speed U through air of density ρ, the lift per unit span is, according to the Kutta-Joukowski theorem (2.5.4),

$$L' = \rho U \Gamma \qquad (4.2.1)$$

The circulation is defined as

$$\Gamma = \oint \mathbf{V} \cdot d\mathbf{s} \qquad (4.2.2)$$

in which $\mathbf{V}$ is the fluid velocity and $d\mathbf{s}$ is a line element along any closed curve enclosing the vortex sheet. The integration is performed in the clockwise direction. Thus, as far as the sectional lift is concerned, the wing may be treated as an infinitely long *bound vortex* of circulation Γ whose axis is parallel to the span. A bound vortex is one that does not move with the fluid like a free vortex.

When a bound vortex is used to approximate a finite wing, the vortex cannot end at the wing tips according to Helmholtz vortex laws; its ends are carried downstream by the fluid motion to form a pair of oppositely revolving vortices trailing behind the wing. Such a form of the vortex line, with the middle portion fixed to the wing and the remaining parts extending to infinity in the downstream direction, is called a *horseshoe vortex*.

The lift per unit span on a finite wing actually varies along the span, so that the circulation of the bound vortex is a function of the spanwise location. A finite wing symmetrical about its midplane is usually constructed by a super-position of horseshoe vortex elements of various strengths, as sketched schematically in Fig. 4.2.1. An infinite number of these vortices lead to a continuous distribution of circulation $\Gamma(x)$ and therefore of the lift per unit span $\rho U \Gamma(x)$ on a wing of span $2a$. The infinite number of trailing vortices then form a vortex sheet on the x-z plane of strength $\gamma(x)$, which is defined

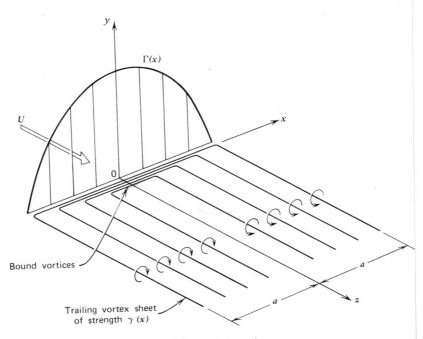

FIGURE 4.2.1 Vortex system for a finite wing.

as the circulation of vortices contained within unit spanwise length of the sheet. From Fig. 4.2.1 we have

$$\gamma(x) = \frac{d\Gamma}{dx} \qquad (4.2.3)$$

The trailing vortices induce at the wing a velocity field pointing in the negative y direction, called the downwash, causing an induced drag that is a characteristic of a wing of finite span. Detailed analyses of finite wings appear in Chapter 6 of Kuethe and Chow (1976) and are not elaborated on here.

The trailing vortex sheet shown in Fig. 4.2.1 cannot always remain on the x-z plane, as assumed for the convenience of theoretical analyses. Each line vortex moves under the influence of the others, and the motion differs from one vortex to another, depending on the instantaneous geometry of the sheet. In reality the vortex sheet is found to roll up, forming two well-defined column vortices trailing behind an airplane. Figure 4.2.2 shows the rolling up of the vortex sheet behind a delta wing, whose horseshoe-vortex arrangement is somewhat different from that shown in Fig. 4.2.1.

The roll-up of a vortex sheet is a three-dimensional phenomenon, as exhibited by Fig. 4.2.2. Instead of dealing with this realistic but extremely cumbersome situation, we examine a simplified two-dimensional problem by considering the deformation of a vortex sheet of width $2a$ that extends to infinity in either direction along the z-axis. In this way the constraint that one end of the sheet must be attached to the bound vortices inside the wing is removed. The configurations of the two-dimensional vortex sheet at various times closely resemble those at different downstream sections along the deforming three-dimensional sheet.

For an elliptically loaded wing with circulation Γ_0 in the plane of symmetry, the variation of circulation with spanwise location is written

$$\Gamma = \frac{(a^2 - x^2)^{1/2}}{a} \Gamma_0 \qquad (4.2.4)$$

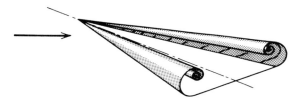

FIGURE 4.2.2 Roll-up of trailing vortex sheet behind a delta wing.

Rolling Up of the Trailing Vortex Sheet Behind a Finite Wing

and, from (4.2.3), the strength of the initially planar trailing vortex sheet is

$$\gamma = -\frac{\Gamma_0 x}{a(a^2 - x^2)^{1/2}} \tag{4.2.5}$$

It can easily be shown that the total circulation of the vortices of positive strength on the left half of the sheet is Γ_0, and that of the vortices of negative strength on the right is $-\Gamma_0$. To study the evolution of this continuous vortex distribution, the sheet is discretized by subdividing it into spanwise intervals and replacing each subdivision by a point vortex of circulation equal to the circulation of all vortices within that subdivision. The velocity of each point vortex can be obtained as a function of the position of the others, from which the displacement of that vortex at a short time period later is computed. The deformation of the sheet with respect to time is thus traced by repeatedly integrating forward in time.

Suppose that at time $t = 0$ the initially flat vortex sheet is divided into m equally spaced intervals, as shown in Fig. 4.2.3, m being a reasonably large even integer. If each subdivision is replaced by a discrete vortex at the middle, the distance between any two neighboring vortices is $2a/m$, and each of the two outermost vortices is at a distance of a/m from the wing tip. These m discrete vortices are named in an ascending order starting from the left end, as marked in the figure. By substitution from (4.2.5) for γ, the circulation or the strength of the ith vortex located at (x_i, y_i) is

$$g_i = \int_{x_i - a/m}^{x_i + a/m} \gamma\, dx$$

$$= \Gamma_0 \left[\sqrt{1 - \left(\frac{x_i}{a} + \frac{1}{m}\right)^2} - \sqrt{1 - \left(\frac{x_i}{a} - \frac{1}{m}\right)^2} \right] \tag{4.2.6}$$

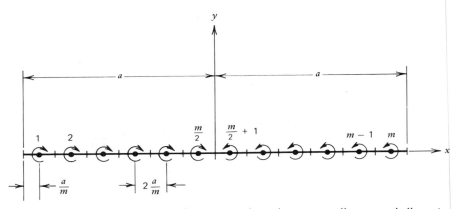

FIGURE 4.2.3 Replacement of a vortex sheet by m equally spaced discrete point vortices.

where x_i is the initial abscissa of the point vortex and $y_i = 0$ at the same time.

The Biot-Savart law (2.5.1) is applied to compute the velocity W_i of the ith vortex induced by the jth vortex at (x_j, y_j), with both vortices at two arbitrary positions, as depicted in Fig. 4.2.4. Vector algebra could still be used here, as in Section 2.5, for computing the motion of a trailing vortex pair above the runway. However, a different approach will be taken in solving the problem at hand in order to avoid the large errors that would result from the previous method when the distance between vortices becomes small. If $R[= \sqrt{(x_i - x_j)^2 + (y_i - y_j)^2}]$ is the distance between the two concerned vortices, the magnitude of the induced velocity is $W_i = g_j/(2\pi R)$. When it is decomposed into x and y directions, this velocity has the components

$$u_i = W_i \sin \theta = \frac{y_i - y_i}{R} W_i$$

$$v_i = -W_i \cos \theta = -\frac{x_i - x_j}{R} W_i$$

Adding the contributions from all vortices other than the ith vortex itself and replacing u by dx/dt and v by dy/dy, we obtain

$$\frac{dx_i}{dt} = \sum_{j \neq i}^{m} \frac{g_j}{2\pi} \frac{y_i - y_j}{(x_i - x_j)^2 + (y_i - y_j)^2} \tag{4.2.7}$$

$$\frac{dy_i}{dt} = -\sum_{j \neq i}^{m} \frac{g_j}{2\pi} \frac{x_i - x_j}{(x_i - x_j)^2 + (y_i - y_j)^2} \tag{4.2.8}$$

It is more convenient to work with dimensionless quantities. In terms of dimensionless variables defined by

$$X = \frac{x}{a}, \qquad Y = \frac{y}{a}, \qquad G = \frac{g}{\Gamma_0}, \qquad T = \frac{t}{2\pi a^2/\Gamma_0}$$

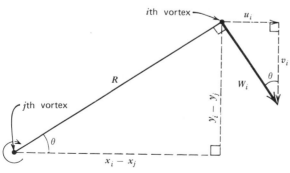

FIGURE 4.2.4 Velocity of the ith vortex induced by the jth vortex.

Rolling Up of the Trailing Vortex Sheet Behind a Finite Wing

(4.2.6) to (4.2.8) become

$$G_i = \sqrt{1 - \left(X_i + \frac{1}{m}\right)^2} - \sqrt{1 - \left(X_i - \frac{1}{m}\right)^2} \tag{4.2.9}$$

$$\frac{dX_i}{dT} = \sum_{j \neq i}^{m} G_j \frac{Y_i - Y_j}{(X_i - X_j)^2 + (Y_i - Y_j)^2} \tag{4.2.10}$$

$$\frac{dY_i}{dT} = -\sum_{j \neq i}^{m} G_j \frac{X_i - X_j}{(X_i - X_j)^2 + (Y_i - Y_j)^2} \tag{4.2.11}$$

The dimensionless distance between any two neighboring vortices at the initial instant is then $2/m$.

(4.2.10) and (4.2.11) form a system of two simultaneous first-order ordinary differential equations. If the system is written in the general form that

$$\frac{dx}{dt} = F_1(x, y), \qquad \frac{dy}{dt} = F_2(x, y) \tag{4.2.12}$$

the solution can be obtained using the fourth-order Runge-Kutta method. The formulas for computing the solution x' and y' at $t + h$ based on the solution x and y at t are, simplified from (1.4.3) for solving simultaneous second-order equations,

$$\begin{aligned}
\Delta_1 x &= hF_1(x, y) \\
\Delta_1 y &= hF_2(x, y) \\
\Delta_2 x &= hF_1(x + \tfrac{1}{2}\Delta_1 x, y + \tfrac{1}{2}\Delta_1 y) \\
\Delta_2 y &= hF_2(x + \tfrac{1}{2}\Delta_1 x, y + \tfrac{1}{2}\Delta_1 y) \\
\Delta_3 x &= hF_1(x + \tfrac{1}{2}\Delta_2 x, y + \tfrac{1}{2}\Delta_2 y) \\
\Delta_3 y &= hF_2(x + \tfrac{1}{2}\Delta_2 x, y + \tfrac{1}{2}\Delta_2 y) \\
\Delta_4 x &= hF_1(x + \Delta_3 x, y + \Delta_3 y) \\
\Delta_4 y &= hF_2(x + \Delta_3 x, y + \Delta_3 y) \\
x' &= x + \tfrac{1}{6}(\Delta_1 x + 2\,\Delta_2 x + 2\,\Delta_3 x + \Delta_4 x) \\
y' &= y + \tfrac{1}{6}(\Delta_1 y + 2\,\Delta_2 y + 2\,\Delta_3 y + \Delta_4 y)
\end{aligned} \tag{4.2.13}$$

This numerical integration procedure is programmed in a subroutine named RUNTTA, in which the names XNEXT and YNEXT are used, respectively, for x' and y'. It should be pointed out that in Section 2.5 on the trajectory of a vortex pair, the numerical computation of Program 2.3 is equivalent to that in Euler's method, whose truncation error is $O(h^2)$ in comparison with $O(h^5)$ in the fourth-order Runge-Kutta method. A much higher accuracy is needed in the present problem because of the much stronger induced velocities caused by the shorter distances between neighboring vortices. Of course, the computing time required for the present problem is much longer.

In Program 4.2 the right-hand sides of (4.2.10) and (4.2.11) are respectively called U(XI, YI) and V(XI, YI) for obvious reasons, and are defined in two separate function subprograms.

When progressing in time the step size dT cannot be large in order to avoid numerical instabilities. Moore (1974), in studying the same problem, found that the restriction on the time step is, after being rewritten in our notation,

$$dT \ll \frac{8\pi}{m^2} \qquad (4.2.14)$$

The value $m = 40$ is used in Program 4.2. By choosing the value $1/(25m)$ for dT this inequality is satisfied. The results based on some even smaller step sizes exhibit only minor changes near the wing tips.

Because the vortex sheet is symmetric about the y-axis at all times, computations are needed only for discrete vortices on the left half of the sheet, with i running from 1 to $m/2$. The kth vortex, where $k = m + 1 - i$, on the right half of the sheet is the mirror image of the ith vortex about the y-axis.

At any time step, the newly computed coordinates of the vortices are stored temporarily in the arrays XNEW and YNEW. These values are later assigned to the arrays X and Y after the positions of all the $m/2$ vortices have been updated. This procedure is necessary; when computing the new position of any one vortex, the present locations (X, Y) of all vortices are needed.

NMAX is the maximum number of time steps allowed in a single run, and it is chosen to be 500 in Program 4.2. To avoid showing the large amount of numerical output, the coordinates of the vortices on the left half of the vortex sheet are printed every NFREQ time steps.

For easier visualization we omit the printed output of the program and, instead, plot in Fig. 4.2.5 the shape of the vortex sheet at eight representative time instants. Initially the tip vortex starts to move upward as a result of its peculiarity that vortices are on its one side whereas none on the other. The rest of the vortices move downward at approximately equal speeds. The mutual interactions among vortices in the tip region cause the sheet to curve up and be rolled into a spiral. As more and more vortices are drawn into the ever-expanding spiral region, the vortex sheet is continuously stretched, leaving longer distances between neighboring vortices along the sheet. Finally, the originally flat vortex sheet will be evolved into two tightly rolled spirals revolving in opposite directions. The situation is similar to what is observed behind an aircraft.

Using 40 discrete vortices to replace the vortex sheet does not give accurate results at large values of T, since the constituent vortices are separated far apart. For example, at $T = 0.35$, irregularities start to appear on the sheet in the spiral region. Results at later time steps show that the paths of

```
C                    ***** PROGRAM 4.2 *****

C              STUDYING THE ROLL-UP OF A VORTEX SHEET

      EXTERNAL U,V
      DIMENSION G(40),X(40),Y(40),XNEW(20),YNEW(20)
      COMMON I,M,G,X,Y

      M = 40
      NMAX = 500
      NFREQ = 5
      M2 = M/2

C      ..... SPECIFY DIMENSIONLESS STRENGTHS AND INITIAL
C            COORDINATES OF VORTICES .....
      NDT = 0
      T   = 0.0
      DX = 1.0 / FLOAT(M)
      X(1) = -1.0 + DX
      DO 1   I = 2, M
    1 X(I) = X(I-1) + 2.0*DX
      DO 2   I = 1, M2
      Y(I) = 0.0
    2 G(I) = SQRT(1.0-(X(I)+DX)**2) - SQRT(1.0-(X(I)-DX)**2)
      M2P1 = M2 + 1
      DO 3   I = M2P1, M
      Y(I) = 0.0
    3 G(I) = -G(M+1-I)

C      ..... COMPUTE THE POSITION OF EACH VORTEX ON THE LEFT HALF
C            OF THE SHEET AT TIME DT LATER .....
      DT = 1.0 / (25.*M)
    4 NDT = NDT + 1
      IF( NDT.GT.NMAX )  GO TO 10
      T = T + DT
      DO 5   I = 1, M2
    5 CALL RUNTTA( X(I), Y(I), DT, U, V, XNEW(I), YNEW(I) )

C      ..... ASSIGN VALUES OF XNEW AND YNEW TO X AND Y, RESPECTIVELY ....
      DO 6   I = 1, M2
      X(I) = XNEW(I)
    6 Y(I) = YNEW(I)

C      ..... VORTICES ON THE RIGHT HALF OF THE SHEET ARE THE
C            MIRROR IMAGES OF THOSE ON THE LEFT .....
      DO 7   I = M2P1, M
      X(I) = -X(M+1-I)
    7 Y(I) =  Y(M+1-I)

C      ..... PRINT COORDINATES EVERY NFREQ TIME STEPS .....
      IF( NDT/NFREQ*NFREQ .EQ. NDT )  GO TO 8
      GO TO 4
    8 WRITE(6,9)  NDT, T, ( (X(I),Y(I)), I=1,M2 )
    9 FORMAT( //5X, 5HNDT =, I4, 1H,, 5X, 3HT =, F5.3 / (5(5X,2F9.3)) )
      GO TO 4
   10 STOP
      END
```

```
      SUBROUTINE RUNTTA( X, Y, H, F1, F2, XNEXT, YNEXT )
C     ..... INTEGRATION OF THE SIMULTANEOUS FIRST-ORDER EQUATIONS
C           DX/DT = F1(X,Y)  AND  DY/DT = F2(X,Y)  USING FOURTH-
C           ORDER RUNGE-KUTTA METHOD. XNEXT AND YNEXT ARE THE
C           VALUES OF X AND Y, RESPECTIVELY, AT T+H .....
      D1X = H * F1( X, Y )
      D1Y = H * F2( X, Y )
      D2X = H * F1( X+D1X/2., Y+D1Y/2. )
      D2Y = H * F2( X+D1X/2., Y+D1Y/2. )
      D3X = H * F1( X+D2X/2., Y+D2Y/2. )
      D3Y = H * F2( X+D2X/2., Y+D2Y/2. )
      D4X = H * F1( X+D3X, Y+D3Y )
      D4Y = H * F2( X+D3X, Y+D3Y )
      XNEXT = X + (D1X + 2.*D2X + 2.*D3X + D4X)/6.0
      YNEXT = Y + (D1Y + 2.*D2Y + 2.*D3Y + D4Y)/6.0
      RETURN
      END



      FUNCTION U( XI, YI )
C     ..... IT REPRESENTS THE RIGHT-HAND SIDE OF (4.2.10) .....
      DIMENSION G(40),X(40),Y(40)
      COMMON I,M,G,X,Y
      U = 0.0
      DO 1  J = 1, M
      IF( J.EQ.I )  GO TO 1
      YY = YI - Y(J)
      U = U + G(J)*YY/((XI-X(J))**2+YY**2)
    1 CONTINUE
      RETURN
      END


      FUNCTION V( XI, YI )
C     ..... IT REPRESENTS THE RIGHT-HAND SIDE OF (4.2.11) .....
      DIMENSION G(40),X(40),Y(40)
      COMMON I,M,G,X,Y
      V = 0.0
      DO 1  J = 1, M
      IF( J.EQ.I )  GO TO 1
      XX = XI - X(J)
      V = V - G(J)*XX/(XX**2+(YI-Y(J))**2)
    1 CONTINUE
      RETURN
      END
```

some vortices in that region are contorted, and certain parts of the vortex sheet may become crossed. The computation should actually be terminated before this unrealistic situation has developed. The chaotic motion of the spiral has been found by several authors. A brief description of some previous research works is referred to Moore (1974). To eliminate the chaotic motion, Moore successfully used a concentrated tip vortex to represent the tightly rolled portion of the vortex sheet.

Rolling Up of the Trailing Vortex Sheet Behind a Finite Wing

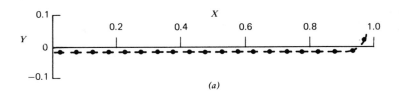

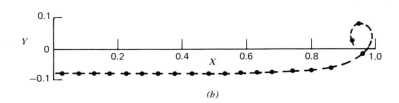

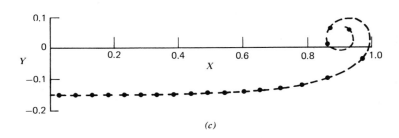

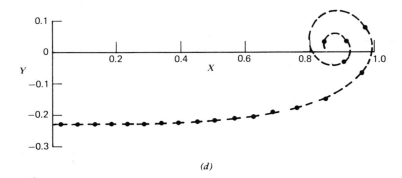

FIGURE 4.2.5 Evolution of an initially flat vortex sheet. (a) $T = 0.005$. (b) $T = 0.025$. (c) $T = 0.050$, (d) $T = 0.075$.

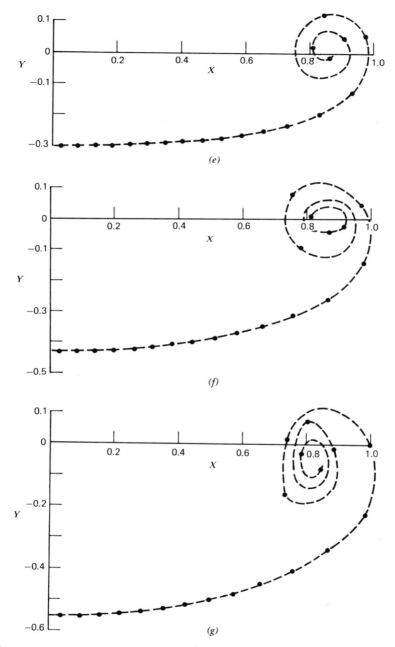

FIGURE 4.2.5 (cont.)　(e) $T = 0.10$. (f) $T = 0.15$. (g) $T = 0.20$.

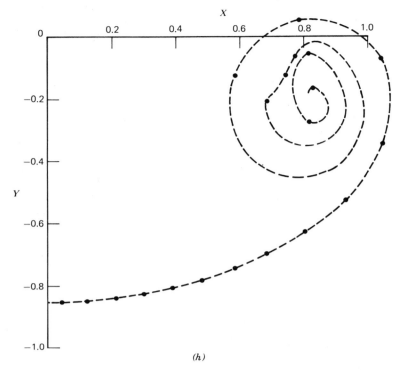

(h)

FIGURE 4.2.5 (cont.) (h) T = 0.35.

Results of higher accuracies can be obtained by choosing larger values for *m*. However, the computing time, which is 225 s on a CDC 6400 computer for *m* = 40, increases roughly with m^2.

Let us now apply our result to a realistic problem. For the Boeing 747 considered previously in Section 2.5 with half-span $a = 30$ m, the circulation averaged along the span is 815 m²/s during take-off. Assuming an elliptical distribution of circulation along the span, as described by (4.2.4), we require that

$$2a \cdot (815) = \int_{-a}^{a} \frac{\Gamma_0}{a} (a^2 - x^2)^{1/2} \, dx$$

$$= \frac{1}{2} \pi a \Gamma_0$$

giving $\Gamma_0 = 1038$ m²/s. The characteristic time $2\pi a^2/\Gamma_0$ is about 5.45 s, and therefore the last plot for $T = 0.35$ in Fig. 4.2.5 shows the vortex sheet configuration at $t = 1.91$ s. For a take-off speed of 67 m/s, this is the approximate configuration of the three-dimensional trailing vortex sheet at a distance of 128 m behind the airplane. The plot also shows that after the sheet has been rolled up, the centers of the two spirals are separated by a distance shorter than the wing span b. The assumption made in Section 2.5 that the initial distance between trailing vortex pair is b is only approximately correct.

Figure 4.2.5 also reveals that the tip region of the vortex sheet descends slower than the middle region. The vertical position of the midpoint of the sheet is plotted in Fig. 4.2.6 as a function of time. It descends initially with speed $\Gamma_0/2a$, and slows down toward a constant speed approximately equal to half of that value.

The analysis on which Program 4.2 was based may be improved. In that analysis we subdivide a vortex sheet into small spanwise intervals and replace each one by a point vortex at the center. Since the sheet strength γ is not uniform, the vortices within each subdivision will generally induce a velocity at its center. Thus the discrete vortex placed there will move under its own influence. This motion was not taken into account, although its magnitude

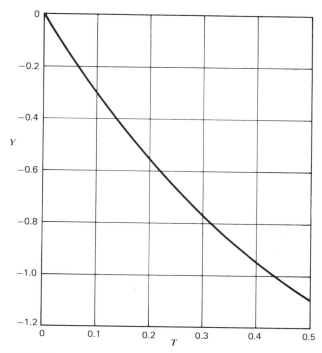

FIGURE 4.2.6 Position of the center of the vortex sheet as a function of time.

Rolling Up of the Trailing Vortex Sheet Behind a Finite Wing

may be small compared with the sum of those induced there by other point vortices. To be more realistic, when a vortex sheet segment is replaced by a discrete vortex, it should be placed at a point on that segment at which the self-induced velocity is zero. The computation for determining the position of such a point for a straight segment can be carried out without any difficulty. However, after the segment is deformed into a curve, the location of the point of vanishing self-induced velocity changes. The determination of its position on an arbitrary curve becomes too involved to be considered practical.

Instead of being replaced by initially equispaced vortices, as demonstrated here, the vortex sheet may also be replaced by discrete vortices of equal strength. Moore's results based on these two approximations show remarkable agreement.

Problem 4.2 Replace the vortex sheet considered in Program 4.2 by 40 discrete vortices of equal strength. Plot the initial configuration of the sheet, and compare the shapes of the sheet with those shown in Fig. 4.2.5 at the same time instants.

4.3 Helmholtz Instability of a Velocity Discontinuity

The method of discretization of a continuous vortex sheet described in Section 4.2 is now applied to study the stability of an inviscid, incompressible, parallel flow containing a velocity discontinuity. In a coordinate system moving with the mean speed of the flow and with the x-axis coinciding with the interface of the velocity discontinuity, the velocity field appears to be that shown in Fig. 4.3.1. The fluid above the x-axis moves at a uniform velocity U, and that below it moves at another uniform velocity of equal magnitude but in the opposite direction. The x-axis, or the interface of the velocity discontinuity, is thus the site of a vortex sheet of uniform circulation $2U$ per unit width according to the definition (4.2.2).

Such a vortex sheet of infinite span is unstable in that if any part of the sheet is given a displacement, it will keep going away and will not return to its original position. The theoretical work on the stability of a velocity discontinuity was initiated by Helmholtz (1868). The analysis based on the assumption of small sinusoidal disturbances, which can be found in many books on fluid mechanics (e.g., in Section 7.1 of Batchelor, 1967), shows that the interface is unstable to a disturbance of any wavelength. Linearized analyses determine the onset of instabilities, but do not show what will happen after the vortex sheet becomes unstable. The study of the latter nonlinear phenomenon is made possible by using the numerical method of the

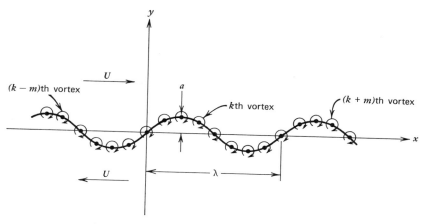

FIGURE 4.3.1 Approximation of velocity discontinuity by discrete vortices.

last section to replace the vortex sheet by a row of discrete vortices. Furthermore, when solving the problem numerically, the amplitudes of the displacement are no longer restricted to be small, as they usually are in the analytical work. The first study along this line was that of Rosenhead (1931) using the Euler integration scheme to compute the time development of a plane vortex sheet. His result was later improved by Birkhoff and Fisher (1959) and by Hama and Burke (1960) using the Runge-Kutta algorithm with the availability of digital computers.

Let us divide the vortex sheet into segments of equal length λ, and approximate each segment by m equispaced discrete vortices, m being a reasonably large, even integer. Because the circulation of the continuous vortex sheet is $2U$ per unit length, each discrete vortex has a circulation of $2U\lambda/m$. Assume that at the initial instant these vortices are displaced vertically in such a way that the coordinates (x_j, y_j) of the jth vortex are related through the equation

$$y_j = a \sin\left(\frac{2\pi x_j}{\lambda}\right) \qquad \text{at } t = 0 \qquad (4.3.1)$$

where the index j runs from $-\infty$ to $+\infty$, with $j = 1$ denoting the vortex at the origin of the coordinate system, and λ is the wavelength of the deformation. The initial configuration is sketched in Fig. 4.3.1.

Consider a row of vortices consisting of the kth vortex, the $(k \pm m)$th vortex, the $(k \pm 2m)$th vortex, and so forth, as shown in Fig. 4.3.1. From the discussions in Section 2.7, the complex potential of the flow induced by this

Helmboltz Instability of a Velocity Discontinuity

row of vortices is

$$w_k(z) = \sum_{n=-\infty}^{\infty} i\,\frac{2U\lambda}{2\pi m}\log{(z - z_k - n\lambda)}$$

$$= i\,\frac{U\lambda}{m\pi}\log\left[\sin\frac{\pi(z - z_k)}{\lambda}\right] \qquad (4.3.2)$$

in which i is the imaginary unit $\sqrt{-1}$, and $z(= x + iy)$ and $z_k(= x_k + iy_k)$ represent the coordinates of an arbitrary point and those of the kth vortex, respectively. The deduction from the first to the second expression for w_k, which is referred to in Art. 156 of Lamb (1932), greatly simplifies the analysis to that without performing the infinite summation. Thus the complex potential of the m rows of vortices that make up the entire sheet is

$$w(z) = \sum_{k=1}^{m} w_k(z) = \sum_{k=1}^{m} i\,\frac{U\lambda}{m\pi}\log\left[\sin\frac{\pi(z - z_k)}{\lambda}\right] \qquad (4.3.3)$$

Substituting this expression into (2.7.9) and separating the final result into real and imaginary parts after differentiation, we obtain the velocity components u and v at the point (x, y). Then, letting this point be in coincidence with the jth vortex, we have

$$\frac{dx_j}{dt} = \frac{U}{m}\sum_{k\neq j}^{m}\frac{\sinh{[2\pi(y_j - y_k)/\lambda]}}{\cosh{[2\pi(y_j - y_k)/\lambda]} - \cos{[2\pi(x_j - x_k)/\lambda]}} \qquad (4.3.4)$$

$$\frac{dy_j}{dt} = -\frac{U}{m}\sum_{k\neq j}^{m}\frac{\sin{[2\pi(x_j - x_k)/\lambda]}}{\cosh{[2\pi(y_j - y_k)/\lambda]} - \cos{[2\pi(x_j - x_k)/\lambda]}} \qquad (4.3.5)$$

After introducing the following dimensionless variables

$$X = \frac{x}{\lambda}, \qquad Y = \frac{y}{\lambda}, \qquad A = \frac{a}{\lambda}, \qquad T = \frac{t}{\lambda/U}$$

(4.3.1), (4.3.4), and (4.3.5) become

$$Y_j = A\sin{(2\pi X_j)} \qquad \text{at } T = 0 \qquad (4.3.6)$$

$$\frac{dX_j}{dT} = \frac{1}{m}\sum_{k\neq j}^{m}\frac{\sinh{[2\pi(Y_j - Y_k)]}}{\cosh{[2\pi(Y_j - Y_k)]} - \cos{[2\pi(X_j - X_k)]}} \qquad (4.3.7)$$

$$\frac{dY_j}{dT} = -\frac{1}{m}\sum_{k\neq j}^{m}\frac{\sin{[2\pi(X_j - X_k)]}}{\cosh{[2\pi(Y_j - Y_k)]} - \cos{[2\pi(X_j - X_k)]}} \qquad (4.3.8)$$

It can be shown either mathematically or physically that both velocity components vanish at all times at the nodal points where $j = 1$, $(m/2) + 1$, or $m + 1$, and that the form of the wave is always antisymmetric about a nodal point. Thus computation for the deformation of the vortex sheet is needed only for the range of j varying from 2 to $m/2$ within half a wavelength. After figuring out the antisymmetric geometry of the remaining half of the wave, we can obtain the configuration of the entire sheet by repeating this representative wave pattern for every wavelength along the x-axis.

Coordinates (X_j, Y_j) at various time instants are to be computed by integrating (4.3.7) and (4.3.8) again using the fourth-order Runge-Kutta method. With the right-hand sides of these two differential equations defined as functions U(XJ, YJ) and V(XJ, YJ), respectively, the subroutine RUNTTA, written for Program 4.2, is borrowed directly to carry out the required numerical integrations. The form of Program 4.3, constructed for the present problem, is very similar to that of the previous program. Two function subprograms SINH(Z) and COSH(Z) are added, respectively, for the evaluation of hyperbolic functions $\sinh z$ and $\cosh z$. The computation is performed based on $m = 40$, an initial amplitude $A = 0.05$, and a time step size of 0.002.

Without presenting the output of Program 4.3, we plot the result in Fig. 4.3.2 showing the sheet configuration at various time instants. During the initial period all discrete vortices within the range $0 < X < 1$ are moving away from the X-axis and in the meantime toward the midpoint at $X = 0.5$. This process continues until kinks have developed near the central portion of the wave when $T \sim 0.2$. The sheet then starts to roll up into clusters of vortices, as revealed by the succeeding plots. While moving toward the midpoint, these clusters draw more and more vortices into their expanding territories. Irregular sheet configurations are found in the later stages of the computation, similar to what was observed in the spiral region discussed in the preceding section. Hama and Burke (1960) reported that a much smoother rolling up could be achieved by starting with a row of unequally spaced vortices.

A test has been made with the time step size reduced to half the previous value. Except with some insignificant differences in the coordinates of certain cluster vortices, the result is practically the same as before. The present program takes about eight times more computing time than Program 4.2 does to finish the computation for one time step. The extra time is spent on the evaluation of the functions U and V, which are much more involved than the functions defined in Program 4.2.

Program 4.3 is run once more for $A = 0.1$ and $dT = 0.002$ to study the effect of increasing amplitude of the initial deformation on the evolution of the vortex sheet. The result, plotted in Fig. 4.3.3, reveals a similar behavior but a faster roll-up process than that for $A = 0.05$. Furthermore, irregularities are found to appear earlier in the cluster region of the vortex sheet. The pictures in Fig. 4.3.3 may also be visualized as the dimensionless plots of a vortex sheet subject to an initial deformation having the same amplitude as that shown in Fig. 4.3.2a but a wavelength only half as long.

Rolling Up of the Trailing Vortex Sheet Behind a Finite Wing

```
C                    ***** PROGRAM 4.3 *****

C           NONLINEAR INSTABILITY OF AN INFINITE VORTEX SHEET

      EXTERNAL U,V
      DIMENSION X(40),Y(40),XNEW(20),YNEW(20)
      COMMON J,M,TWOPI,X,Y

C     ..... SPECIFY INPUT DATA.  THE INITIAL AMPLITUDE OF THE WAVY
C           DEFORMATION IS FIVE PERCENT OF THE WAVELENGTH .....
      M = 40
      NMAX = 200
      NFREQ = 5
      M2 = M / 2
      M2P2 = M2 + 2
      A = 0.05
      PI = 4.0 * ATAN(1.0)
      TWOPI = 2.0 * PI

C     .....SPECIFY AND PRINT INITIAL CONDITIONS .....
      NDT = 0
      T   = 0.0
      DX = 1.0 / FLOAT(M)
      X(1) = 0.0
      Y(1) = 0.0
      DO 1  J = 2, M
      X(J) = X(J-1) + DX
    1 Y(J) = A * SIN(TWOPI*X(J))
      WRITE(6,2)  NDT, T, ( (X(J),Y(J)), J=1,M2 )
    2 FORMAT( //5X, 5HNDT =, I4, 1H,, 5X, 3HT =, F5.3/(5(5X,2F9.3)) )

C     ..... COMPUTE SHEET CONFIGURATION FOR ONE-HALF
C           OF A WAVELENGTH AT TIME DT LATER .....
      DT = 0.002
    3 NDT = NDT + 1
      IF( NDT.GT.NMAX )  GO TO 8
      T = T + DT
      DO 4  J = 2, M2
    4 CALL RUNTTA( X(J), Y(J), DT, U, V, XNEW(J), YNEW(J) )

C     ..... ASSIGN VALUES OF XNEW AND YNEW TO X AND Y, RESPECTIVELY ....
      DO 5  J = 2, M2
      X(J) = XNEW(J)
    5 Y(J) = YNEW(J)

C     ..... VORTICES ON THE OTHER HALF OF THE WAVE ARE LOCATED
C           FROM THE ANTISYMMETRIC GEOMETRY .....
      DO 6  J = M2P2, M
      X(J) = 1.0 - X(M+2-J)
    6 Y(J) = -Y(M+2-J)

C     ..... PRINT COORDINATES EVERY NFREQ TIME STEPS .....
      IF( NDT/NFREQ*NFREQ .EQ. NDT )  GO TO 7
      GO TO 3
    7 WRITE(6,2)  NDT, T, ( (X(J),Y(J)), J=1,M2 )
      GO TO 3
    8 STOP
      END
```

```
      SUBROUTINE RUNTTA( X, Y, H, F1, F2, XNEXT, YNEXT )
C     ..... BORROWED FROM PROGRAM 4.2 .....



      FUNCTION U( XJ, YJ )
C     ..... IT REPRESENTS THE RIGHT-HAND SIDE OF (4.3.7) .....
      DIMENSION X(40),Y(40)
      COMMON J,M,TWOPI,X,Y
      U = 0.0
      DO 1  K = 1, M
      IF( K.EQ.J )  GO TO 1
      XX = TWOPI * (XJ-X(K))
      YY = TWOPI * (YJ-Y(K))
      U = U + SINH(YY)/(COSH(YY)-COS(XX))
    1 CONTINUE
      U = U / FLOAT(M)
      RETURN
      END



      FUNCTION V( XJ, YJ )
C     ..... IT REPRESENTS THE RIGHT-HAND SIDE OF (4.3.8) .....
      DIMENSION X(40),Y(40)
      COMMON J,M,TWOPI,X,Y
      V = 0.0
      DO 1  K = 1, M
      IF( K.EQ.J )  GO TO 1
      XX = TWOPI * (XJ-X(K))
      YY = TWOPI * (YJ-Y(K))
      V = V - SIN(XX)/(COSH(YY)-COS(XX))
    1 CONTINUE
      V = V / FLOAT(M)
      RETURN
      END



      FUNCTION SINH( Z )
C     ..... HYPERBOLIC SINE OF Z .....
      SINH = (EXP(Z)-EXP(-Z)) / 2.0
      RETURN
      END



      FUNCTION COSH( Z )
C     ..... HYPERBOLIC COSINE OF Z .....
      COSH = (EXP(Z)+EXP(-Z)) / 2.0
      RETURN
      END
```

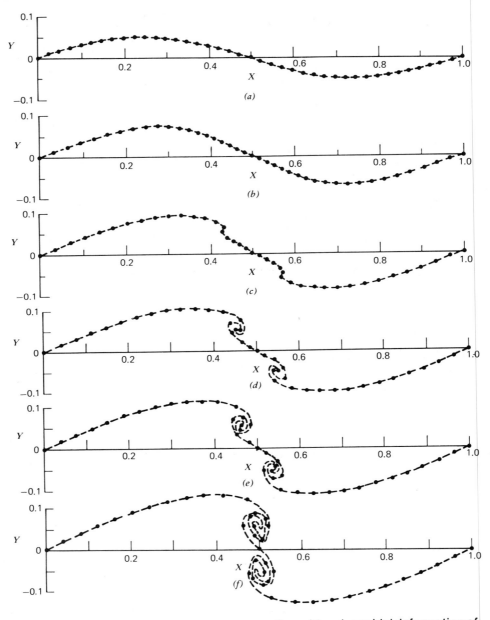

FIGURE 4.3.2 Evolution of a vortex sheet starting with a sinusoidal deformation of amplitude $A = 0.05$ (a) $T = 0$. (b) $T = 0.15$. (c) $T = 0.20$. (d) $T = 0.23$. (e) $T = 0.25$. (f) $T = 0.30$.

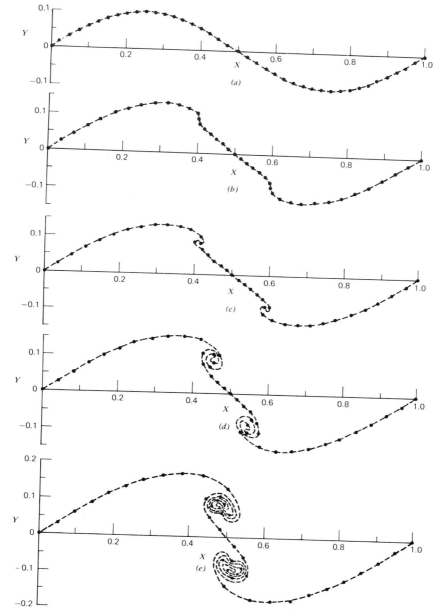

FIGURE 4.3.3 Evolution of a vortex sheet starting with a sinusoidal deformation of amplitude $A = 0.1$. (*a*) $T = 0$. (*b*) $T = 0.14$. (*c*) $T = 0.15$. (*d*) $T = 0.20$. (*e*) $T = 0.25$.

Problem 4.3 The formation of vortex streets in the wake of two-dimensional bluff bodies can be explained by considering the nonlinear interaction of two infinite vortex sheets, initially a fixed distance h apart, in an inviscid incompressible fluid. The vortex sheets are formed because of the two velocity discontinuities shown in Fig. 4.3.4. They may be approximated by two rows of discrete vortices. If m is the number of vortices in a length λ in each row, the circulation of a vortex in the upper row is $2U\lambda/m$ in the clockwise direction and that in the lower row is $-2U\lambda/m$ in the opposite direction. The stability of the vortex sheets is to be examined subject to an antisymmetric periodic disturbance of wavelength λ as sketched in Fig. 4.3.4.

Let (x_k, y_k) and (x'_k, y'_k) be, respectively, the coordinates of the kth vortex in the upper row and those of the kth vortex in the lower row at time t. If the kth vortex in the lower row is designated as the one that is one-half wavelength advanced from the kth vortex in the upper row, for this particular mode of disturbance we have

$$x'_k = x_k + \tfrac{1}{2}\lambda, \qquad k = 1, 2, \ldots, m \qquad (4.3.9)$$

$$y'_k = -y_k, \qquad k = 1, 2, \ldots, m \qquad (4.3.10)$$

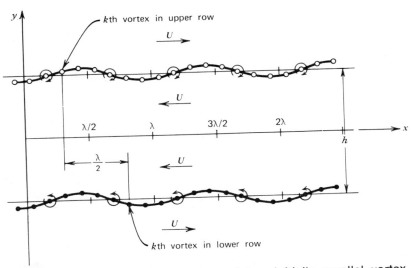

FIGURE 4.3.4 Periodic perturbation of two initially parallel vortex sheets.

Secondary Flows and Flow Instabilities

These relations enable us to simplify the analysis. The velocity induced at (x_j, y_j) has two contributions. The part caused by the vortices in the upper row is already described by the right-hand sides of (4.3.4) and (4.3.5); the other, caused by the vortices in the lower row, gives velocity components

$$-\frac{U}{m}\sum_{k=1}^{m}\frac{\sinh\,[2\pi(y_j + y_k)/\lambda]}{\cosh\,[2\pi(y_j + y_k)/\lambda] + \cos\,[2\pi(x_j - x_k)/\lambda]} \qquad (4.3.11)$$

$$-\frac{U}{m}\sum_{k=1}^{m}\frac{\sin\,[2\pi(x_j - x_k)/\lambda]}{\cosh\,[2\pi(y_j + y_k)/\lambda] + \cos\,[2\pi(x_j - x_k)/\lambda]} \qquad (4.3.12)$$

which are derived by changing the sign of U in (4.3.4) and (4.3.5), adding a prime to x_k and y_k, and then using the relations (4.3.9) and (4.3.10). Furthermore, in the absence of a deformation, these two flat vortex sheets will induce on each other a velocity $-U$, so that both will move along the negative x-axis at that speed. A term U is to be added to (4.3.11) if the velocity is measured in a coordinate system that is stationary with respect to the undeformed sheets.

Thus we obtain two simultaneous equations of the form

$$\frac{dx_j}{dt} = f_1(x_j, y_j), \qquad \frac{dy_j}{dt} = f_2(x_j, y_j)$$

The function f_1 is the sum of U, the right-hand side of (4.3.4) and (4.3.11), whereas f_2 is the sum of the right-hand side of (4.3.5) and (4.3.12).

Write a computer program for studying the nonlinear interaction of the two vortex sheets for various ratios of h to the wavelength λ.

The formulation shown here is based on that by Abernathy and Kronauer (1962). Their computed result shows that the strong interaction between the vortex sheets leads to the formation of concentrated clouds of vorticity, or the Kármán vortex streets, downstream in the wake of a bluff body.

4.4 Upwind Differencing and Artificial Viscosity

One of the main difficulties encountered in solving fluid dynamic problems is caused by the nonlinear substantial derivative that appears, for example, in the Euler equation (2.1.5), the Navier-Stokes equation (3.1.2), and the energy equation (3.1.3). Special care must be taken in approximating a substantial derivative by a finite difference expression. To prepare for the numerical computations of the next two chapters in which the nonlinear terms in the Navier-Stokes equation are retained, we derive here a numerical scheme suitable for handling such terms by considering a simplified one-dimensional model equation

$$\frac{\partial \zeta}{\partial t} + u \frac{\partial \zeta}{\partial x} = 0 \tag{4.4.1}$$

where ζ is the vorticity. This equation is obtained by taking the curl of the Euler equation (2.1.5) and then assuming a one-dimensional configuration. It describes the conservation of vorticity in an inviscid fluid moving parallel to the x-axis. For illustration let us assume the velocity u to be a positive constant.

To replace the differential equation by a finite-difference one, the time-space plane is subdivided into small meshes of size $h \times \tau$. Using the forward-differencing formula (2.2.6) for the time derivative and the backward-differencing formula (2.2.7) for the spatial derivative, (4.4.1) at grid point (x_i, t_j) is approximated by

$$\frac{\zeta_{i,j+1} - \zeta_{i,j}}{\tau} = -u \frac{\zeta_{i,j} - \zeta_{i-1,j}}{h} \tag{4.4.2}$$

having truncation errors of $O(\tau, h)$. Because of its special appearance [i.e., the spatial difference is on the upwind side of the point indexed (i, j)], the numerical scheme (4.4.2) is said to be in the *upwind differencing form*. (See Fig. 4.4.1.) Rearranging terms in (4.4.2) gives

$$\zeta_{i,j+1} = (1 - C)\zeta_{i,j} + C \zeta_{i-1,j} \tag{4.4.3}$$

in which $C = u\tau/h$ is the Courant number defined in (2.12.5). (4.4.3) enables us to compute vorticity at any time level based on the distribution of vorticity at the previous time step.

It is important to know whether the numerical scheme so constructed is

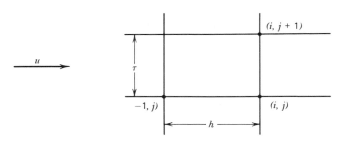

FIGURE 4.4.1 Upwind differencing scheme.

Secondary Flows and Flow Instabilities

computationally stable. Its behavior will be examined here using a method called the *discrete perturbation stability analysis*, which is conceptually different from von Neumann's stability analysis, introduced in Section 2.12. The method was first used by Thom and Apelt (1961), and it was later further developed by Thoman and Szewczyk (1966). To apply this method, a steady-state solution that $\zeta_{i,j} = 0$ for all i is assumed at time level t_j. If a disturbance ϵ is introduced at x_m, its influence at the next time step will be shown at two points located at x_m and x_{m+1}. The disturbances there are computed according to (4.4.3), and have the following form.

$$\zeta_{m,j+1} = (1 - C)\epsilon \qquad (4.4.4)$$

$$\zeta_{m+1,j+1} = C\epsilon \qquad (4.4.5)$$

For stability is is required that both

$$|\zeta_{m,j+1}/\epsilon| \leqslant 1 \quad \text{and} \quad |\zeta_{m+1,j+1}/\epsilon| \leqslant 1 \qquad (4.4.6)$$

so that these disturbances will not grow in time. From the first of (4.4.6) we have

$$-1 \leqslant (1 - C) \leqslant +1 \qquad (4.4.7)$$

The inequality on the right-hand side is automatically satisfied by the upwind differencing scheme in which C is positive. To satisfy the left-hand inequality, a restriction on the magnitude of C is obtained that

$$C \leqslant 2 \qquad (4.4.8)$$

However, the second part of (4.4.6) puts a more restrictive condition on C that

$$C \leqslant 1 \qquad (4.4.9)$$

Thus, for a positive C satisfying (4.4.9), the computation at the first time step is stable. In a similar manner, the disturbances at the second time step can be calculated. They are

$$\zeta_{m,j+2} = (1 - C)^2\epsilon \qquad (4.4.10)$$

$$\zeta_{m+1,j+2} = 2C(1 - C)\epsilon \qquad (4.4.11)$$

$$\zeta_{m+2,j+2} = C^2\epsilon \qquad (4.4.12)$$

It can easily be shown that the previous requirements for C also insure the computational stability at the second time step, and that the amplitudes of disturbances will decrease with increasing time steps if the perturbations at succeeding time steps are explicitly written out. Therefore the condition that

$$0 < C \leqslant 1 \qquad (4.4.13)$$

is the stability criterion for the numerical scheme (4.4.3).

Upwind Differencing and Artificial Viscosity

Problem 4.4 Show that the same stability criterion (4.4.13) results if von Neumann's stability analysis is used.

If the spatial derivative in (4.4.1) were approximated by the forward (or downwind) differencing form, a finite-difference equation would be obtained having the same appearance as that of (4.4.3), but with a negative value of C. Such a numerical scheme can never satisfy one of the stability requirements, as shown on the right side of the inequality (4.4.7). Thus a downwind differencing scheme is numerically unstable.

One of the salient features of the upwind differencing scheme is that it possesses the *transportive property* that a perturbation is advected only in the direction of the fluid motion. Suppose at x_m a disturbance ϵ is introduced at time t_j, when the trivial solution that $\zeta_{i,j} = 0$ for all i is assumed. The change of vorticity at x_m during a time period τ is, according to (4.4.2),

$$\zeta_{m,j+1} - \zeta_{m,j} = -\frac{u\tau}{h}(\epsilon - 0) = -\frac{u\tau}{h}\epsilon \qquad (4.4.14)$$

The change at x_{m+1}, which is a point one grid downstream from x_m, during the same period is calculated also from (4.4.2) as

$$\zeta_{m+1,j+1} - \zeta_{m+1,j} = -\frac{u\tau}{h}(0 - \epsilon) = +\frac{u\tau}{h}\epsilon \qquad (4.4.15)$$

The result shows that the perturbation is transported, as it should be, from one point to another in the downstream direction. It is also concluded that the numerical solution to the differential equation (4.4.1) based on (4.4.3) is exact if $u\tau/h$, or C, is chosen to be unity.

Problem 4.5 Using an approximation with forward differencing in time and central differencing in space, (4.4.1) is replaced by the finite-difference equation

$$\zeta_{i,j+1} = \zeta_{i,j} - C(\zeta_{i+1,j} - \zeta_{i-1,j}) \qquad (4.4.16)$$

Use the discrete perturbation stability analysis to show that stability of this numerical scheme can be achieved only when $C \to 0$. Show also that (4.4.16) does not possess the transportive property.

The upwind differencing scheme (4.4.2) is now reexamined from a different point of view. Expanding its first and last terms in the form of a Taylor series about the value of $\zeta_{i,j}$, we obtain

$$\zeta_{i,j+1} = \zeta_{i,j} + \tau\left(\frac{\partial\zeta}{\partial t}\right)_{i,j} + \tfrac{1}{2}\tau^2\left(\frac{\partial^2\zeta}{\partial t^2}\right)_{i,j} + O(\tau^3) \qquad (4.4.17)$$

Secondary Flows and Flow Instabilities

$$\zeta_{i-1,j} = \zeta_{i,j} - h\left(\frac{\partial \zeta}{\partial x}\right)_{i,j} + \tfrac{1}{2}h^2\left(\frac{\partial^2 \zeta}{\partial x^2}\right)_{i,j} + O(h^3) \tag{4.4.18}$$

Substituting into (4.4.2) and dropping the subscripts i and j gives

$$\frac{\partial \zeta}{\partial t} + \frac{\tau}{2}\frac{\partial^2 \zeta}{\partial t^2} + O(\tau^2) = -u\frac{\partial \zeta}{\partial x} + \tfrac{1}{2}uh\frac{\partial^2 \zeta}{\partial x^2} + O(h^2) \tag{4.4.19}$$

The second term on the left side can be evaluated for constant u from (4.4.1) as

$$\frac{\partial^2 \zeta}{\partial t^2} = \frac{\partial}{\partial t}\left(-u\frac{\partial \zeta}{\partial x}\right) = -u\frac{\partial}{\partial x}\left(\frac{\partial \zeta}{\partial t}\right) = u^2\frac{\partial^2 \zeta}{\partial x^2} \tag{4.4.20}$$

Upon substitution of this expression in (4.4.19) and with terms up to the first order in τ and h retained, the differential equation

$$\frac{\partial \zeta}{\partial t} + u\frac{\partial \zeta}{\partial x} = \mu_e\frac{\partial^2 \zeta}{\partial x^2} \tag{4.4.21}$$

is obtained in which

$$\mu_e = \tfrac{1}{2}uh(1 - C) \tag{4.4.22}$$

is a constant determined by the grid size τ and h. Because the term on the right-hand side of (4.4.21) is caused by the errors inherited when (4.4.1) is replaced by (4.4.2), and because this term is analogous to the diffusive viscous force term in the Navier-Stokes equation (3.1.7), the coefficient μ_e is called the *artificial viscosity*. Its effect is to introduce artificial damping and diffusion in the numerical solution. μ_e vanishes when either $C = 1$ or both τ and h approach zero. Under that condition (4.4.21) reverts to the original differential equation (4.4.1).

Hirt (1968) argued that the effect of a diffusion term is to smear out a disturbance in ζ so that a negative diffusion coefficient is physically impossible, or otherwise the opposite would happen. The requirement that $\mu_e \geqslant 0$ results in $C \leqslant 1$, which is the stability criterion (4.4.13) derived previously for the upwind numerical scheme (4.4.3). The present procedure for determining the stability, based on a differential equation constructed from the finite-difference equation through Taylor series expansion, is called *Hirt's stability analysis*.

Problem 4.6 Applying Hirt's stability analysis to the explicit formula (3.6.3) for solving hyperbolic equation (3.6.1), show that the resulting differential equation is

$$\frac{\partial u}{\partial t} = \nu\frac{\partial^2 u}{\partial y^2} + \tfrac{1}{2}\nu h^2(\tfrac{1}{6} - R)\frac{\partial^4 u}{\partial y^4} \tag{4.4.23}$$

by retaining terms up to $O(\tau, h^2)$, in which $R = \nu\tau/h^2$. Positive terms of even derivatives, like the second-order derivative term, cause damping. Thus the stability condition $R \leqslant \frac{1}{6}$ so obtained is more restrictive than the condition $R \leqslant \frac{1}{2}$ obtained in Section 3.6 using von Neumann's stability analysis.

Show also that when Hirt's stability analysis is applied to the implicit formula (3.7.1), the same conclusion is deduced that the numerical scheme is stable for any positive value of R.

Problem 4.7 Show that by using the discrete perturbation stability analysis, the condition $R \leqslant \frac{1}{2}$ is obtained for (3.6.3) in agreement with von Neumann's method, but no conclusion can be reached for the stability of the implicit formula (3.7.1).

The results stated in Problems 4.6 and 4.7 indicate that a stability criterion deduced from the discrete perturbation stability analysis or from Hirt's stability analysis may or may not agree with that deduced from von Neumann's analysis. The reason for the disagreement is that the condition for computational stability in each analysis is derived based on a different physical argument. Both von Neumann's and the discrete perturbation analyses require that the amplitude of a disturbance should not grow with time. But the former is concerned with a distribution of disturbances that can be synthesized by a Fourier series, whereas the latter traces only the influence of a point perturbation. Hirt's stability condition is based on the fact that a diffusion coefficient cannot be negative. Despite the fact that von Neumann's stability analysis is most commonly used, each method has its own advantages as well as limitations. A comparison and evaluation of these three methods can be found in Chapter 3, Section A-5, of Roache (1972).

By using a model equation (4.4.1), we have shown that the substantial derivative may be approximated by the upwind differencing scheme, which is computationally stable if the ratio τ/h is chosen appropriate for the local fluid velocity. Furthermore, the upwind differencing scheme introduces, in addition to the truncation errors, an artificial diffusion whose effect is to smear out perturbations in the numerical solution. For thorough referencing and a more detailed discussion on the subject and its related topics, Sections A-8 to A-11 in Chapter 3 of Roache (1972) are recommended.

4.5 Secondary Flows Caused by a Rotational Electromagnetic Force Field

The development of secondary flows is a salient feature of a fluid flow in the presence of rotational force fields. Examples of rotational forces are the shear force in a viscous fluid, the electromagnetic force in an electrically conducting fluid, the pressure force in a stratified fluid, and the Coriolis force experienced by a flow in rotation. An example of the secondary flows caused by the viscous force in the wake of a spherical body has already been given in Section 4.1, and the one to be studied in this section concerns the effect of the electromagnetic force. The numerical techniques learned from these two examples can easily be extended to solve the stratified or rotating flow problems. When a secondary flow occurs, the perturbed fluid motion has a magnitude of the same order as that of the primary flow. Therefore, the secondary flow is a nonlinear phenomenon, which is contained only in the solution of the governing equations whose nonlinear terms are retained.

When an electric current is passed through a conducting fluid, a magnetic field is generated. In the laboratory under a steady state, even if the fluid medium is in motion, the magnetic field $\mathbf{H}$ and current density $\mathbf{J}$ are governed approximately by the following equations.

$$\mathbf{J} = \nabla \times \mathbf{H} \qquad (4.5.1)$$

$$\nabla \times \nabla \times \mathbf{H} = 0 \qquad (4.5.2)$$

The governing equations cannot be independent of the fluid velocity $\mathbf{V}$ in problems of astronomical scales. See Section 1.2 of Cowling (1957) for details. In the presence of a current and a magnetic field, the conducting fluid is subject to an electromagnetic body force per unit volume of the form

$$\mathbf{f} = \mathbf{J} \times \mu_m \mathbf{H} \qquad (4.5.3)$$

where μ_m is the magnetic permeability of the fluid medium. Unless the current is uniform or is in some simple geometry, such as a two-dimensional configuration, this body force is in general rotational; that is, $\nabla \times \mathbf{f} \neq 0$. This rotational force causes many interesting phenomena in a conducting flow that are not commonly found in the ordinary fluid flows.

The equation of motion for an incompressible conducting fluid is obtained by simply adding the electromagnetic force term, expressed by (4.5.3), to the right-hand side of the Navier-Stokes equation (3.1.7). Rewriting the equation in the form of (4.1.2) with

$$\nabla \times \mathbf{V} = \zeta \qquad (4.5.4)$$

being the vorticity, we have

$$\frac{\partial \mathbf{V}}{\partial t} - \mathbf{V} \times \zeta = -\nabla\left(\frac{p}{\rho} + \tfrac{1}{2}V^2\right) - \nu\nabla \times \zeta + \frac{\mu_m}{\rho}\mathbf{J} \times \mathbf{H}$$

in which $\nu = \mu/\rho$ is the kinematic viscosity. Taking the curl of this vector equation and substituting $\mathbf{J}$ from (4.5.1) yields

$$\frac{\partial \zeta}{\partial t} - \nabla \times (\mathbf{V} \times \zeta) = -\nu\nabla \times \nabla \times \zeta + \frac{\mu_m}{\rho}\nabla \times [(\nabla \times \mathbf{H}) \times \mathbf{H}] \qquad (4.5.5)$$

Secondary Flows Caused by a Rotational Electromagnetic Force Field

337

This equation describes the transportation and production of vorticity in a conducting viscous incompressible fluid. The continuity equation for such a fluid remains the same form as that for a nonconducting fluid.

$$\nabla \cdot \mathbf{V} = 0 \qquad (4.5.6)$$

The procedure for solving a magnetohydrodynamic problem in the laboratory scale is outlined as follows. The steady electromagnetic fields are first determined from (4.5.1) and (4.5.2) utilizing their appropriate boundary conditions specified by the problem. Upon substitution of $\mathbf{H}$ into the last term of (4.5.5), the effect of the electromagnetic force is computed as a known function of spatial coordinates. The solution $\mathbf{V}$ to the equation (4.5.5), satisfying the specified fluid boundary conditions, is then sought under the constraint (4.5.6).

As a specific example let us compute the motion of a conducting fluid flowing through a circular tube of radius r_0 in the presence of a converging-diverging current flow. Such a current distribution is obtained by applying potential differences across a small screen electrode of radius r_1 and two large screen electrodes covering the ends of the nonconducting tube. The electrodes, through which the fluid can freely move, are separated by the same distance z_0, as shown in Fig. 4.5.1. In the presence of a conducting fluid, electric current will flow along the tube to form a constriction at the central electrode. When it interacts with the associated magnetic field in the azimuthal direction, this current produces a rotational force field that will modify the originally uniform motion of the fluid. In addition, the electromagnetic force, $\mathbf{J} \times \mu_m \mathbf{H}$, has a component pointing toward the tube axis that causes a "*pinch effect*" to build up the pressure near the axis. An analysis of the magnetohydrostatic pinch due to a uniform current can be found in Section 6.13 of Hughes and Young (1966). In our problem the pinch is nonuniform; it has a stronger effect in the neighborhood of the central electrode. Thus the fluid is moving through a pressure hump created electromagnetically. The situation is somewhat analogous to the situation in which a plasma is blowing toward an electric arc.

Because of the axisymmetric geometry of the flow, the stream function introduced in Section 2.6 can again be used, so that the continuity equation is satisfied and the axial and radial velocity components are expressed as

$$u_z = \frac{1}{r}\frac{\partial \psi}{\partial r}, \qquad u_r = -\frac{1}{r}\frac{\partial \psi}{\partial z} \qquad (4.5.7)$$

The magnetic field has only one component, h_θ, in the azimuthal direction if the current flow is axisymmetric.

Let $\pi r_0^2 u_0$ and $\pi r_0^2 J_0$ be the volume flow rate and the total current, respectively, passing through the tube. By using r_0, u_0, and J_0 as the reference

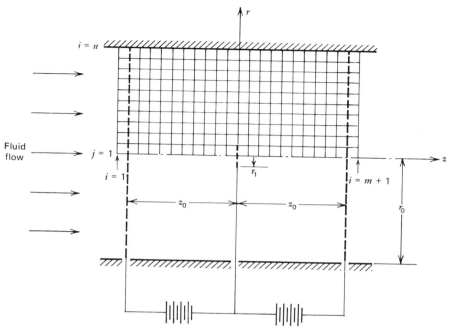

FIGURE 4.5.1 Flow through a nonconducting tube in the presence of a converging-diverging electric current field.

quantities, the following dimensionless variables can be introduced.

$$R = \frac{r}{r_0}, \qquad Z = \frac{z}{r_0}, \qquad U = \frac{u_z}{u_0}, \qquad V = \frac{u_r}{u_0}$$

$$T = \frac{t}{r_0/u_0}, \qquad \Psi = \frac{\psi}{r_0^2 u_0}, \qquad \Omega = \frac{-\zeta}{u_0/r_0}, \qquad H = \frac{h_\theta}{r_0 J_0} \qquad (4.5.8)$$

In terms of these new variables, the governing equations for an axisymmetric configuration become

$$\left(\frac{\partial^2}{\partial R^2} - \frac{1}{R}\frac{\partial}{\partial R} + \frac{\partial^2}{\partial Z^2}\right)RH = 0 \qquad (4.5.9)$$

$$U = \frac{1}{R}\frac{\partial \Psi}{\partial R} \qquad (4.5.10)$$

$$V = -\frac{1}{R}\frac{\partial \Psi}{\partial Z} \qquad (4.5.11)$$

$$\left(\frac{\partial^2}{\partial R^2} - \frac{1}{R}\frac{\partial}{\partial R} + \frac{\partial^2}{\partial Z^2}\right)\Psi = R\Omega \qquad (4.5.12)$$

Secondary Flows Caused by a Rotational Electromagnetic Force Field **339**

$$\frac{\partial \Omega}{\partial T} = -\frac{\partial(U\Omega)}{\partial Z} - \frac{\partial(V\Omega)}{\partial R} + C\frac{H}{R}\frac{\partial H}{\partial Z}$$

$$+ \frac{1}{Re}\left(\frac{\partial^2}{\partial R^2} + \frac{1}{R}\frac{\partial}{\partial R} - \frac{1}{R^2} + \frac{\partial^2}{\partial Z^2}\right)\Omega \tag{4.5.13}$$

The two dimensionless parameters in the last equation are $C(\equiv \mu_m r_0^2 J_0^2/\frac{1}{2}\rho u_0^2)$, the magnetic pressure number, and $Re(\equiv u_0 r_0/\nu)$, the Reynolds number of the flow. A comparison between (4.5.9) and (4.5.12) reveals that the combined variable RH is analogous to the stream function Ψ of an irrotational flow, so that the constant-RH curves are the analog of the streamlines for the current flow.

Simple boundary conditions are chosen for our analysis. For example, instead of spending time on computing the growth of the boundary layer along the tube wall, which is a phenomenon not essential to the present problem, a slip boundary condition is assumed there that states that the wall coincides with the outermost streamline and that along this particular streamline vorticity vanishes. The boundary conditions for the electromagnetic field are that the current must be tangent to the nonconducting tube wall and the radial component of the current flow vanishes at both the end and central electrodes.

Since the electromagnetic field is symmetric about the central electrode at $z = 0$, we need only to find the solution in the region $-z_0 \leqslant z \leqslant 0$ to the left of the central electrode. The mirror image of this solution gives that in the region to the right. The boundary conditions in terms of RH are listed below.

$$R = 0 \quad \text{and} \quad -\frac{z_0}{r_0} \leqslant Z \leqslant 0: \quad RH = 0 \tag{4.5.14}$$

$$R = 1 \quad \text{and} \quad -\frac{z_0}{r_0} \leqslant Z \leqslant 0: \quad RH = \tfrac{1}{2} \tag{4.5.15}$$

$$0 < R < \frac{r_1}{r_0} \quad \text{and} \quad Z = 0: \quad \frac{\partial RH}{\partial Z} = 0 \tag{4.5.16}$$

$$\frac{r_1}{r_0} \leqslant R < 1 \quad \text{and} \quad Z = 0: \quad RH = \tfrac{1}{2} \tag{4.5.17}$$

$$0 < R < 1 \quad \text{and} \quad Z = -\frac{z_0}{r_0}: \quad \frac{\partial RH}{\partial Z} = 0 \tag{4.5.18}$$

When there is a net flow through the tube, the flow pattern, unlike the electromagnetic field, is not symmetric about the central electrode, so the whole region between the two end electrodes must be considered. By assuming that the flow is purely axial when passing through the end screen electrodes, the fluid boundary conditions are as follows.

$$R = 0 \quad \text{and} \quad -\frac{z_0}{r_0} \leqslant Z \leqslant \frac{z_0}{r_0}: \quad \Psi = \Omega = 0 \tag{4.5.19}$$

$$R = 1 \quad \text{and} \quad -\frac{z_0}{r_0} \leqslant Z \leqslant \frac{z_0}{r_0}: \quad \Psi = \tfrac{1}{2}, \Omega = 0 \tag{4.5.20}$$

$$0 < R < 1 \quad \text{and} \quad Z = \pm\frac{z_0}{r_0}: \quad \frac{\partial \Psi}{\partial Z} = \frac{\partial \Omega}{\partial Z} = 0 \tag{4.5.21}$$

The system of equations, including the nonlinear equation (4.5.13), is to be solved numerically for flows whose Reynolds numbers are not small. These problems generally cannot be solved analytically by using linearization techniques. The solution will be obtained at a finite number of grid points having coordinates $Z_i = (i-2)\Delta Z - z_0/r_0$, $R_j = (j-1)\Delta R$ and at discrete time steps separated by constant intervals ΔT. In the present work ΔZ and ΔR are chosen to have the same value h; this is not, however, generally necessary.

To replace the governing differential equations by finite-difference equations, evaluated at (Z_i, R_j), we approximate all linear spatial derivatives by using the central-difference formulas (2.2.8) and (2.2.9). Thus, the first four equations (4.5.9) to (4.5.12) become

$$(RH)_{i,j} = \tfrac{1}{4}[(RH)_{i+1,j} + (RH)_{i-1,j} + \alpha_j(RH)_{i,j+1} + \beta_j(RH)_{i,j-1}] \tag{4.5.22}$$

$$U_{i,j} = \frac{\Psi_{i,j+1} - \Psi_{i,j-1}}{2hR_j} \tag{4.5.23}$$

$$V_{i,j} = -\frac{\Psi_{i+1,j} - \Psi_{i-1,j}}{2hR_j} \tag{4.5.24}$$

$$\Psi'_{i,j} = \tfrac{1}{4}(\Psi'_{i+1,j} + \Psi'_{i-1,j} + \alpha_j\Psi'_{i,j+1} + \beta_j\Psi'_{i,j-1} - h^2R_j\Omega'_{i,j}) \tag{4.5.25}$$

where

$$\alpha_j = 1 - \frac{h}{2R_j} = \frac{j-1.5}{j-1}$$

$$\beta_j = 1 + \frac{h}{2R_j} = \frac{j-0.5}{j-1}$$

(4.5.22) is in the same form as (2.11.14) derived previously. The (i, j) subscripts denote quantities evaluated at the grid point (Z_i, R_j), while the notation $(RH)_{i,j}$ is specifically used to replace $R_jH_{i,j}$. The unprimed and primed flow variables represent, respectively, those at two consecutive time levels T and $T + \Delta T$. Along the tube axis the velocity is calculated according to the formula

$$U_{i,1} = \frac{2\Psi_{i,2}}{h^2} \tag{4.5.26}$$

Secondary Flows Caused by a Rotational Electromagnetic Force Field **341**

This relation is derived from (4.5.10) by assuming constant velocity within a small radial distance h from the axis.

Care must be taken in approximating the equation of motion (4.5.13). The first two nonlinear terms on the right-hand side are similar to the substantial derivative in (4.4.1); they will be replaced by the upwind-difference scheme considered in the previous section. That scheme must be modified, however, to accommodate the arbitrary local flow direction, which may vary from one grid point to another. We adopt here a scheme developed by Torrance (1968) that has been successfully used in solving natural convection (Torrance and Rockett, 1969) and rotating flow (Kopecky and Torrance, 1973) problems.

We first define U_f and U_b as the average axial velocities evaluated, respectively, at half a grid forward and backward from the point (Z_i, R_j) in the Z direction, or,

$$U_f = \tfrac{1}{2}(U_{i+1,j} + U_{i,j})$$
$$U_b = \tfrac{1}{2}(U_{i,j} + U_{i-1,j})$$

(4.5.27)

Similarly, we define

$$V_f = \tfrac{1}{2}(V_{i,j+1} + V_{i,j})$$
$$V_b = \tfrac{1}{2}(V_{i,j} + V_{i,j-1})$$

(4.5.28)

as the radial velocities averaged at half a grid forward and backward from that point in the R direction, respectively. It can easily be verified that the upwind differencing form is automatically preserved when the following numerical formulas are used.

$$\left[\frac{\partial(U\Omega)}{\partial Z}\right]_{i,j} = \frac{1}{2h}\left[(U_f - |U_f|)\Omega_{i+1,j}\right.$$
$$+ (U_f + |U_f| - U_b + |U_b|)\Omega_{i,j}$$
$$\left.- (U_b + |U_b|)\Omega_{i-1,j}\right]$$

(4.5.29)

$$\left[\frac{\partial(V\Omega)}{\partial R}\right]_{i,j} = \frac{1}{2h}\left[(V_f - |V_f|)\Omega_{i,j+1}\right.$$
$$+ (V_f + |V_f| - V_b + |V_b|)\Omega_{i,j}$$
$$\left.- (V_b + |V_b|)\Omega_{i,j-1}\right]$$

(4.5.30)

The remaining terms in (4.5.13) are approximated by using forward differencing in time and central differencing in space. Their individual expressions follow.

$$\left(\frac{\partial\Omega}{\partial T}\right)_{i,j} = \frac{\Omega'_{i,j} - \Omega_{i,j}}{\Delta T}$$

(4.5.31)

$$\left(\frac{H}{R}\frac{\partial H}{\partial Z}\right)_{i,j} = \left(\frac{RH}{R^3}\frac{\partial RH}{\partial Z}\right)_{i,j} = (RH)_{i,j}\frac{(RH)_{i+1,j} - (RH)_{i-1,j}}{2hR_j^3}$$

(4.5.32)

$$\left(\frac{\partial^2 \Omega}{\partial R^2} + \frac{1}{R}\frac{\partial \Omega}{\partial R} - \frac{\Omega}{R^2} + \frac{\partial^2 \Omega}{\partial Z^2}\right)_{i,j} = \frac{1}{2hR_j}(\Omega_{i,j+1} - \Omega_{i,j-1})$$

$$+ \frac{1}{h^2}\left[\Omega_{i,j+1} + \Omega_{i,j-1} + \Omega_{i+1,j} + \Omega_{i-1,j} - \left(4 + \frac{h^2}{R_j^2}\right)\Omega_{i,j}\right] \qquad (4.5.33)$$

Substitution of (4.5.29) to (4.5.33) into (4.5.13) results in an equation that enables us to compute Ω' at $T + \Delta T$ based on the dimensionless vorticity distribution Ω at the previous time step T.

In this way the fluid equations governing an unsteady flow have been formulated for numerical solution. To find the steady-state solution for the tube flow passing through a current constriction, a uniform axial fluid flow is first assumed. The finite-difference equations are then used repeatedly to compute the flow conditions at increasing time steps, until the difference between the solutions at two consecutive steps becomes negligibly small. When this happens the result is considered to be the steady-state solution for which we are looking. Thus we replace the original problem by an unsteady flow problem concerning the development of a uniform flow after a converging-diverging current field is suddenly turned on. The marching procedure described here is a powerful technique that is quite commonly employed in fluid dynamics to find the steady-state solution.

However, numerical instability may occur in the process of approaching the steady-state solution if the size of the time increment is not properly chosen. For a given grid size h, the constraints on ΔT are to be derived from a quasilinear analysis of Lax and Richtmyer (1956) for linear equations instead of from Hirt's stability analysis. The latter is not adequate for the present case. As a first step, the difference representation of (4.5.13) is written in the form

$$\Omega'_{i,j} = a_1 \Omega_{i+1,j} + a_2 \Omega_{i-1,j} + a_3 \Omega_{i,j} + a_4 \Omega_{i,j+1} + a_5 \Omega_{i,j-1} + b \qquad (4.5.34)$$

in which the coefficients a_k, expressed in terms of velocity components at time T, are considered constant over a period ΔT in the computation, and b is a constant determined by the magnetic field. According to Lax and Richtmyer, the numerical scheme is stable if the coefficients a_k are all positive. It can be shown that this requirement is automatically satisfied for all the coefficients except a_3. To require that a_3 be greater than or equal to zero, a condition is derived that specifies the upper limit for the time increment.

$$\Delta T \leq \left[\frac{1}{2h}(U_f + |U_f| - U_b + |U_b| + V_f + |V_f| - V_b + |V_b|)\right.$$

$$\left. + \frac{1}{Re\,h^2}\left(4 + \frac{h^2}{R_j^2}\right)\right]^{-1} \qquad (4.5.35)$$

Although the current density does not appear explicitly in this relation, its effect is shown indirectly through the velocities U_f, U_b, V_f, and V_b defined in (4.5.27) and (4.5.28). In general, a higher current density causes higher velocities in the neighborhood of the central electrode and, therefore, needs a smaller ΔT in the computation.

The procedure as adopted in Program 4.4 for solving this problem is outlined in the following steps.

1. Values of the magnetic pressure number C and the Reynolds number Re are assigned according to the specification of the problem. In this case we assume $C = 0.3$ and $Re = 50$, and let the tube configuration be described by $z_0/r_0 = 1$ and $r_1/r_0 = 0.1$. By choosing $h = 0.1$ for the dimensionless grid size, there will be 21 vertical and 11 horizontal grid lines in the upper half of the fluid domain bounded by the end electrodes. Two additional vertical grid lines are needed outside this domain to accommodate the derivative boundary conditions (4.5.18) and (4.5.21). If indices i and j are named according to the grid system laid out in Fig. 4.5.1, we have the values $m = 22$ and $n = 11$. The coordinates of all grid points as well as the coefficients α and β, which appeared in (4.5.22) and (4.5.25), are then computed. The time step size is chosen to be $\Delta T = 0.01$.

2. As mentioned earlier the constant-RH curves represent the streamlines of the current flow, so that it is more convenient to compute RH instead of the magnetic field itself. For finding the solution to (4.5.9) satisfying boundary conditions (4.5.14) to (4.5.18), the subroutine LAPLCE is constructed. This subroutine is similar to the subroutine LIEBMN defined in Program 3.7, except that in the present case the boundary conditions are specified within the subprogram instead of in the main program. Because of the symmetry of the magnetic field about the central electrode, the solution for RH in the left half of the tube is first computed by using the iterative formula (4.5.22) and is then appropriately transmitted to the right half of the tube. Similar to LIEBMN, a small positive quantity ERRMAX is used here to control the accuracy of the solution. An integer variable ITER is inserted as an argument of LAPLCE, whose numerical value when returned to the main program represents the number of iterations that have been performed to obtain the solution of desired accuracy.

3. Based on the values of RH, the term $(CH/R)\,\partial H/\partial Z$ caused by the electromagnetic force is computed at all interior grid points according to (4.5.32). These values, stored in the two-dimensional array EM, do not vary with time. The points on several current streamlines whose RH values are specified in the array PSIA are located by calling the subroutine SEARCH, borrowed from Program 2.2. Their coordinates together with the RH and EM distributions are printed in the output.

4. Having determined the electromagnetic effect, we can now proceed to compute the flow variables. In Program 4.4 an integer variable **NSTEP** is introduced as the time step counter. The computation will be terminated when **NSTEP** exceeds a specified number **MAXSTP** no matter whether the steady state is reached or not. Thus, if an error were made in the program that cause the solution not to converge to a steady state, this limitation would prevent an excessive waste of the computer time.

At the initial instant, $\text{NSTEP} = 0$, a uniform flow of unit dimensionless speed is assumed whose stream function is $\frac{1}{2}R^2$, obtained by integrating (4.5.10). The initial vorticity is zero everywhere in such a flow. From the known stream function, the velocity components at all interior grid points are computed using (4.5.23) and (4.5.24), and the velocity along the tube axis, which is the lower boundary of the grid system, is obtained using (4.5.26). The velocities at the left and right ends of the grid system where $i = 1$ and $m + 1$, respectively, can be deduced from the boundary conditions (4.5.21). The first condition states that

$$V_{2,j} = V_{m,j} = 0 \qquad \text{for } j = 2, 3, \ldots, n - 1 \qquad (4.5.36)$$

Keeping that condition in terms of Ψ and differentiating it with respect to R, we have, for the same range of j,

$$\left(\frac{\partial U}{\partial Z}\right)_{2,j} = \left(\frac{\partial U}{\partial Z}\right)_{m,j} = 0 \qquad (4.5.37)$$

The second condition in (4.5.21) can be rewritten in terms of velocity components. Substitution of Ω from (4.5.12) with use of the conditions (4.5.36) yields

$$\left(\frac{\partial^2 V}{\partial Z^2}\right)_{2,j} = \left(\frac{\partial^2 V}{\partial Z^2}\right)_{m,j} = 0 \qquad (4.5.38)$$

After the derivatives are replaced by central difference approximations (2.2.8) and (2.2.9), these two velocity boundary conditions just derived become, for $j = 2, 3, \ldots, n - 1$,

$$U_{1,j} = U_{3,j}, \qquad U_{m+1,j} = U_{m-1,j} \qquad (4.5.39)$$

$$V_{1,j} = -V_{3,j}, \qquad V_{m+1,j} = -V_{m-1,j} \qquad (4.5.40)$$

To compute the velocity along the tube wall, we use the boundary conditions (4.5.20). When expressed in terms of velocity components, they state that $V = 0$ and $\partial U/\partial R = 0$ at the wall. It can easily be shown that with a truncation error $O(h^2)$ the latter can be approximated by

$$\frac{-3U_{i,n} + 4U_{i,n-1} - U_{i,n-2}}{2h} = 0$$

from which the axial velocity at the wall is computed.

Secondary Flows Caused by a Rotational Electromagnetic Force Field **345**

$$U_{i,n} = \frac{4U_{i,n-1} - U_{i,n-2}}{3} \tag{4.5.41}$$

The one-sided finite-difference approximation is used for $\partial U/\partial R$ because there are no grid points beyond the wall.

5. Based on the present vorticity and velocity distributions, the vorticity distribution at time ΔT later is computed from

$$\Omega'_{i,j} = \Omega_{i,j} + \Delta T \text{ [right-hand side of (4.5.13)]} \tag{4.5.42}$$

where the indices (i, j) apply to any interior grid point and the terms contained within the brackets are computed according to (4.5.27) to (4.5.33). In this part of the computation there is a frequently used parameter $\Delta T/(2h)$. Its value is to be calculated early in Program 4.4 and is then assigned to the variable name CONST.

The boundary values of the vorticity are assigned in the following way. Along the tube axis and the wall, vorticity always remains the constant value of zero. But on the upstream and downstream sides it must be updated according to the second condition in (4.5.21) that $\partial\Omega/\partial Z = 0$ at both electrodes, or

$$\Omega_{1,j} = \Omega_{3,j}, \qquad \Omega_{m+1,j} = \Omega_{m-1,j} \tag{4.5.43}$$

At each time level the differences $(\Omega'_{i,j} - \Omega_{i,j})$ are evaluated at all interior grid points. The sum of their absolute values is assigned to the name SUM, which will be used later. The name OMEGA is used in the program for both Ω and Ω'.

6. The distribution of stream function corresponding to the newly computed vorticity is obtained by calling the subroutine POISSN, which solves (4.5.12) using the iterative formula (4.5.25) until the sum of absolute errors at all grid points becomes less than or equal to ERRMAX. The boundary values of Ψ are described in (4.5.19) to (4.5.21). Again, ITER in this subroutine is the total number of iterations executed when the final result is returned to the main program. In Program 4.4 Ψ is named PSI.

7. At this stage we check to see if the steady state is reached by examining whether the value of SUM obtained in Step 5 is less than or equal to EPSLON, a prescribed, small, positive quantity. In the meantime the time step counter NSTEP is checked to see if it equals the allowed number MAXSTP. When either one of the two conditions is satisfied, we go to Step 8 to print out the result. Otherwise we go back to Step 4 to compute the velocity field based on the present distribution of the stream function. Steps 4 to 7 are iterated, each time NSTEP is increased by one, until we are able to proceed to the next step.

8. The values of Ψ, U, V, and Ω at all grid points between the end electrodes are printed for the last time step. Subroutine SEARCH is called once more to find the coordinates of points along several streamlines.

In Program 4.4 the output data for field variables are printed directly in the form of arrays to show the detailed structure of the solution. The coordinates

```
C                        ***** PROGRAM 4.4 *****
C          TUBE FLOW IN THE PRESENCE OF AN ELECTROMAGNETIC FORCE
       DIMENSION ALPHA(11),BETA(11),Z(23),R(11),RH(23,11),EM(23,11),
     A           PSI(23,11),U(23,11),V(23,11),OMEGA(23,11),
     B           PSIA(8),ZZ(50),RR(50)
       COMMON ALPHA,BETA,R,M,MM1,MP1,N,NM1,H
       DATA PSIA/ 0.0, 0.02, 0.05, 0.1, 0.15, 0.2, 0.3, 0.4 /
       DATA C,RE,DT,EPSLON,MAXSTP/ 0.3, 50., 0.01, 0.01, 1000 /
       H = 0.1
       CONST = DT / (2.0*H)
       M = 22
         MM1 = M - 1
         MP1 = M + 1
       N = 11
         NM1 = N - 1
C      ..... COMPUTE Z(I),R(J),ALPHA(J), AND BETA(J) .....
       Z(1) = -1.0 - H
       DO 1 I = 2, M
     1 Z(I) = Z(I-1) + H
       R(1) = 0.0
       DO 2 J = 2, N
       R(J) = R(J-1) + H
       REALJ = FLOAT(J)
       ALPHA(J) = (REALJ-1.5) / (REALJ-1.0)
     2 BETA(J)  = (REALJ-0.5) / (REALJ-1.0)
C      ..... COMPUTE AND PRINT RH(I,J) .....
       CALL LAPLCE( RH, 0.0001, ITER )
       WRITE(6,50)
       WRITE(6,51)  ITER
       DO 3 J = 1, N
       K = N - J + 1
     3 WRITE(6,52)   ( PH(I,K), I=2,M )
C      ..... COMPUTE AND PRINT ELECTROMAGNETIC FORCE TERM .....
       DO 4 I = 2, M
       EM(I,1) = 0.0
       EM(I,N) = 0.0
       DO 4 J = 2, NM1
     4 EM(I,J) = C * RH(I,J) * (RH(I+1,J)-RH(I-1,J)) / (2.*H*R(J)**3)
       WRITE(6,53)
       DO 5 J = 1, N
       K = N - J + 1
     5 WRITE(6,54)   ( EM(I,K), I=2,M )
C      ..... FIND AND PRINT COORDINATES OF POINTS ALONG CURRENT
C            STREAMLINES WHOSE RH VALUES ARE SPECIFIED IN PSIA .....
       WRITE(6,50)
       DO 6 L = 1, 8
       CALL SEARCH( Z, R, RH, PSIA(L), ZZ, RR, KMAX )
     6 WRITE(6,55)   ( PSIA(L), ((ZZ(K),RR(K)),K=2,KMAX) )
```

Secondary Flows Caused by a Rotational Electromagnetic Force Field

```
C       ..... ASSIGN INITIAL VALUES TO PSI AND OMEGA .....
        NSTEP = 0
        DO 7  I = 1, MP1
        DO 7  J = 1, N
        PSI(I,J) = R(J)**2 / 2.0
      7 OMEGA(I,J) = 0.0


C       ..... COMPUTE VELOCITY COMPONENTS .....
      8 DO 9  I = 2, M
        DO 9  J = 2, NM1
        U(I,J) =  (PSI(I,J+1)-PSI(I,J-1)) / (2.*H*R(J))
      9 V(I,J) = -(PSI(I+1,J)-PSI(I-1,J)) / (2.*H*R(J))
        DO 10  I = 1, MP1
        U(I,1) = 2.0 * PSI(I,2) / H**2
     10 V(I,1) = 0.0
        DO 11  J = 1, N
        U(1,J) =  U(3,J)
     11 V(1,J) = -V(3,J)
        DO 12  I = 2, M
        U(I,N) = (4.0*U(I,NM1)-U(I,N-2)) / 3.0
     12 V(I,N) = 0.0
        DO 13  J = 1, N
        U(MP1,J) =  U(MM1,J)
     13 V(MP1,J) = -V(MM1,J)


C       ..... FOR THE NEXT TIME STEP COMPUTE OMEGA AT THE INTERIOR POINTS.
C             THE ABSOLUTE VALUES OF THE CHANGE OF LOCAL VORTICITY ARE
C             ACCUMULATED IN THE NAME SUM.  BOUNDARY VALUES OF OMEGA ARE
C             THEN UPDATED .....
        NSTEP = NSTEP + 1
        SUM = 0.0
        DO 14  I = 2, M
        DO 14  J = 2, NM1
        UF = ( U(I+1,J)+U(I,J) ) / 2.0
        UFA = ABS( UF )
        UB = ( U(I-1,J)+U(I,J) ) / 2.0
        UBA = ABS( UB )
        VF = ( V(I,J+1)+V(I,J) ) / 2.0
        VFA = ABS( VF )
        VB = ( V(I,J-1)+V(I,J) ) / 2.0
        VBA = ABS( VB )
        OM1 = (UF-UFA)*OMEGA(I+1,J) + (UF+UFA-UB+UBA)*OMEGA(I,J)
     A      - (UB+UBA)*OMEGA(I-1,J)
        OM2 = (VF-VFA)*OMEGA(I,J+1) + (VF+VFA-VB+VBA)*OMEGA(I,J)
     A      - (VB+VBA)*OMEGA(I,J-1)
        OM3 = OMEGA(I+1,J) + OMEGA(I-1,J) + OMEGA(I,J+1) + OMEGA(I,J-1)
     A      - (4.0+(H/R(J))**2)*OMEGA(I,J)
        OM4 = OMEGA(I,J+1) - OMEGA(I,J-1)
        OMDIFF = - CONST*(OM1+OM2) + DT/(H*H*RE)*OM3
     A           + CONST/(R(J)*RE)*OM4 + DT*EM(I,J)
        OMEGA(I,J) = OMEGA(I,J) + OMDIFF
        SUM = SUM + ABS(OMDIFF)
     14 CONTINUE
        DO 15  J = 2, NM1
     15 OMEGA(1,J) = OMEGA(3,J)
```

```
      DO 16  J = 2, NM1
 16  OMEGA(MP1,J) = OMEGA(MM1,J)

C    ..... COMPUTE STREAM FUNCTION FROM THE NEW VORTICITY FIELD .....
     CALL POISSN( PSI, OMEGA, 0.0005, ITER )

C    ..... CHECK TO SEE IF PROGRESSION IN TIME IS TO BE TERMINATED ....
     IF( SUM.LE.EPSLON .OR. NSTEP.EQ.MAXSTP )  GO TO 17
     GO TO 8
 17  WRITE(6,50)
     WRITE(6,56)  NSTEP, ITER, SUM
     DO 18  J = 1, N
     K = N - J + 1
 18  WRITE(6,52)  ( PSI(I,K), I=2,M )
     WRITE(6,57)
     DO 19  J = 1, N
     K = N - J + 1
 19  WRITE(6,52)  ( U(I,K), I=2,M )
     WRITE(6,50)
     WRITE(6,58)
     DO 20  J = 1, N
     K = N - J + 1
 20  WRITE(6,52)  ( V(I,K), I=2,M )
     WRITE(6,59)
     DO 21  J = 1, N
     K = N - J + 1
 21  WRITE(6,52)  ( OMEGA(I,K), I=2,M )
     WRITE(6,50)
     DO 22  L = 1, 8
     CALL SEARCH( Z, R, PSI, PSIA(L), ZZ, RR, KMAX )
 22  WRITE(6,60)  ( PSIA(L), ((ZZ(K),RR(K)), K=1,KMAX) )

C    ..... FORMATS ARE LISTED BELOW .....
 50  FORMAT( 1H1 )
 51  FORMAT( /// 20X, 7HITER = , I3, 20X, 7HRH(I,J) )
 52  FORMAT( / 2X, 21F6.3 )
 53  FORMAT( ////// 50X, 7HEM(I,J) )
 54  FORMAT( / 2X, 21F6.2 )
 55  FORMAT( // 10X43HTHE POINTS (Z,R) ON CURRENT STREAMLINE RH =,
    A         F5.2 / (6(4X, F5.2, 2H ,, F5.2)) )
 56  FORMAT( /// 20X, 8HNSTEP = , I4, 10X, 7HITER = , I3,
    A         10X, 6HSUM = , F7.4 // 5X, 8HPSI(I,J) )
 57  FORMAT( ////// 5X, 6HU(I,J) )
 58  FORMAT( / 5X, 6HV(I,J) )
 59  FORMAT( ////// 5X, 10HOMEGA(I,J) )
 60  FORMAT( // 10X40HTHE POINTS (Z,R) ON THE STREAMLINE PSI =,
    A         F6.3 / (6(4X, F5.2, 2H ,, F5.2)) )

     END




     SUBROUTINE LAPLCE( RH, ERRMAX, ITER )
C    ..... FINDING RH IN THE LEFT HALF OF THE TUBE SATISFYING
C          EQUATION (4.5.9) AND BOUNDARY CONDITIONS (4.5.14 - 18).
C          ITERATIVE SCHEME (4.5.22) IS USED REPEATEDLY UNTIL THE SUM
C          OF ABSOLUTE ERRORS BECOMES ≤ ERRMAX.  ITER IS THE TOTAL
C          NUMBER OF ITERATIONS EXECUTED.  THE SOLUTION IN THE OTHER
C          HALF OF THE TUBE IS OBTAINED BY CONSIDERING ITS SYMMETRY
C          ABOUT THE CENTRAL ELECTRODE .....

     DIMENSION RH(23,11),ALPHA(11),BETA(11),R(11)
     COMMON ALPHA,BETA,R,M,MM1,MP1,N,NM1,H
```

```
C      ..... ASSUME A UNIFORM CURRENT FLOW AT THE BEGINNING WHICH ALREADY
C            SATISFIES BOUNDARY CONDITIONS (4.5.14 AND 15).  THEN ASSIGN
C            VALUES OF RH ACCORDING TO (4.5.17).  FOR THE CENTRAL
C            ELECTRODE USED IN THE PRESENT PROBLEM HAVING A RADIUS EQUAL
C            TO THE GRID SIZE H, (4.5.16) IS NOT NEEDED .....
       ITER = 0
       IM = M/2 + 1
       IMM1 = IM - 1
       IMP1 = IM + 1
       DO 1   I = 2, IM
       DO 1   J = 1, N
     1 RH(I,J) = R(J)**2 / 2.0
       DO 2   J = 2, NM1
     2 RH(IM,J) = 0.5

C      ..... BEFORE EACH ITERATION, ADD ONE TO ITER
C            AND LET ERROR EQUAL ZERO .....
     3 ITER = ITER + 1
       ERROR = 0.0

C      ..... ASSIGN VALUES TO RH(1,J) ACCORDING TO (4.5.18) .....
       DO 4   J = 2, NM1
     4 RH(1,J) = RH(3,J)

C      ..... COMPUTE RH(I,J) AT INTERIOR POINTS USING (4.5.22) .....
       DO 5   I = 2, IMM1
       DO 5   J = 2, NM1
       RHOLD = RH(I,J)
       RH(I,J) = 0.25*( RH(I+1,J) + RH(I-1,J) + ALPHA(J)*RH(I,J+1)
      A                 + BETA(J)*RH(I,J-1) )
     5 ERROR = ERROR + ABS( RH(I,J)-RHOLD )

C      ..... CHECK TO SEE IF THE REQUIRED ACCURACY IS REACHED.  IF SO,
C            THE BOUNDARY VALUES ARE UPDATED AND THE SOLUTION IS
C            TRANSMITTED TO THE RIGHT HALF OF THE TUBE .....
       IF( ERROR.GT.ERRMAX )  GO TO 3
       DO 6   J = 2, NM1
     6 RH(1,J) = RH(3,J)
       DO 7   I = IMP1, MP1
       DO 7   J = 1, N
     7 RH(I,J) = RH(M-I+2,J)
       RETURN
       END

       SUBROUTINE POISSN( PSI, OMEGA, ERRMAX, ITER )
C      ..... FINDING PSI SATISFYING EQUATION (4.5.12) AND
C            BOUNDARY CONDITIONS (4.5.19 - 21).  ITERATIVE SCHEME
C            (4.5.25) IS USED REPEATEDLY UNTIL THE SUM OF ABSOLUTE
C            ERRORS BECOMES ≤ ERRMAX.  ITER IS THE TOTAL NUMBER
C            OF ITERATIONS EXECUTED .....

       DIMENSION PSI(23,11),OMEGA(23,11),ALPHA(11),BETA(11),R(11)
       COMMON ALPHA,BETA,R,M,MM1,MP1,N,NM1,H
C      ..... ASSUME A UNIFORM FLOW AT THE BEGINNING .....
       ITER = 0
       DO 1   I = 2, M
       DO 1   J = 1, N
     1 PSI(I,J) = R(J)**2 / 2.0

C      ..... BEFORE EACH ITERATION, ADD ONE TO ITER
C            AND LET ERROR EQUAL ZERO .....
     2 ITER = ITER + 1
       ERROR = 0.0
```

```
C      ..... ASSIGN VALUES TO PSI AT I = 1 AND M+1 .....
       DO 3  J = 2, NM1
       PSI(1,J) = PSI(3,J)
     3 PSI(MP1,J) = PSI(MM1,J)

C      ..... COMPUTE PSI(I,J) AT INTERIOR POINTS USING (4.5.25) .....
       DO 4  I = 2, M
       DO 4  J = 2, NM1
       PSIOLD = PSI(I,J)
       PSI(I,J) = 0.25*( PSI(I+1,J) + PSI(I-1,J) + ALPHA(J)*PSI(I,J+1)
      A            + BETA(J)*PSI(I,J-1) - H*H*R(J)*OMEGA(I,J) )
     4 ERROR = ERROR + ABS( PSI(I,J)-PSIOLD )

C      ..... CHECK TO SEE IF THE REQUIRED ACCURACY IS REACHED.  IF SO,
C            THE BOUNDARY VALUES OF PSI ARE UPDATED AND THE SOLUTION
C            IS RETURNED TO THE MAIN PROGRAM .....
       IF( ERROR.GT.ERRMAX )  GO TO 2
       DO 5  J = 2, NM1
       PSI(1,J) = PSI(3,J)
     5 PSI(MP1,J) = PSI(MM1,J)
       RETURN
       END




       SUBROUTINE SEARCH( X, Y, PSI, PSIA, XX, YY, KMAX )
C      ..... TAKEN FROM PROGRAM 2.2 WITH CHANGES IN THE DIMENSION
C            STATEMENT, AND WITH THE REPLACEMENT OF THE COMMON
C            STATEMENT BY A DATA STATEMENT .....
       DIMENSION X(23),Y(11),PSI(23,11),XX(50),YY(50)
       DATA M,N,DY/ 22, 11, 0.1 /
```

of points along streamlines are printed but are not shown in the book. An alternative method of presenting the output is illustrated in Program 4.5, in which the results at various time steps are displayed graphically instead.

Program 4.4 is run for EPSLON = 0.01 and MAXSTP = 1000. To obtain the printed result on a CDC 6400 computer, approximately 6 min of central processing time has been used. Notice that the maximum allowable error used in subroutine LAPLCE is 0.0001, while that used in subroutine POISSN is 0.0005. The reason for having a less accurate result from POISSN is that a considerable amount of computer time can be saved after it is called several hundred times in the program. You may verify that practically the same results will be obtained even if higher accuracies are required in POISSN. On the other hand, reducing the size of EPSLON will cause an increase in the number of time steps required to reach the approximate steady state, yet the flow pattern so obtained does not differ significantly from the original one. If data were printed out at some intermediate time steps, they would show that the stream function reached a steady state long before the vorticity field did. The value assigned to EPSLON is satisfactory for the present problem.

A graphical presentation of some of the numerical output is shown in Fig.

Secondary Flows Caused by a Rotational Electromagnetic Force Field   **351**

ITER = 98

RH(I,J)

```
.500  .500  .500  .500  .500  .500  .500  .500  .500  .500  .500  .500  .500  .500  .500  .500  .500  .500  .500  .500  .500
.415  .416  .418  .421  .426  .433  .442  .454  .468  .483  .500  .483  .468  .454  .442  .433  .426  .421  .418  .416  .415
.338  .339  .343  .349  .358  .371  .389  .411  .437  .468  .500  .468  .437  .411  .389  .371  .358  .349  .343  .339  .338
.268  .270  .274  .282  .295  .313  .337  .368  .407  .451  .500  .451  .407  .368  .337  .313  .295  .282  .274  .270  .268
.204  .206  .211  .221  .235  .256  .285  .325  .374  .434  .500  .434  .374  .325  .285  .256  .235  .221  .211  .206  .204
.148  .149  .154  .164  .179  .201  .233  .278  .338  .414  .500  .414  .338  .278  .233  .201  .179  .164  .154  .149  .148
.098  .099  .104  .112  .125  .146  .178  .226  .294  .387  .500  .387  .294  .226  .178  .146  .125  .112  .104  .099  .098
.057  .058  .061  .067  .078  .094  .122  .166  .238  .349  .500  .349  .238  .166  .122  .094  .078  .067  .061  .058  .057
.026  .027  .028  .032  .038  .048  .066  .100  .165  .288  .500  .288  .165  .100  .066  .048  .038  .032  .028  .027  .026
.007  .007  .007  .008  .010  .014  .021  .036  .075  .180  .500  .180  .075  .036  .021  .014  .010  .008  .007  .007  .007
0.000 0.000 0.000 0.000 0.000 0.000 0.000 0.000 0.000 0.000 0.000 0.000 0.000 0.000 0.000 0.000 0.000 0.000 0.000 0.000 0.000
```

EM(I,J)

```
0.00   0.00   0.00   0.00   0.00   0.00   0.00   0.00   0.00   0.00   0.00   0.00   0.00   0.00   0.00   0.00   0.00   0.00   0.00   0.00   0.00
0.00    .00    .01    .01    .02    .02    .03    .03    .03    .03   0.00   -.03   -.03   -.03   -.03   -.02   -.02   -.01   -.01   -.00   0.00
0.00    .00    .01    .02    .02    .04    .06    .07    .09    .09   0.00   -.09   -.07   -.06   -.04   -.03   -.02   -.02   -.01   -.00   0.00
0.00    .01    .02    .03    .04    .06    .08    .11    .15    .18   0.00   -.18   -.15   -.11   -.08   -.06   -.04   -.03   -.02   -.01   0.00
0.00    .01    .02    .04    .06    .09    .14    .20    .28    .38   0.00   -.38   -.28   -.20   -.14   -.09   -.06   -.04   -.02   -.01   0.00
0.00    .01    .03    .05    .08    .13    .22    .35    .55    .80   0.00   -.80   -.55   -.35   -.22   -.13   -.08   -.05   -.03   -.01   0.00
0.00    .01    .03    .06    .11    .18    .33    .61   1.11   1.87   0.00  -1.87  -1.11   -.61   -.33   -.18   -.11   -.06   -.03   -.01   0.00
0.00    .02    .04    .08    .15    .28    .55   1.11   2.42   5.08   0.00  -5.08  -2.42  -1.11   -.55   -.28   -.15   -.08   -.04   -.02   0.00
0.00    .02    .06    .11    .22    .49   1.08   2.42   5.80  18.08   0.00 -18.08  -5.80  -2.42  -1.08   -.49   -.22   -.11   -.06   -.02   0.00
0.00    .02    .08    .22    .71   2.94  16.04  14.62   0.00 -14.62 -16.04  -2.94   -.71   -.22   -.08   -.02   0.00
0.00   0.00   0.00   0.00   0.00   0.00   0.00   0.00   0.00   0.00   0.00   0.00   0.00   0.00   0.00   0.00   0.00   0.00   0.00   0.00   0.00
```

NSTEP = 230   ITER = 56   SUM = .0099

PSI(I,J)

.500	.500	.500	.500	.500	.500	.500	.500	.500	.500	.500	.500	.500	.500	.500	.500	.500
.404	.404	.404	.404	.404	.405	.404	.403	.403	.403	.403	.403	.404	.404	.405	.405	.405
.319	.319	.318	.318	.317	.318	.318	.317	.316	.316	.316	.316	.317	.318	.319	.320	.320
.243	.243	.242	.242	.241	.242	.243	.241	.240	.239	.239	.240	.240	.241	.244	.245	.245
.178	.178	.177	.176	.175	.177	.178	.175	.174	.173	.173	.173	.174	.176	.179	.180	.181
.123	.123	.122	.121	.120	.122	.123	.120	.119	.117	.117	.117	.117	.119	.125	.126	.126
.079	.079	.077	.077	.074	.078	.079	.074	.072	.071	.072	.073	.074	.077	.081	.081	.082
.044	.044	.043	.042	.040	.041	.040	.038	.037	.036	.037	.040	.043	.044	.046	.047	.047
.020	.020	.019	.018	.016	.018	.016	.014	.012	.012	.014	.017	.020	.021	.022	.021	.021
.005	.005	.005	.005	.003	.004	.003	.002	.000	.000	.002	.005	.006	.006	.006	.006	.006
0.000	0.000	0.000	0.000	0.000	0.000	0.000	0.000	0.000	0.000	0.000	0.000	0.000	0.000	0.000	0.000	0.000

U(I,J)

1.007	1.007	1.008	1.010	1.012	1.014	1.017	1.020	1.022	1.023	1.022	1.020	1.017	1.014	1.010	1.007	1.005	1.003	1.002	1.001	1.001
1.007	1.007	1.008	1.009	1.011	1.014	1.016	1.019	1.021	1.023	1.022	1.020	1.016	1.013	1.010	1.007	1.004	1.002	1.001	1.000	1.000
1.005	1.006	1.006	1.008	1.010	1.012	1.015	1.018	1.021	1.022	1.022	1.019	1.015	1.011	1.008	1.005	1.002	1.001	1.000	.999	.999
1.003	1.003	1.004	1.005	1.007	1.009	1.012	1.016	1.020	1.022	1.021	1.017	1.012	1.007	1.004	1.001	.999	.998	.997	.996	.996
1.000	1.000	1.001	1.002	1.004	1.008	1.012	1.017	1.021	1.021	1.019	1.012	1.006	1.000	.997	.995	.994	.993	.993	.993	.993
.996	.996	.996	.997	.997	1.000	1.005	1.012	1.017	1.013	1.003	.994	.989	.986	.986	.987	.988	.989	.990	.990	.990
.991	.991	.990	.988	.985	.986	.990	1.000	1.009	1.001	.982	.972	.970	.972	.977	.981	.985	.988	.990	.990	.991
.987	.986	.984	.980	.975	.969	.961	.983	.967	.940	.937	.948	.962	.974	.984	.992	.998	1.001	1.002	1.001	1.000
.982	.981	.978	.972	.962	.947	.927	.903	.898	.880	.871	.910	.951	1.001	1.014	1.022	1.027	1.029	1.030		
.979	.978	.973	.964	.949	.923	.879	.807	.701	.689	.871	.987	1.045	1.070	1.080	1.081	1.078	1.075	1.071	1.069	
.977	.975	.970	.959	.938	.900	.827	.682	.405	.036	.392	1.030	1.225	1.252	1.231	1.200	1.172	1.148	1.129	1.115	1.109

353

```
V(I,J)

0.000  0.000  0.000  0.000  0.000  0.000  0.000  0.000  0.000  0.000  0.000  0.000  0.000  0.000  0.000  0.000  0.000  0.000
-0.000  .001  .002   .003   .003  -.003  -.003  -.002  -.001  -.001 -0.000
-0.000  .001  .004   .006   .006  -.006  -.006  -.005  -.004  -.002 -.001 -0.000
-0.000  .002  .006   .010   .011  -.010  -.011  -.008  -.007  -.004 -.002 -0.000
-0.000  .003  .008   .015   .017   .012  -.013  -.013  -.010  -.005 -.004 -0.000
-0.000  .003  .011   .020   .025   .019  -.018  -.017  -.014  -.007 -.002 -0.000
-0.000  .004  .013   .027   .035   .033  -.031  -.023  -.017  -.009 -.005 -0.000
-0.000  .004  .013   .033   .048   .057  -.052  -.029  -.020  -.008 -.005 -0.000
-0.000  .003  .012   .035   .062   .099  -.090  -.039  -.028  -.013 -.002 -0.000
-0.000  .003  .012   .035   .089  -.149  -.072  -.048  -.029  -.017 -.005 -0.000
-0.000  .002  .008   .028   .058  -.208  -.087  -.042  -.017  -.005 -.003 -0.000
-0.000  .004  .015   .055   .106  -.248  -.055  -.002  -.005  -.001 -.003 -0.000
0.000  0.000  0.000  0.000   .161  -.131   .013   .015   .013   .003  .004 -0.000
0.000  0.000  0.000  0.000  0.000  0.000  0.000   .011   .008  .005 -0.000
                                                  0.000  0.000  0.000  0.000  0.000

OMEGA(I,J)

0.000  0.000  0.000  0.000  0.000  0.000  0.000  0.000  0.000  0.000  0.000  0.000  0.000  0.000  0.000  0.000  0.000  0.000
.003   .004   .006   .007   .009   .012   .015   .018   .018   .013   .011   .009   .005   .006   .005   .004   .004  0.000
.007   .007   .012   .015   .020   .026   .035   .043   .043   .029   .023   .019   .016   .013   .010   .009   .009  0.000
.010   .011   .018   .024   .033   .046   .065   .081   .081   .051   .041   .033   .027   .022   .019   .016   .012  0.000
.013   .014   .025   .034   .050   .074   .113   .146   .146   .086   .066   .052   .032   .025   .019   .014   .011  0.000
.015   .016   .031   .046   .072   .117   .196   .271   .271   .140   .099   .070   .049   .032   .018   .007  -.008  0.000
.016   .017   .023   .057   .100   .188   .534   .547   .547   .203   .115   .056   .015  -.014  -.036  -.052  -.070  0.000
.015   .016   .025   .067   .136   .308  1.124  1.239  1.239   .339   .161  -.017  -.112  -.164  -.191  -.203  -.197  0.000
.012   .012   .020   .068   .168   .487  2.392  3.220  2.392   .504  -.537  -.699  -.648  -.580  -.511  -.447  -.350  0.000
.006   .008   .012   .051   .156   .596  9.341  4.379-3.809-3.212-2.310-1.652-1.199 -.887 -.670 -.516 -.406 -.340  0.000
0.000  0.000  0.000  0.000  0.000  0.000  0.000  0.000  0.000  0.000  0.000  0.000  0.000  0.000  0.000  0.000  0.000  0.000
```

354

4.5.2. In the upper half of the tube in that figure, several current streamlines and several fluid streamlines are plotted using dashed and solid lines, respectively. Three axial velocity profiles are plotted in the lower half of the tube. A plot of the dimensionless speed along the tube axis is shown below the tube. Figure 4.5.2 clearly shows that when moving toward a current constriction created by the small central electrode, the flow decelerates continuously in the region near the tube axis. The minimum speed, whose magnitude is close

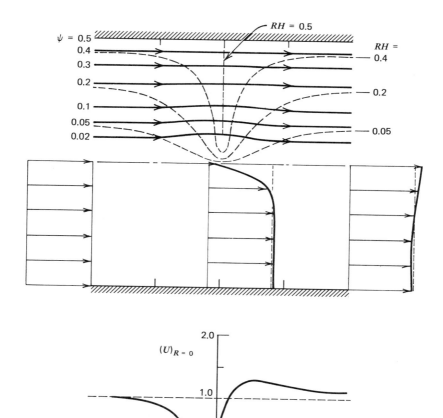

FIGURE 4.5.2 Streamlines, axial velocity profiles, and the velocity along the tube axis for $C = 0.3$ and $Re = 50$. Dashed lines shown in the upper half of the tube are the streamlines of the electric current.

Secondary Flows Caused by a Rotational Electromagnetic Force Field   **355**

to zero, occurs on the axis at a short distance ahead of the small electrode. After passing that location, the central flow accelerates to speeds higher than the entering speed and then decelerates toward that value after a maximum value is attained.

Similarly, the result for $C = 0.5$ is obtained and is plotted in Fig. 4.5.3. This higher magnetic pressure number can be achieved by increasing the potential differences across the electrodes while keeping the other conditions the same

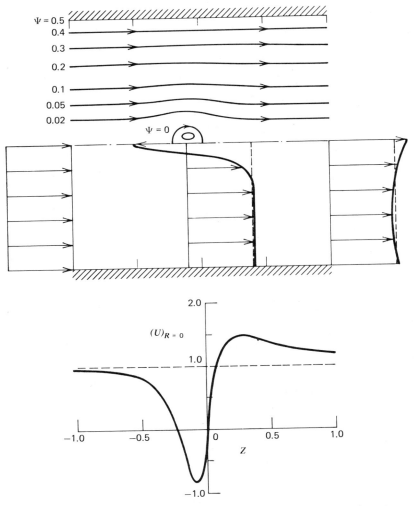

Figure 4.5.3 Streamlines, axial velocity profiles, and velocity along the tube axis for $C = 0.5$ and $Re = 50$.

Secondary Flows and Flow Instabilities

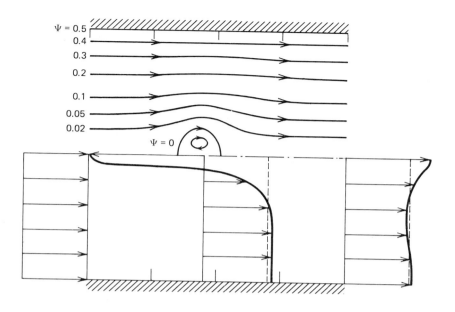

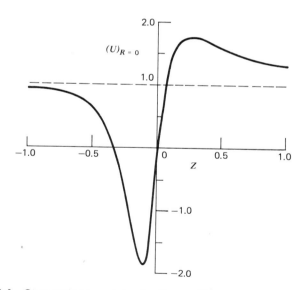

FIGURE 4.5.4 Streamlines, axial velocity profiles, and velocity along the tube axis for $C = 0.7$ and $Re = 50$.

as before. In this case the central flow does not have enough kinetic energy to go through the steep pressure gradient, so that it becomes stagnant on the tube axis before it reaches the small electrode. When this happens, a secondary flow is formed upstream from the middle section. Further computations indicate that the size of the secondary flow grows with increasing C. One example plotted for $C = 0.7$ is shown in Fig. 4.5.4.

In the three cases mentioned here, a constant time step size of $\Delta T = 0.01$ is used. The size of ΔT is determined after a few trials. With that chosen value of ΔT, the inequality (4.5.35) is satisfied and computational stability is maintained for all three cases. Program 4.4 shows that the required computing time increases rapidly with increasing C. Additional discussions on this class of magnetohydrodynamic flows can be found in Uberoi and Chow (1977).

Problem 4.8   If the two large screen electrodes in Fig. 4.5.1 are replaced by solid conducting plates, an enclosed chamber is formed without a net fluid flow passing through it, whereas the current flow still remains the same as before. The originally stationary fluid cannot maintain stationary in the presence of the rotational electromagnetic body force, and a motion is set up in the chamber.

In this case an appropriate characteristic speed, $r_0 J_0 (2\mu_m/\rho)^{1/2}$, is used to replace u_0 in (4.5.8). The nondimensionalized equations are in the same form as (4.5.9) to (4.5.13), except that C now has a fixed value of 1 in the last equation. Since the flow streamline pattern is symmetric about the central electrode, only the left half of the chamber needs to be considered. The fluid boundary conditions are changed from (4.5.19) to (4.5.21) to

$$R = 0 \text{ or } 1, \quad \text{and} \quad -\frac{z_0}{r_0} \leqslant Z \leqslant 0: \quad \Psi = 0$$

$$0 < R < 1, \quad \text{and} \quad Z = -\frac{z_0}{r_0} \text{ or } 0: \quad \Psi = 0$$

The no-slip condition is assumed for velocity at all solid walls.

In this program an improvement on Program 4.4 is to be made such that the size of ΔT is determined by the stability criterion (4.5.35). At every time step the right-hand side of (4.5.35) is first calculated at all grid points, and the lowest value among them, obtained through comparison, is the ΔT to be used in the flow computation for that time step. This procedure is then repeated again for each succeeding step.

For the same geometry and Reynolds number as specified in Program 4.4, compute and plot the steady-state solution for the flow.

4.6 Bénard and Taylor Instabilities

When a horizontal layer of fluid is heated from below in a gravitational field, the density of the fluid at any location becomes smaller than the fluid just above it. If a fluid parcel is displaced slightly upward into the region of higher density, a buoyant force will assist it to move further upward. Similarly, if the fluid parcel is displaced downward into a region of smaller density, it will keep moving in the same direction. Without a sufficiently large, viscous, retarding force, this situation is said to be unstable, and the instability appears in the form of a net of hexagonal convection cells.

Lord Rayleigh (1916) made the first theoretical analysis of the so-called *Bénard problem* concerning the stability of a fluid layer in the presence of a temperature gradient parallel to the gravitational force. An extensive treatment on this subject based on linearized theories can be found in the book by Chandrasekhar (1961). A linearized theory predicts only the onset of instabilities. Once the flow becomes unstable, the initially small disturbances will grow with time, and the subsequent fluid motion will be governed by the nonlinear Navier-Stokes equation. Now that we already have a successful upwind-difference numerical scheme for approximating the nonlinear terms in that equation, the method used in Section 4.5 will be adopted here to study the Bénard problem numerically on the computer.

The governing equations for the fluid motion are (3.1.1) to (3.1.3), the continuity, the Navier-Stokes (adding the buoyant force), the energy equations and, also, the equation of state. For the last equation we use, instead of (3.1.5), the following form:

$$\rho = \rho_0[1 - \alpha(T - T_0)] \tag{4.6.1}$$

where α is the coefficient of volume expansion and T_0 is the temperature at which the fluid density is ρ_0. For ordinary gases or liquids, α is of the order of 10^{-3} or 10^{-4}. Based on this fact, considerable simplifications can be made by using the *Boussinesq approximation* that, if temperature variations are not too large, ρ can be considered constant everywhere except in the buoyant force term. Under further assumption of constant physical properties of the fluid, (3.1.1) and (3.1.2) become

$$\nabla \cdot \mathbf{V} = 0 \tag{4.6.2}$$

$$\rho_0 \frac{D\mathbf{V}}{Dt} = -\mathbf{j}(\rho - \rho_0)g - \nabla p + \mu \nabla^2 \mathbf{V} \tag{4.6.3}$$

in which $\mathbf{j}$ is the unit vector along the y-axis opposite to the direction of the gravitational acceleration g. It can be shown (see Section II.8, Chandrasekhar, 1961) that the energy equation (3.1.3) reduces to

$$\frac{DT}{Dt} = \mathcal{K}\nabla^2 T \tag{4.6.4}$$

where $\mathcal{K}(= k/\rho_0 c_v)$ is the coefficient of thermometric conductivity, and c_v is the constant-volume specific heat. Note that the dissipation function drops out completely from the energy equation in the present approximation.

We consider a two-dimensional flow in the x-y plane contained between two flat plates at $y = 0$ and $y = H$, respectively. The fluid originally at a uniform temperature T_0 is heated from below by increasing the temperature of the lower plate suddenly to T_1. Being perpendicular to the fluid motion, the vorticity is in the z direction and its magnitude is designated ζ. Again, a stream function ψ can be introduced to satisfy (4.6.2).

For this problem H is a reference length, $\mu/\rho_0 H$ is a reference velocity, and $T_1 - T_0$ is a reference temperature difference. Based on these reference quantities, the following dimensionless variables are constructed.

$$X = \frac{x}{H}, \qquad Y = \frac{y}{H}, \qquad T = \frac{t}{\rho_0 H^2/\mu}$$

$$U = \frac{u}{\mu/\rho_0 H}, \qquad V = \frac{v}{\mu/\rho_0 H}, \qquad \Psi = \frac{\psi}{\mu/\rho_0},$$

$$\Omega = \frac{\zeta}{\mu/\rho_0 H^2}, \qquad \theta = \frac{T - T_0}{T_1 - T_0}. \tag{4.6.5}$$

From now on T is used to designate the dimensionless time, and θ is used to designate the dimensionless temperature difference.

Density difference in (4.6.3) is first replaced by temperature difference, with substitution from (4.6.1). Then the pressure gradient term is eliminated after taking the curl of the resulting equation of motion. When expressed in dimensionless form, the governing equations are

$$U = \frac{\partial \Psi}{\partial Y} \tag{4.6.6}$$

$$V = -\frac{\partial \Psi}{\partial X} \tag{4.6.7}$$

$$\Omega = -\left(\frac{\partial^2}{\partial X^2} + \frac{\partial^2}{\partial Y^2}\right)\Psi \tag{4.6.8}$$

$$\frac{\partial \Omega}{\partial T} + \frac{\partial(U\Omega)}{\partial X} + \frac{\partial(V\Omega)}{\partial Y} = Gr\frac{\partial \theta}{\partial X} + \left(\frac{\partial^2}{\partial X^2} + \frac{\partial^2}{\partial Y^2}\right)\Omega \tag{4.6.9}$$

$$\frac{\partial \theta}{\partial T} + \frac{\partial (U\theta)}{\partial X} + \frac{\partial (V\theta)}{\partial Y} = Pr^{-1}\left(\frac{\partial^2}{\partial X^2} + \frac{\partial^2}{\partial Y^2}\right)\theta \qquad (4.6.10)$$

in which $Gr = \alpha g H^3 (T_1 - T_0)/\nu^2$ is the *Grashof number* and $Pr = \nu/\mathcal{K}$ is the Prandtl number, $\nu(=\mu/\rho_0)$ being the kinematic viscosity coefficient. In some analyses the Rayleigh number instead of the Grashof number is used, which is defined as the product of *Gr* and *Pr*. The nonlinear terms in (4.6.9) and (4.6.10) are written in the same form, so that both equations can be solved by using the same numerical technique.

In Program 4.5 we consider water ($Pr = 6.75$) enclosed by a rectangular box, shown schematically in Fig. 4.6.1. Because of the symmetry about the Y-axis, only the left half of the flow needs to be computed, thus resulting in large savings in computational efforts. Square meshes of size $h \times h$ are chosen for the numerical work, and the concerned fluid region is covered with m vertical and n horizontal grid lines. Since the dimensionless distance between parallel plates is unity, the grid size h has the value of $1/(n-1)$. The values $m = 21$ and $n = 9$ are assumed for the present example.

At time $T = 0$, the temperature distribution is such that $\theta = 0$ everywhere, except $\theta = 1$ on the lower plate. As time increases fluid temperature changes, but the values at solid walls are kept always at the initial condition. The boundary condition along the Y-axis is $\partial \theta / \partial X = 0$, as required by the symmetry of the temperature field. The temperature boundary conditions are indicated in Fig. 4.6.1.

Except the nonlinear terms, all spatial derivatives in the governing differential equations are approximated at the interior grid points using the

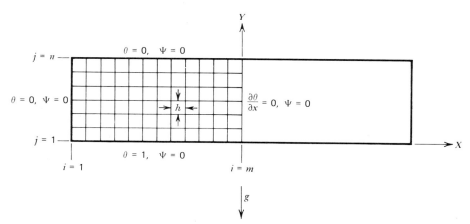

FIGURE 4.6.1 Schematic representation of the problem considered in Program 4.5.

Bénard and Taylor Instabilities

central-difference formula. The finite-difference forms of (4.6.6) and (4.6.7) are

$$U_{i,j} = \frac{\Psi_{i,j+1} - \Psi_{i,j-1}}{2h} \tag{4.6.11}$$

$$V_{i,j} = - \frac{\Psi_{i+1,j} - \Psi_{i-1,j}}{2h} \tag{4.6.12}$$

(4.6.8) conforms with the generalized Poisson equation (2.10.1) if Ψ is replaced by f and $-\Omega$ by q. The successive overrelaxation method represented by the iterative scheme (2.10.13) is programmed in the subroutine SORLX, which is used to find the stream function for a certain known vorticity distribution. The maximum error ERRMAX allowed in our program for the S.O.R. method is 0.0001. The boundary conditions for Ψ stated in this subroutine are those shown in Fig. 4.6.1, specified particularly for the present problem. The condition that $\Psi = 0$ on the three solid walls comes from the fact that there is no net flow across these boundaries. Ψ must also vanish along the Y-axis in order to make the fluid motion to its right the mirror image of that to its left. When the same subroutine is used elsewhere, care must be taken to check whether the boundary conditions need to be modified accordingly.

The symmetric distribution of θ and the antisymmetric distribution of Ψ about the Y-axis result in the following boundary conditions for all values of j.

$$\theta_{m+1,j} = \theta_{m-1,j} \tag{4.6.13}$$

$$\Psi_{m,j} = 0 \tag{4.6.14}$$

$$\Psi_{m+1,j} = - \Psi_{m-1,j} \tag{4.6.15}$$

$$U_{m+1,j} = - U_{m-1,j} \tag{4.6.16}$$

$$U_{m,j} = 0 \tag{4.6.17}$$

$$V_{m,j} = \frac{\Psi_{m-1,j}}{h} \tag{4.6.18}$$

$$\Omega_{m,j} = 0 \tag{4.6.19}$$

The last three conditions are deduced from (4.6.11), (4.6.12), and (4.6.8), respectively, by using (4.6.14) and (4.6.15).

To solve both (4.6.9) and (4.6.10), a function subprogram named PNEW is constructed for handling a generalized equation of the form

$$\frac{\partial P}{\partial T} = - \frac{\partial (UP)}{\partial X} - \frac{\partial (VP)}{\partial Y} + A \frac{\partial Q}{\partial X} + B \left(\frac{\partial^2 P}{\partial X^2} + \frac{\partial^2 P}{\partial Y^2} \right) \tag{4.6.20}$$

The computationally stable upwind-differencing scheme of the previous section is used to approximate the first two terms on the right-hand side of this equation. With the forward and backward velocity components defined in the same forms as those shown in (4.5.27) and (4.5.28), we further define

$$P1 = (U_f - |U_f|)P_{i+1,j} + (U_f + |U_f| - U_b + |U_b|)P_{i,j} - (U_b + |U_b|)P_{i-1,j}$$

$$(4.6.21)$$

$$P2 = (V_f - |V_f|)P_{i,j+1} + (V_f + |V_f| - V_b + |V_b|)P_{i,j} - (V_b + |V_b|)P_{i,j-1}$$

$$(4.6.22)$$

The terms multiplied by A and B are approximated by central-differencing schemes. For them we let

$$P3 = Q_{i+1,j} - Q_{i-1,j} \qquad (4.6.23)$$

$$P4 = P_{i+1,j} + P_{i-1,j} + P_{i,j+1} + P_{i,j-1} - 4P_{i,j} \qquad (4.6.24)$$

Finally, a forward-differencing scheme is used to approximate the time derivative, so that

$$\left(\frac{\partial P}{\partial T}\right)_{i,j} = \frac{1}{\Delta T}(P'_{i,j} - P_{i,j}) \qquad (4.6.25)$$

in which ΔT is the size of time increment and, as in the previous section, a prime is used to denote the value of a variable evaluated at time $T + \Delta T$. Thus, after rearranging terms, (4.6.20) becomes

$$P'_{i,j} = P_{i,j} + \frac{\Delta T}{2h}\left(-P1 - P2 + A \cdot P3 + 2B\frac{P4}{h}\right) \qquad (4.6.26)$$

When (4.6.26) is used to integrate (4.6.9) at an interior grid point, we replace P by Ω, Q by θ, and let $A = Gr$ and $B = 1$. The value evaluated from the right-hand side of (4.6.26) is stored temporarily as an element of a two-dimensional array named **OMNEW**. Similarly, for integrating (4.6.10), $A = 0$, $B = Pr^{-1}$, and P is replaced by θ. Since Q is multiplied by zero in this case, it cannot have any influence on the computation. Q is arbitrarily replaced by θ in Program 4.5. The newly computed value for θ is assigned to the array **THNEW**. After computations have been done at all grid points, **OMNEW** and **THNEW** represent the updated vorticity and temperature distributions at $T + \Delta T$. Their elements are then assigned respectively back to **OMEGA** and **THETA**, which are the variable names used in the program for Ω and θ. The data may be printed or plotted before proceeding to the next time step.

This procedure is a little different from that followed in Program 4.4. There we are interested in the steady-state solution instead of the solution at every time step. Replacing the value of vorticity stored in the array **OMEGA** by the newly computed value speeds up the convergence toward the final state. That

Bénard and Taylor Instabilities

method cannot be used in the present program, in which the development of the convective flow at various stages is the main concern.

The function PNEW is used to compute Ω and θ at all interior points except those on the Y-axis, along which vorticity vanishes according to (4.6.19). Temperature along that axis is computed separately by again applying (4.6.26), but with the symmetric boundary condition (4.6.13) incorporated.

Vorticities at solid walls are computed according to (4.6.8). On the vertical surface at $i = 1$, $U = 0$ and $\partial U/\partial Y = 0$ so that

$$\Omega_{1,j} = \left(\frac{\partial V}{\partial X}\right)_{1,j} = \frac{1}{2h}(4V_{2,j} - V_{3,j}) \tag{4.6.27}$$

This approximation is obtained by using a three-point forward differencing scheme having a truncation error $O(h^2)$, and the fact that $V_{1,j} = 0$. Similarly, on the top and bottom walls where $V = 0$, $\partial V/\partial X = 0$, and $U = 0$, the boundary values of vorticity are approximated by

$$\Omega_{i,1} = \frac{1}{2h}(-4U_{i,2} + U_{i,3}) \tag{4.6.28}$$

$$\Omega_{i,n} = \frac{1}{2h}(4U_{i,n-1} - U_{i,n-2}) \tag{4.6.29}$$

In summary, the procedure for our numerical computations is outlined as follows. At any time instant the vorticity and temperature distributions are obtained from the conditions at the previous time step; however, at the initial instant they are prescribed by the initial conditions. Stream function is computed based on the vorticity distribution by solving (4.6.8) with the help of the subroutine SORLX. Velocity components are calculated from (4.6.11) and (4.6.12) once Ψ becomes known. By using the function subprogram PNEW, (4.6.9) and (4.6.10) are integrated to find the vorticity and temperature in the interior region at the next time step. Their boundary values are either fixed by the boundary conditions or updated appropriately in the way just described. The same process is repeated for each of the following time steps until the time step counter NSTEP reaches a specified value MAXSTP.

Problem 4.9 The computational stability of the numerical scheme (4.6.26) is to be examined. When it is applied to integrate the energy equation (4.6.10), it can be written as

$$\theta'_{i,j} = a_1\theta_{i+1,j} + a_2\theta_{i-1,j} + a_3\theta_{i,j} + a_4\theta_{i,j+1} + a_5\theta_{i,j-1} \tag{4.6.30}$$

Show that except for a_3, all the coefficients are positive, no matter what the flow direction is. According to the quasilinear analysis of Lax and Richtmyer (1956), the scheme is stable if every coefficient in (4.6.30) is positive or, equivalently, if $a_3 \geq 0$. Show that this requirement gives the stability criterion that

$$\Delta T \leq \left[\frac{1}{2h} (U_f + |U_f| - U_b + |U_b| + V_f + |V_f| - V_b + |V_b|) + \frac{4}{h^2 Pr} \right]^{-1}$$

(4.6.31)

A similar inequality may be derived from (4.6.9), which puts an additional constraint on the time step size ΔT. These relations, however, show that ΔT is dependent on the velocity field as well as on h, Pr, and Gr; its size will therefore vary as fluid motion gradually develops. A program can be written in such a way that the maximum allowable size of ΔT at every time step is determined to satisfy all stability criteria at every grid point. To make it simple, we use a constant value of 0.0025 for ΔT in Program 4.5, which is obtained after several trial runs with larger and smaller step sizes for $Pr = 6.75$ and $Gr = 1000$.

Having determined the appropriate time step size, we first run the program for a long period of time with MAXSTP = 2000 and print out numerical results at some selected time instants. The output reveals that no drastic change in flow pattern occurs after the step counter NSTEP is beyond 400. In Program 4.5, as shown, we finally use MAXSTP = 400, costing approximately 5 min on a CDC 6400 computer. Also, to avoid printing a large amount of data and for better visualization of the flow, in the output of this program only some representative stream patterns are plotted at several critical stages of the flow development.

A subroutine named CONTUR is constructed for the sole purpose of plotting the flow. Here we use the method described in Section 2.11 to print different symbols in the fluid domain, depending on the local value of the stream function. The final picture displays a rough contour plot of the flow pattern. Most of the variable names in this subroutine are kept in conformity with those in Program 2.8, whose definitions can be found in Section 2.11. There is a slight modification, however, in this subroutine. Since the same subroutine is called at different stages involving both slow and fast fluid motions, it is not convenient in the present case to choose a constant interval size δ as before to subdivide the large range of stream function. For example, at an earlier stage Ψ is of order of ± 0.001, whereas its peak values increase to around ± 1.0 at later times. This difficulty is avoided by dividing Ψ into a number of subdivisions of any desired span, whose end values are specified as the elements of a one-dimensional array called VALUE. The printing symbols corresponding to these subdivisions are specified as the elements of the array SYMBOL, just the same as before.

Flow pattern is plotted only at particular time steps when NSTEP equals one of the selected values stored in the array NPLOT. These values, specified

Bénard and Taylor Instabilities

in a data statement in the main program, are selected after inspecting the printed data output. We use NFIG in the main program to indicate the order of the figures to be plotted. Since an output page of the computer can contain only two figures of our specified size, we start each of the odd-numbered figures on top of a new page. The total number of figures to be printed in Program 4.5 is eight.

```
C                       ***** PROGRAM 4.5 *****
C          INSTABILITY OF A LAYER OF LIQUID HEATED FROM BELOW
      DIMENSION PSI(21,9),U(21,9),V(21,9),OMEGA(21,9),THETA(21,9),
     A           OMNEW(21,9),THNEW(21,9),RHS(21,9),NPLOT(8)
      COMMON H,DT,I,J, UF,UFA, UB,UBA, VF,VFA, VB,VBA
      DATA GR,PR,MAXSTP / 1000.0, 6.75, 400 /
      DATA NPLOT / 10, 20, 50, 80, 100, 200, 300, 400 /
      PRINVS = 1.0 / PR
      DT = 0.0025
      M = 21
        MM1 = M-1
      N = 9
        NM1 = N-1
      H = 1.0 / (N-1.)
C     ..... SPECIFY INITIAL FLOW AND TEMPERATURE FIELDS.  THETA,
C           U, V AT SOLID WALLS, AND U, OMEGA AT I=M WILL BE
C           KEPT THE SAME FOR ALL TIMES .....
      NSTEP = 0
      NFIG = 0
      T = 0.0
      DO 1  I = 1, M
      DO 1  J = 1, N
      U(I,J) = 0.0
      V(I,J) = 0.0
      OMEGA(I,J) = 0.0
      THETA(I,J) = 0.0
    1 IF( J.EQ.1 )  THETA(I,J) = 1.0

C     ..... COMPUTE STREAM FUNCTION BASED ON VORTICITY FIELD .....
    2 DO 3  I = 1, M
      DO 3  J = 1, N
    3 RHS(I,J) = -OMEGA(I,J)
      CALL SORLX( PSI, RHS, M, N, H, 0.0001, ITER )

C     ..... COMPUTE U AND V AT INTERIOR POINTS .....
      DO 4  I = 2, MM1
      DO 4  J = 2, NM1
      U(I,J) = (PSI(I,J+1)-PSI(I,J-1)) / (2.*H)
    4 V(I,J) = -(PSI(I+1,J)-PSI(I-1,J)) / (2.*H)
      DO 5  J = 2, NM1
    5 V(M,J) = PSI(MM1,J) / H

C     ..... INCREASE TIME STEP COUNTER BY ONE.  TERMINATE COMPUTATION
C           IF NSTEP EXCEEDS MAXSTP, OTHERWISE COMPUTE VORTICITY AND
C           TEMPERATURE AT ALL INTERIOR POINTS.  THEY ARE STORED
C           TEMPORARILY AS THE ELEMENTS OF ARRAYS OMNEW AND THNEW .....
      NSTEP = NSTEP + 1
      IF( NSTEP.GT.MAXSTP )  STOP
      T = T + DT
      DO 6  I = 2, MM1
      DO 6  J = 2, NM1
```

```
      UF   = (U(I+1,J)+U(I,J)) / 2.0
      UFA  = ABS( UF )
      UB   = (U(I-1,J)+U(I,J)) / 2.0
      UBA  = ABS( UB )
      VF   = (V(I,J+1)+V(I,J)) / 2.0
      VFA  = ABS( VF )
      VB   = (V(I,J-1)+V(I,J)) / 2.0
      VBA  = ABS( VB )
      OMNEW(I,J) = PNEW( OMEGA, THETA, M, N, GR, 1.0 )
    6 THNEW(I,J) = PNEW( THETA, THETA, M, N, 0.0, PRINVS )
      DO 7  J = 2, NM1
      UB   = U(M-1,J) / 2.0
      UBA  = ABS( UB )
      VF   = (V(M,J+1)+V(M,J)) / 2.0
      VFA  = ABS( VF )
      VB   = (V(M,J-1)+V(M,J)) / 2.0
      VBA  = ABS( VB )
      P1 = -2.*(UB+UBA)*THETA(MM1,J) + 2.*(-UB+UBA)*THETA(M,J)
      P2 = (VF-VFA)*THETA(M,J+1) + (VF+VFA-VB+VBA)*THETA(M,J)
     A         - (VB+VBA)*THETA(M,J-1)
      P4 = 2.*THETA(MM1,J) + THETA(M,J+1) + THETA(M,J-1) - 4.*THETA(M,J)
      THNEW(M,J) = THETA(M,J) + DT/(2.*H)*( -P1 - P2 + 2.*PRINVS*P4/H )
    7 OMNEW(M,J) = 0.0

C     ..... THE ELEMENTS OF OMNEW AND THNEW ARE THEN
C           ASSIGNED TO OMEGA AND THETA, RESPECTIVELY .....
      DO 8  I = 2, M
      DO 8  J = 2, NM1
      OMEGA(I,J) = OMNEW(I,J)
    8 THETA(I,J) = THNEW(I,J)

C     ..... OMEGA AT SOLID WALLS ARE UPDATED .....
      DO 9  I = 1, M
      OMEGA(I,1) = (-4.*U(I,2)+U(I,3)) / (2.*H)
    9 OMEGA(I,N) = (4.*U(I,N-1)-U(I,N-2)) / (2.*H)
      DO 10  J = 2, NM1
   10 OMEGA(1,J) = (4.*V(2,J)-V(3,J)) / (2.*H)

C     ..... PLOT FLOW PATTERN WHEN NSTEP EQUALS ONE OF THE VALUES
C           SPECIFIED IN NPLOT.  IN EITHER CASE UPDATE STREAM
C           FUNCTION AND THEN REPEAT THE SAME PROCESS .....
      DO 11  K = 1, 8
      IF( NSTEP-NPLOT(K) )  11, 12, 11
   11 CONTINUE
      GO TO 2
   12 NFIG = NFIG + 1
      IF( NFIG/2*2 .EQ. NFIG )  GO TO 14
      WRITE(6,13)
   13 FORMAT( 1H1 )
   14 WRITE(6,15)
   15 FORMAT( // )
      CALL CONTUR( PSI, 2.5, 1.0, 100, H )
      WRITE(6,16)  T
   16 FORMAT( // 66X, 3HT =, F5.3 )
      GO TO 2
      END
```

```
      SUBROUTINE SORLX( F, Q, M, N, H, ERRMAX, ITER )
C     ..... SUCCESSIVE OVERRELAXATION SCHEME (2.10.13) IS USED TO
C           SOLVE POISSON EQUATION (2.10.1),   LAP( F ) = Q,   IN A
C           RECTANGULAR DOMAIN OF SIZE  (M-1)H * (N-1)H.  ITERATION
C           IS STOPPED WHEN THE SUM OF ABSOLUTE ERRORS AT ALL GRID
C           POINTS BECOMES LESS THAN OR EQUAL TO ERRMAX.  ITER IS
C           THE TOTAL NUMBER OF ITERATIONS PERFORMED.  BOUNDARY
C           CONDITIONS USED HERE ARE THAT F=0 ON EACH OF THE FOUR
C           BOUNDING SURFACES .....

      DIMENSION F(M,N),Q(M,N)
      MM1 = M - 1
      NM1 = N - 1

C     ..... CALCULATE THE OPTIMUM VALUE OF OMEGA
C           ACCORDING TO (2.10.14) .....
      PI = 4.0 * ATAN(1.0)
      ALPHA = COS(PI/M) + COS(PI/N)
      OPTOM = (8.0-4.*SQRT(4.-ALPHA**2)) / ALPHA**2

C     ..... AT THE BEGINNING ASSUME A CONSTANT VALUE OF ZERO FOR F,
C           THE BOUNDARY VALUES ARE THUS AUTOMATICALLY INCLUDED .....
      ITER = 0
      DO 1  I = 1, M
      DO 1  J = 1, N
    1 F(I,J) = 0.0

C     ..... BEFORE EACH ITERATION, ADD ONE TO ITER
C           AND SET ERROR TO ZERO .....
    2 ITER = ITER + 1
      ERROR = 0.0

C     ..... COMPUTE F(I,J) AT INTERIOR POINTS USING (2.10.13) .....
      DO 3  I = 2, MM1
      DO 3  J = 2, NM1
      FOLD = F(I,J)
      F(I,J) = F(I,J) + 0.25*OPTOM*( F(I-1,J) + F(I+1,J) + F(I,J-1)
     A                  + F(I,J+1) - 4.*F(I,J) - H*H*Q(I,J) )
    3 ERROR = ERROR + ABS(F(I,J)-FOLD)

C     ..... ITERATE UNTIL ERROR IS ≤ ERRMAX .....
      IF( ERROR.GT.ERRMAX )  GO TO 2
      RETURN
      END




      SUBROUTINE CONTUR( PSI, XRANGE, YRANGE, NX, H )
C     ..... GENERATING A CONTOUR PLOT OF PSI IN A RECTANGULAR DOMAIN
C           OF SIZE  (XRANGE)*(YRANGE).  NX IS THE NUMBER OF SYMBOLS
C           TO BE PRINTED IN A HORIZONTAL LINE, AND H IS THE SIZE
C           OF THE SQUARE GRIDS BASED ON WHICH VALUES OF PSI(I,J)
C           ARE COMPUTED.  SYMBOLS SPECIFIED IN THE ARRAY *SYMBOL*
C           ARE THOSE USED FOR PSI FALLING WITHIN THE INTERVALS
C           BOUNDED BY VALUES SPECIFIED IN THE ARRAY *VALUE*.  SEE
C           PROGRAM 2.8 FOR DEFINITION OF VARIABLE NAMES .....
```

368

```
      DIMENSION PSI(21,9),VALUE(17),SYMBOL(18),GRAPH(100)
      DATA VALUE / -.9, -.5, -.3, -.2, -.05, -.01, -.003, -.001,
     A              0., .001, .003, .01, .05, .2, .3, .5, .9 /
      DATA SYMBOL / 1HD, 1H , 1HC, 1H , 1HB, 1H , 1HA, 1H , 1H-,
     A              1H+, 1H , 1H1, 1H , 1H2, 1H , 1H3, 1H , 1H4 /
      REAL LX,LY
      NY = 0.6 * NX * YRANGE / XRANGE
      DELX = XRANGE / NX
      DELY = YRANGE / NY
      DO 5  JSYMBL = 1, NY
      Y = YRANGE - (JSYMBL-0.5)*DELY
      J = 1 + Y/H
         LY = Y - (J-1)*H
         HMLY = H - LY
      DO 4  ISYMBL = 1, NX
      X = (ISYMBL-0.5) * DELX
      I = 1 + X/H
         LX = X - (I-1)*H
         HMLX = H - LX
         A1 = HMLX * HMLY
         A2 = HMLX * LY
         A3 = LX * LY
         A4 = LX * HMLY
      PSIC = ( A1*PSI(I,J) + A2*PSI(I,J+1) + A3*PSI(I+1,J+1)
     A         + A4*PSI(I+1,J) ) / H**2
C
C     ..... DETERMINE NRANGE BASED ON PSIC.  LET NRANGE EQUAL 18
C           IF PSIC IS GREATER THAN VALUE(17).  CHARACTERS STORED
C           IN THE ARRAY *GRAPH* ARE THEN PRINTED ROW BY ROW .....
      DO 2  K = 1, 17
      IF( PSIC-VALUE(K) )  1, 1, 2
1     NRANGE = K
      GO TO 3
2     CONTINUE
      NRANGE = 18
3     GRAPH( ISYMBL ) = SYMBOL( NRANGE )
4     CONTINUE
5     WRITE(6,6)  ( GRAPH(I), I=1,NX )
6     FORMAT( 20X, 100A1 )
      RETURN
      END




      FUNCTION PNEW( P, Q, M, N, A, B )
C     ..... INTEGRATING THE FOLLOWING EQUATION WITH RESPECT TO T :
C           DP/DT = -D(UP)/DX - D(VP)/DY + A*DQ/DX + B*LAP(P)
C           IN WHICH D AND LAP REPRESENT PARTIAL DERIVATIVE AND THE
C           LAPLACIAN OPERATOR, RESPECTIVELY .....
      DIMENSION P(M,N),Q(M,N)
      COMMON H,DT,I,J, UF,UFA, UB,UBA, VF,VFA, VB,VBA
      P1 = (UF-UFA)*P(I+1,J) + (UF+UFA-UB+UBA)*P(I,J)
     A     - (UB+UBA)*P(I-1,J)
      P2 = (VF-VFA)*P(I,J+1) + (VF+VFA-VB+VBA)*P(I,J)
     A     - (VB+VBA)*P(I,J-1)
      P3 = Q(I+1,J) - Q(I-1,J)
      P4 = P(I+1,J) + P(I-1,J) + P(I,I+1) + P(I,J-1) - 4.*P(I,J)
      PNEW = P(I,J) + DT/(2.*H)*( -P1 - P2 + A*P3 + 2.*B*P4/H )
      RETURN
      END
```

```
T = .025
```

```
T = .050
```

```
T = .125
```

370

T = .200

T = .250

T = .500

```
  111                                    11  A                              AAA  1111                              1111  •
1          222222222222222222222       1A    BBBBBBBBAABBABBBABBBBBBBBBBA      A1                2222222222                    1
1       222222222222222222222222222     A  BBBBBBBBBBB       ABBBBBBBBBBBBBB    A1     222222222222222222222222         1
1    2222222                      2?222   •BBBBB     CCCCCCC         BBBBAABB   A    2222222222222222222222222222     1
      22222      333333333333333   222  •BBBB   CCCCCCCCCCCCCCCCC    BBBBB   A  2222222222        2222222222222     1
      2222  3333333        333333 22   •BB  CCCCC        CCCCCCC   BBBB A 222222            222222?
      2222  33333          3333 22? AB   CCC              CCCC  BBB A 2222    133333333       222222
      222  3333             137  22AAB CCC                CCCC  BBB 12222   3333333333333333       2222
      222  333              333 22AAB CCC                CCCC  BB  222   3333333333333333       2222
      222 333              13 22 AB CC                   CCC  BB  222  333333333333333333      2222
      222 333                •   333 2  AB CC    DDDDDD    CCC  BA  22   333333      13333333   2222
      22  333     444444     33 ? AB CC   DDDDDD    CCC BBA222 333333      3333333     222
      22  333    4444444     33 ? AB CC   DDDDDD    CCC BBB1222 33333      333333     222
      22  333    444444      33 ? AB CC    DDDDD    CCC BB  22  333333      3333333    2222
      222 333     444        33 ? AB CC              CCC  BB  22  333333      3333333   2222
      222 333               133 ?? B CCC            cCCC BAB  22  33333333     13333333   2222
      222  333              133 ?? BB  CCC          CCCC  BBB  22  333333333333333333    2222
      222  3333             333 ?? BB  CCC          CCCCC RBB  2222  333333333333333333   2222
      2222  33333          3333 ??? BB  CCCCC       CCCCCCC  BBBA 2222   333333333333    222222?
       2222  33333333333333333333  222 •ABBB  CCCCCCCCCCCCC    BBBA A 222222              22222222?
       222222   333333333333      2222 • BBBBB          BBABBA  •  2222222222        2222222222222?    1
1      22222222                 22222?22    •  BBBBBBBBB BBBAABBBBBBBABBA  1  222222222?222222222222222   1
1      2222222222222222222222?2?22        •  ABBBBBBABBBABBBABBBBBBB    A    2222222222?2222?2222222     11
   11                                  1•AA                            AA •111                            1111
```

                              T = .750

```
  111                                    11•AA          •        BBBBBABBBBBBABABBBBBBBBBBBA      A1  ?2?22222?22222??222?22??222222?     1
1          222222222222222222222222       •   BBBBBBBBBBB       ABBBBBBBBBBB    A  222222222          22222222     •
1       222222222222222 22222222222     •  BBBBB       CCCCCCC     BBAAB A 22222   133333333333?    22222?
      2222222                  2?2222  •  BBB   CCCCCCCCCCCCCCCC    BBBBA 222   333333333333333333333       222
      22222   333333333333333   222  BBB  CCCCC            CCCCC  BBBA 222      33333333333333        222
      2222  333333        3333333  3333333       ??? BBB  CCCC    CCCCC  BBHA 22  33333       3333  ??
      2222  3333            333 22•BB  CCC         CCC  BBA222 3333       3333 ??
      222  3333             333  221BB  CCC         CCC  BBA22  333       33 ?
      222? 3333            133 22?BB  CCC          CCC  BBA22 33        33 ?
      22   333            33 22?BB  CC             CC  BBA22 33    444444    33 ??
      22   333            33  2?BB  CC    DD          CC  BBA22 33   444444444   33 ??
      22   333    44      333 ? BB  CC    DDDDD       CC  BB•22 33  44444444444   13 ??
      22   333   444444   333 ? BB  CC   DDDDDDDD     CC  BB 22 33  444444444444   13 ??
      22   333   4444444  333 ? BB  CC   DDDDDDDD     CC  BB122 33  4444444444    13 ??
      22   333   444444   333 ? BB  CC    DDDDDD      CCC  BB122 33  44444444     13 ??
      22   333    44      333 ? BB  CC     DDDN      CCC  BA122 333    444444     13 ??
      222 333             33  ? BB  CCC            CCC  BA122 333               333 ?
      222  333            133 ?? BB  CCC          CCC  BB122 333              133 ??
      2222  3333          3133 ?? BBB  CCCC       CCCCC  BBB122  333            1333 ??
      2222  333333       33333 ??? BB  CCCCC     CCCCCCC  BBB 1 222  333333333    333?333? ??
      222222   333333333333333333333  22? ABB  CCCCCCCCCCCCCC    BHB  2222   3333333333333333    22????
       222222   333333333333      22222  BBBBB  CCCCCCCCC     HBABBABB   2222222       2222222?
1     22222222                22222222   BBBBBBBBB        HBABBBABBH    2222222          222222??
1     22222222222222222222222222222  1A   BBBBBBBBBBBBABBAABBBBBBBBH    •  22222222222?22222222?2222222?    1
   11                                  1  A                          AA•11                            2
```

                              T =1.000

Program 4.5 computes for the originally stationary flow confined within a rectangular region bounded by cold top and side walls and a hot bottom plate. Shown in the output is the time history of the flow pattern in the left half of the fluid region.

At an early stage when $T = 0.025$, a weak convective motion develops in the fluid at the corner where the cold and hot walls intersect. Elsewhere the fluid is practically motionless. The plotting symbols indicate that the stream function is positive everywhere at this moment. Along a closed streamline assuming a positive value of Ψ, the fluid motion is in the counterclockwise direction. Conversely, the motion is clockwise along closed streamlines of negative Ψ. Thus the fluid descends along the cold vertical wall and then rises after flowing over the hot surface. This motion becomes stronger at $T = 0.05$, but two bubbles containing clockwise fluid motions are forming on the top and bottom plates at the midplane.

Very soon these two bubbles are connected. The region of negative Ψ expands from the middle and pushes gradually toward the side wall, as revealed by the plot showing two convective cells at $T = 0.125$. In the meantime, the motion near the side wall is intensified. While this trend continues, a small bubble containing counterclockwise motion starts to appear at the center of the bottom plate, as shown at $T = 0.2$. The plots for $T = 0.25$, 0.5, and 0.75 describe the expansion of this bubble and the continuing intensification of the motion in all three cells.

At $T = 1.0$, when NSTEP = 400, the sizes of all three cells are approximately equal. Printed numerical data disclose that at this instant the strongest motion occurs in the cell on the right, and the cell on the left is the weakest. This situation is not always true, however. Subsequent data show that as time increases, the motion in the left cell is strengthened whereas those in the other two are weakened. At $T = 5.0$ the cell on the left becomes the strongest and the middle one the weakest. The differences in peak velocities among the cells are within 20% of the value in the left cell.

After $T = 1.0$ the changes in stream pattern are not drastic. At $T = 5.0$ the left boundary of the middle cell has shifted to the right from its position shown in the plot for $T = 1.0$. The displacement of the lower portion of that boundary is a little longer than that of the upper portion, resulting in a curved interface. Although a steady state has not been reached yet at $T = 5.0$, the flow does not seem to have any further significant changes. Additional computations with MAXSTP greater than 2000 have not been attempted because of the tremendous amount of computer time required for such work.

Temperature profiles are not plotted in the output because of their slow variation with respect to time. In general, the isothermal lines, except those for $\theta = 0$ and 1, have the shape of the cutaway view of an opened umbrella.

Since the pictures shown in the output describe only the behavior of half of the flow, the total number of convective cells to appear in the channel under the present arrangement is two at the beginning, changes to four later, and finally becomes six as the flow continuously evolves. Thus we start from an unstable situation of the fluid system having a density inversion; the numerical solution of the governing equations leads us to a final state, at which the system may be said to be most stable under the imposed temperature boundary conditions. Once the computer program has been written, we can easily change the dimensions of the channel, try different fluid media by varying the Prandtl number, or examine the effect of temperature gradient by assigning various values to the Grashof number, and we observe the results of our numerical experiments on the computer.

There are many advantages to using numerical methods to examine the stability of a fluid flow. If a linearized analytic method were used to study the problem considered in this section, a certain number of cells might be predicted for the onset of instability. However, this prediction may not be valid at later stages, when the ignored inertial force becomes important, as revealed by our numerical result that the number of cells is changing with time. It seems there is no guarantee that the most unstable condition predicted by a linearized theory will show up in the final state. Furthermore, the inequality in cell size and the curved cell boundaries are all nonlinear phenomena and cannot be predicted using linearized theories. Higher-order analyses are generally tedious and are usually formulated based on the linearized result. The validity of the result so obtained is also uncertain.

The numerical solution in Program 4.5 is obtained under the assumption of a two-dimensional flow. The result is unrealistic by virtue of the fact that the observed convection cells are hexagonal when looking from the top, so that the actual fluid motion is three-dimensional.

Problem 4.10 Perform a numerical experiment on the fluid system shown in Fig. 4.6.1, but with some of the boundary conditions changed. It is now assumed that at the initial instant, $\theta = 0$ everywhere, except $\theta = 1$ on the upper surface. Thereafter the temperatures at all bounding surfaces will be kept at their initial values. The upper surface is considered to be an undeformable free surface where both vorticity and shear stress vanish. Assume that there is no initial motion in the water layer.

The situation stated in this problem is somewhat similar to that in an ocean region bounded by two large icebergs, when the upper surface is heated by the sun.

Problem 4.11 The general circulation of the earth's atmosphere is the result of temperature gradients parallel to the earth's surface caused by differential solar heating between polar and equatorial regions, even if the vertical stratification of the atmosphere is considered stable. The global horizontal temperature gradients may cause convective motions in the core of the earth as well as in the ocean.

To study the effect of a horizontal temperature gradient, we consider a two-dimensional flow contained between two infinitely long parallel plates whose surfaces are normal to the direction of the gravitational acceleration. Using the distant H between plates as the reference length and introducing dimensionless variables defined in (4.6.5), we obtain the same set of governing equations as shown in (4.6.6) to (4.6.10).

For the initial condition we assume that $\theta = 0$ everywhere except that on the upper plate a periodic temperature distribution

$$\theta = 1 + 0.5 \sin \left(2\pi \frac{X}{\lambda} \right)$$

is imposed and is kept unchanged thereafter. The temperature on the bottom plate is also maintained always at $\theta = 0$.

Because the fluid motion driven by this periodic temperature gradient is also periodical, we only need to consider the region $0 \leqslant X \leqslant \lambda$. *Periodic boundary conditions* are applied at these two ends. More specifically, if these two vertical surfaces are represented in index notation by $i = 1$ and m, respectively, and if we let S be any scalar variable, the periodic condition for S requires that

$$S_{1,j} = S_{m,j}$$
$$S_{0,j} = S_{m-1,j}$$
$$S_{2,j} = S_{m+1,j}$$

Thus, in computing the vertical velocities on the left surface by a central-differencing scheme, for example, we write

$$V_{1,j} = -\frac{1}{2h} (\Psi_{2,j} - \Psi_{0,j})$$

$$= -\frac{1}{2h} (\Psi_{2,j} - \Psi_{m-1,j})$$

so that it can be evaluated from the stream function within the concerned fluid domain.

For numerical computation let $Pr = 6.75$, $Gr = 1000$, and $\lambda = 2.5$. Plot flow pattern and isothermal contours at several representative time steps.

A flow may become unstable when going along a curved path. This phenomenon can be observed, for example, in the fluid contained between two concentric cylinders rotating at different speeds. When the two speeds are in a right combination, the flow cannot maintain its purely angular motion and becomes unstable. The instability, shown in the form of a series of donut-shaped ring vortices, was first examined by Taylor (1923) using a linearized analysis and is usually referred to as the *Taylor instability*. It turns out that the governing equations and therefore the analysis for this problem are very similar to those for the Bénard problem just considered.

The governing equations for an incompressible fluid were shown at the beginning of Section 4.5, with the omission of the electromagnetic terms for the present case. Let r_i and ω_i be the radius and angular speed of the inner cylinder and r_0 and ω_0 those of the outer. By choosing ω_i^{-1} as the reference time, r_i as the reference length, and $r_i\omega_i$ as the reference speed, a set of dimensionless variables similar to those shown in (4.5.8) can be defined, in which Ω is the θ component of the nondimensionalized vorticity. In addition, because of the presence of the angular velocity u_θ, a new variable is defined as

$$W = \frac{ru_\theta}{r_i^2\omega_i} \tag{4.6.32}$$

which is the angular momentum expressed in dimensionless form. If the motion is still assumed to be axisymmetric, the system of nondimensionalized governing equations consists of (4.5.10) to (4.5.12), and

$$\frac{\partial \Omega}{\partial T} = -\frac{\partial(U\Omega)}{\partial Z} - \frac{\partial(V\Omega)}{\partial R} - 2\frac{W}{R^3}\frac{\partial W}{\partial Z} + \frac{1}{Re}\left(\frac{\partial^2}{\partial R^2} + \frac{1}{R}\frac{\partial}{\partial R} - \frac{1}{R^2} + \frac{\partial^2}{\partial Z^2}\right)\Omega \tag{4.6.33}$$

$$\frac{\partial W}{\partial T} = -\frac{\partial(UW)}{\partial Z} - \frac{\partial(VW)}{\partial R} - \frac{VW}{R} + \frac{1}{Re}\left(\frac{\partial^2}{\partial R^2} - \frac{1}{R}\frac{\partial}{\partial R} + \frac{\partial^2}{\partial Z^2}\right)W \tag{4.6.34}$$

in which $Re = r_i^2\omega_i/\nu$ is the characteristic Reynolds number. (4.6.33) differs from (4.5.13) in that the term containing W has replaced the electromagnetic force term. (4.6.34) is the θ component of the Navier-Stokes equation, which was not needed previously. The nonlinear terms in these two equations are expressed purposely in the same form.

A computational procedure can be set up for solving the rotating flow problem that is analogous to that used for solving the convection problem. You are then ready to make numerical studies of the stability and other nonlinear features of rotating flows through the following suggested problems.

Problem 4.12 Consider an already established steady flow between infinitely long concentric cylinders, whose boundary conditions are

$$W = 1 \qquad \text{at } R = 1 \tag{4.6.35}$$

$$W = \frac{r_0^2\omega_0}{r_i^2\omega_i} \qquad \text{at } R = \frac{r_0}{r_i} \tag{4.6.36}$$

Since there is no motion in the meridian plane, $U = V = 0$, so the angular motion is described by

$$W = [(A^2B - 1)R^2 + A^2(1 - B)]/(A^2 - 1) \tag{4.6.37}$$

which is the steady-state solution of (4.6.34) satisfying boun-

Secondary Flows and Flow Instabilities

dary conditions (4.6.35) and (4.6.36). The parameters

$$A = \frac{r_0}{r_i} \quad \text{and} \quad B = \frac{\omega_0}{\omega_i} \quad (4.6.38)$$

are, respectively, the ratio of the radii and that of the angular speeds; they are given constants.

The purely angular flow is then perturbed slightly in the meridian plane at time $T = 0$. The small perturbation is represented by a weak periodic vorticity distribution

$$\Omega = 0.01 \sin \left(2\pi \frac{Z}{\lambda} \right) \quad (4.6.39)$$

where λ is the dimensionless wavelength. The meridian motion associated with this vorticity is computed from (4.5.10) and (4.5.11) after the stream function has been obtained by solving the equation (4.5.12). Consequently, the angular motion will be modified according to (4.6.34), causing Ω to change. The time history of the flow variation is then traced by progressing in T.

On solid walls both U and V vanish for all times. Responding to a periodic disturbance the resultant fluid motion is also periodic, so that only an axial length of λ is needed in the computation. The periodic boundary conditions stated in Problem 4.11 are applied at the ends of this fluid region.

For numerical computation we assume that $A = 2$, $B = 0$ (for a stationary outer cylinder), $\lambda = 4$, and $Re = 1000$. Suggested intervals are $h = 0.125$ and $\Delta T = 0.02$. The result should show the development of Taylor ring vortices between the cylinders. Similar flow patterns can be found in the work by Strawbridge and Hooper (1968), obtained using a slightly different numerical scheme.

Problem 4.13 Find the steady-state solution for a flow through a rotating tube having an abrupt contraction, as shown in Fig. 4.6.2. At the entrance the flow is in rigid body rotation with the same angular speed ω_0 as that of the tube, and the axial motion is uniform of speed u_0.

In this problem it is natural to choose the tube radius r_0 as the reference length and u_0 the reference speed. The dimensionless variables defined in (4.5.8) can be used directly here, except that the magnetic field is no longer needed. Instead of

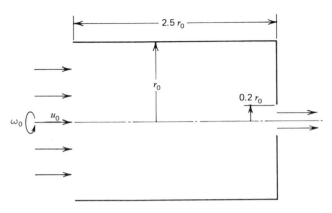

FIGURE 4.6.2 Rotating tube flow through a constriction.

(4.6.32), the dimensionless angular momentum is redefined as

$$W = \frac{ru_\theta}{r_0^2 \omega_0}$$ (4.6.40)

After they are nondimensionalized, the governing equations are the same as those for the previous problem, except that (4.6.33) is now replaced by

$$\frac{\partial \Omega}{\partial T} = -\frac{\partial (U\Omega)}{\partial Z} - \frac{\partial (V\Omega)}{\partial R} - \frac{1}{2R_O^2}\frac{W}{R^3}\frac{\partial W}{\partial Z}$$
$$+ \frac{1}{Re}\left(\frac{\partial^2}{\partial R^2} + \frac{1}{R}\frac{\partial}{\partial R} - \frac{1}{R^2} + \frac{\partial^2}{\partial Z^2}\right)\Omega$$ (4.6.41)

in which $Re = u_0 r_0/\nu$ is the Reynolds number and $Ro = u_0/2r_0\omega_0$ is the *Rossby number*, whose value indicates the relative magnitude between the axial and angular motions of the flow.

Boundary conditions at the entrance have already been specified. For the remaining boundaries we assume no-slip conditions on the rotating tube wall, and we require that $V = 0$ at the exit. Those boundary conditions concerning stream function are specifically stated as follows: $\Psi = 0$ along the tube axis, $\Psi = \frac{1}{2}$ on the wall, $\Psi = \frac{1}{2}R^2$ at the entrance and, finally, $\partial \Psi/\partial Z = 0$ at the exit. Note that (4.5.26) is to be used to calculate the velocity along the tube axis.

For a fixed value $Re = 50$, plot the flow pattern in the meridian plane for $Ro = 2, 1, 0.5$, and 0.2.

Secondary Flows and Flow Instabilities

According to a computation (Yih, 1965, p. 260) for an inviscid rotating flow into a sink (which can be formed by letting the cross-sectional area of our exit approach zero), at small Rossby numbers the flow is found to separate from the axis, forming a secondary flow upstream from the sink.

Problem 4.14   In the book by Scorer (1958), the author used a cartoon as the last problem for discussion. It shows a lady, having arrived at the bottom of a long spiral stairway with a tray, serving coffee to two gentlemen. The caption states: "You'll find it's already stirred."

As the last problem in this book, let us check on the computer to see if what was said by the lady is correct. An idealized coffee cup is shown in Fig. 4.6.3. The motion of the cup can be decomposed into a translation and a rotation. Only the rotation has an effect on the fluid motion inside the cup. At the initial instant when the lady starts to come down the stairway, there is no motion in the coffee, and the cup is suddenly given an angular velocity ω_0. Suppose the lady is moving at a constant pace so that the angular velocity is maintained at the same value. The upper surface of the liquid is assumed to be horizontal and free of shear stresses.

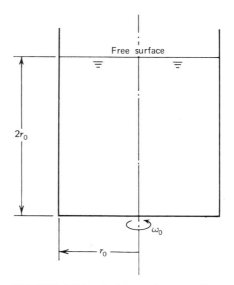

FIGURE 4.6.3 Spin-up in a coffee cup.

Bénard and Taylor Instabilities   **379**

In this problem there is no net axial flow, the dimensionless governing equations are exactly the same as those used in Problem 4.12, except that the reference length and speed are now r_0 and $r_0\omega_0$, respectively, and the dimensionless angular momentum is redefined as $W = ru_\theta/r_0^2\omega_0$. We assume $r_0 = 0.04$ m and $\omega_0 = 0.1 \pi$ rad/s (computed for 3 r/min). In spite of lacking data on the physical properties of hot coffee, we use the value of 10^{-6} m^2/s for kinematic viscosity, which is approximately that of water. The characteristic Reynolds number, $r_0^2\omega_0/\nu$, is 500 after rounded off.

Derive two expressions for ΔT, in a form similar to that of (4.5.35) or (4.6.31), by applying the quasilinear analysis of Lax and Richtmyer on the governing equations (4.6.41) and (4.6.34). In your program, at every time step the lowest value of ΔT, which is obtained by searching through the values at all grid points, is to be used for numerical computations. Selectively plot the flow pattern until the dimensionless time reaches the value 8. The reference time in this problem is $1/\omega_0$.

This is a simplified *spin-up problem*, studying the secondary flows and the development of boundary layers on surfaces in a rotating fluid. The spin-up problems that interest geophysicists and astrophysicists are usually concerned with more complex geometries and may include other effects, such as buoyancy and electromagnetic forces. A general description of the spin-up phenomenon can be found in Greenspan (1968).

APPENDIX

Listing of Some Important Subprograms and Their Usage

---

1. Subprograms for solving algebraic equations:
   TRID (Program 2.1, p. 83) solves a system of linear algebraic equations whose coefficient matrix is in the tridiagonal form.
   DETERM (Program 2.4, p. 113) computes the value of the determinant of a square matrix.
   CRAMER (Program 2.4, p. 114) solves a set of simultaneous linear algebraic equations using Cramer's rule.
   NEWTN1 (Program 4.1, p. 303) finds the root of the equation $F(x) = 0$ using the Newton-Raphson method.
   NEWTN2 (Program 4.1, p. 303) finds the roots of simultaneous equations $G(x, y) = 0$ and $H(x, y) = 0$ using the Newton-Raphson method.

2. Subroutines for solving ordinary differential equations:
   RUNGE (Program 1.2, p. 22) solves a second-order differential equation

$$\frac{d^2x}{dt^2} = F\left(x, \frac{dx}{dt}, t\right)$$

   using the fourth-order Runge-Kutta method.
   KUTTA (Program 1.4, p. 39) solves two simultaneous second-order differential equations

$$\frac{d^2x}{dt^2} = F_1\left(x, y, \frac{dx}{dt}, \frac{dy}{dt}, t\right)$$

$$\frac{d^2y}{dt^2} = F_2\left(x, y, \frac{dx}{dt}, \frac{dy}{dt}, t\right)$$

   using the fourth-order Runge-Kutta method.

381

RK4 (Program 3.1, p. 225) solves a first-order differential equation

$$\frac{dx}{dt} = F(x, t)$$

using the fourth-order Runge-Kutta method.
RUNTTA (Program 4.2, p. 317) solves two simultaneous first-order differential equations

$$\frac{dx}{dt} = F_1(x, y), \qquad \frac{dy}{dt} = F_2(x, y)$$

using the fourth-order Runge-Kutta method.

3. Subprograms for solving partial differential equations:
LIEBMN (Program 3.7, p. 284) solves the Poisson equation

$$\frac{\partial^2 f}{\partial x^2} + \frac{\partial^2 f}{\partial y^2} = q(x, y)$$

within a square domain using Liebmann's iterative method.
LAPLCE (Program 4.4, p. 349) solves the equation

$$\left(\frac{\partial^2}{\partial R^2} - \frac{1}{R}\frac{\partial}{\partial R} + \frac{\partial^2}{\partial Z^2}\right) RH = 0$$

satisfying special boundary conditions (4.5.14) to (4.5.18) using the iterative scheme (4.5.22).
POISSN (Program 4.4, p. 350) solves the equation

$$\left(\frac{\partial^2}{\partial R^2} - \frac{1}{R}\frac{\partial}{\partial R} + \frac{\partial^2}{\partial Z^2}\right) \Psi = R\Omega$$

satisfying boundary conditions (4.5.19) to (4.5.21) using the iterative scheme (4.5.25).
SORLX (Program 4.5, p. 368) solves the Poisson equation

$$\frac{\partial^2 f}{\partial x^2} + \frac{\partial^2 f}{\partial y^2} = q(x, y)$$

in a rectangular domain using the successive overrelaxation method.
PNEW (Program 4.5, p. 369) solves the equation

$$\frac{\partial P}{\partial T} = -\frac{\partial(UP)}{\partial X} - \frac{\partial(VP)}{\partial Y} + A\frac{\partial Q}{\partial X} + B\left(\frac{\partial^2 P}{\partial X^2} + \frac{\partial^2 P}{\partial Y^2}\right)$$

using an upwind-differencing scheme.

4. Subprograms for vector manipulations:

VADD (Program 2.3, p. 102) computes the sum of two vectors.

VSBT (Program 2.3, p. 102) computes the difference of two vectors.

VMAG (Program 2.3, p. 102) computes the magnitude of a vector.

PDSV (Program 2.3, p. 102) computes the product of a scalar and a vector.

VCROSS (Program 2.3, p. 102) computes the cross product of two vectors.

5. Subroutine for conformal mapping:

MAPPNG (Program 2.5, p. 128) maps points from the z' plane onto the z plane through the Joukowski transformation $z = z' + b^2/z'$.

6. Subroutines for finding points along a streamline:

SEARCH (Program 2.2, p. 91) scans through vertical grid lines for such points.

SERCH2 (Program 2.7, p. 165) scans through both vertical and horizontal grid lines for such points.

7. Subroutines for plotting:

PLOTN (Program 1.7, p. 64) prints points in a rectangular domain. It is available at the University of Colorado Computing Center.

EZPLOT (Program 2.5, p. 129) displays points in a rectangular domain on a film. It is available at the University of Colorado Computing Center.

CONTUR (Program 4.5, p. 368) generates a contour plot for stream function in a rectangular domain.

BIBLIOGRAPHY

[The numbers in brackets show the pages on which the references appear]

Abbott, I. H., and von Doenhoff, A. E. (1949), *Theory of Wing Sections*, McGraw-Hill, New York. Paperback edition, Dover, New York, 1959. [117]

Abernathy, F. H., and Kronauer, R. E. (1962), "The Formation of Vortex Streets," *J. Fluid Mech.*, Vol. 13, 1–20. [331]

Batchelor, G. K. (1967), *An Introduction to Fluid Dynamics*, Cambridge University Press, Cambridge, England. [260, 276, 322]

Becker, R. (1922), "Stosswelle und Detonation," *Z. Physik*, Vol. 8, 321–362. Available in English translation as NACA TM 505 and TM 506. [217, 219]

Birkhoff, G. D., and Fisher, J. (1959), "Do Vortex Sheets Roll Up?" *Rendi. Circ. Mat. Palermo*. Ser. 2, Vol. 8, 77–90. [323]

Bisplinghoff, R. L., and Ashley, H. (1962), *Principles of Aeroelasticity*, Wiley, New York. [30]

Blanchard, D. C. (1967), *From Raindrops to Volcanoes*, Doubleday, Garden City, N.Y. [15, 16]

Blasius, H. (1908), "Grenzschichten in Flüssigkeiten mit kleiner Reibung," *Z. Math. u. Physik*, Vol. 56, 1. Available in English translation as NACA TM 1256. [230, 231]

Carnahan, B., Luther, H. A., and Wilkes, J. O. (1969), *Applied Numerical Methods*, Wiley, New York. [4, 115]

Chandrasekhar, S. (1961), *Hydrodynamic and Hydromagnetic Stability*, Oxford University Press, London. [359, 360]

Chigier, N. A. (1974), "Vortexes in Aircraft Wakes," *Scientific American*, Vol. 230, No. 3, 76–83. [97]

Chow, C.-Y., and Halat, J. A. (1969), "Drag of a Sphere of Arbitrary Conductivity in a Current-Carrying Fluid," *Phys. Fluids*, Vol. 12, 2317–2322. [307]

Chow, C.-Y., and Lai, Y.-C. (1972), "Alternating Flow in Trachea," *Respiration Physiology*, Vol. 16, 22–32. [276]

Churchill, R. V. (1948), *Introduction to Complex Variables and Applications*, McGraw-Hill, New York. [118, 121]

Courant, R., Isaacson, E., and Rees, M. (1952), "On the Solution of Nonlinear

Hyperbolic Differential Equations by Finite Differences," *Comm. Pure and Appl. Math.*, Vol. 5, 243–255. [206, 208]

Cowling, T. G. (1957), *Magnetohydrodynamics*, Wiley-Interscience Publishers, New York. [337]

Falkner, V. M., and Skan, S. W. (1930), *Some Approximate Solutions of the Boundary Layer Equations*, British Aeronautical Research Committee Reports & Memo. 1314. [241]

Gilbarg, D., and Paolucci, D. (1953), "The Structure of Shock Waves in the Continuum Theory of Fluids," *Jour. Rational Mech. and Analysis*, Vol. 2, 617–642. [220]

Goldstein, S., editor (1938), *Modern Developments in Fluid Dynamics*, Vol. 1, Clarendon Press, Oxford, England. [6]

Grad, H. (1952), "The Profile of a Steady Plane Shock Wave," *Comm. Pure and Appl. Math.*, Vol. 5, 257–300. [220, 221]

Greenspan, D. (1974), *Discrete Numerical Methods in Physics and Engineering*, Academic Press, New York. [279]

Greenspan, H. P. (1968), *The Theory of Rotating Fluids*, Cambridge University Press, Cambridge, England. [380]

Hama, F. R., and Burke, E. R. (1960), *On the Rolling Up of a Vortex Sheet*, University of Maryland, Tech. Note No. BN-220. [323, 325]

Hamielec, A. E., Storey, S. H., and Whitehead, J. M. (1963), "Viscous Flow Around Fluid Spheres at Intermediate Reynolds Numbers (II)," *Canadian J. Chem. Eng.*, Vol. 41, 246–251. [306]

Hamielec, A. E., Hoffman, T. W., and Ross, L. L. (1967), "Numerical Solution of the Navier-Stokes Equation for Flow Past Spheres: Part 1. Viscous Flow Around Spheres With and Without Radial Mass Efflux," *A.I.Ch.E. Jour.*, Vol. 13, 212–219. [306]

Happel, J., and Brenner, H. (1965), *Low Reynolds Number Hydrodynamics*, Prentice-Hall, Englewood Cliffs, N.J. [17]

Hartree, D. R. (1958), *Numerical Analysis*, second edition, Oxford University Press, London. [208]

von Helmholtz, H. (1868), "Über discontinuirliche Flüssigkeitsbewegungen," *Monatberichte Akad. d. Wiss. Berlin*, Vol. 23, 215–228. [322]

Hess, J. L. (1972), *Calculation of Potential Flow about Arbitrary Three-Dimensional Lifting Bodies*, Report No. MDC J5679-01, McDonnell Douglas Corporation, Long Beach, Calif. [153]

Hess, J. L. (1975), "Review of Integral-Equation Techniques for Solving Potential-Flow Problems with Emphasis on the Surface-Source Method," *Computer Methods in Applied Mechanics and Engineering*, Vol. 5, 145–196. [134]

Hess, J. L., and Smith, A. M. O. (1966), "Calculation of Potential Flow about Arbitrary Bodies," *Progress in Aeronautical Sciences*, Vol. 8, 1–138. [143]

Hirt, C. W. (1968), "Heuristic Stability Theory for Finite-Difference Equations," *J. of Computational Physics*, Vol. 2, 339–355. [335]

Hocking, L. M. (1959), "The Collision Efficiency of Small Drops." *Quart. J. Roy. Meteor. Soc.*, Vol. 85, 44–50. [55]

Howarth, L. (1938), "On the Solution of the Laminar Boundary Layer Equations," *Proc. Roy. Soc. London, A*, Vol. 164, 547–579. [233]

Hughes, W. F., and Gaylord, E. W. (1964), *Basic Equations of Engineering Science*, Schaum Publishing Company, New York. [215, 250, 290, 296, 307]

Hughes, W. F., and Young, F. J. (1966), *The Electromagnetodynamics of Fluids*, Wiley, New York. [338]

von Kármán, T. (1927), *Calculation of Pressure Distribution on Airship Hulls*, NACA TM 574. [105]

Kawaguti, M. (1955), "The Critical Reynolds Number for the Flow Past a Sphere," *Jour. Phys. Soc. Japan*, Vol. 10, 694–699. [293, 300]

Kopecky, R. M., and Torrance, K. E., (1973), "Initiation and Structure of Axisymmetric Eddies in a Rotating Stream," *Computers & Fluids*, Vol. 1, 289–300. [342]

Kuethe, A. M., and Chow, C.-Y. (1976), *Foundations of Aerodynamics: Bases of Aerodynamic Design*, third edition, Wiley, New York. [8, 25, 72, 86, 99, 124, 143, 150, 153, 154]

Kuo, S. S. (1972), *Computer Applications of Numerical Methods*, Addison-Wesley, Reading, Mass. [4]

Lamb, H. (1932), *Hydrodynamics*, sixth edition, Cambridge University Press, Cambridge, England. [6, 61, 104, 324]

Landau, L. D., and Lifshitz, E. M. (1959), *Fluid Mechanics*, Addison-Wesley, Reading, Mass. [309]

Langmuir, I. (1948), "The Production of Rain by a Chain Reaction in Cumulus Clouds at Temperatures above Freezing," *J. Meteor.*, Vol. 5, 175–192. [55]

Lax, P. D., and Richtmyer, R. D. (1956), "Survey of the Stability of Linear Finite Difference Equations," *Comm. Pure Appl. Math.*, Vol. 9, 267–293. [343, 364]

Liepmann, H. W., and Roshko, A. (1957), *Elements of Gasdynamics*, Wiley, New York. [186, 190, 202, 203, 209, 214]

Lin, C. C. (1955), *The Theory of Hydrodynamic Stability*, Cambridge University Press, Cambridge, England. [241]

Lindsey, W. F. (1938), *Drag of Cylinders of Simple Shapes*, NACA Report 619. [33]

McCormick, J. M., and Salvadori, M. G. (1964), *Numerical Methods in FORTRAN*, Prentice-Hall, Englewood Cliffs, N.J. [77]

Moore, D. W. (1974), "A Numerical Study of the Roll-Up of a Finite Vortex Sheet," *J. Fluid Mech.* Vol. 63, 225–235. [315, 317]

Pohlhausen, E. (1921), "Der Wärmeaustausch zwischen festen Körpern und

Flüssigkeiten mit kleiner Reibung und kleiner Wärmeleitung," *Zeit. angew. Math. u. Mech.*, Vol. 1, 115. [249]

Prandtl, L. (1904), "Über Flüssigkeitsbewegung bei sehr kleiner Reibung," *Proc. III Intern. Math. Congr.*, Heidelberg. [70]

Rayleigh, Lord (1911), "On the Motion of Solid Bodies through Viscous Liquid," *Phil. Mag.*, Ser. VI, Vol. 21, 697–711. [260]

Rayleigh, Lord (1916), "On Convective Currents in a Horizontal Layer of Fluid When the Higher Temperature Is on the Under Side," *Phil. Mag.*, Ser. VI, Vol. 32, 529–546. [359]

Richtmyer, R. D., and Morton, K. W. (1967), *Difference Methods for Initial-Value Problems*, second edition, Wiley-Interscience Publishers, New York. [217]

Roache, P. J. (1972), *Computational Fluid Dynamics*, Hermosa Publishers, Albuquerque, N.M. [161, 217, 336]

Rosenhead, L. (1931), "The Formation of Vortices from a Surface of Discontinuity," *Proc. Roy. Soc. London, A*, Vol. 134, 170–192. [323]

Rosenhead, L. (1940), "The Steady Two-Dimensional Radial Flow of Viscous Fluid Between Two Inclined Walls," *Proc. Roy. Soc. London, A*, Vol. 175, 436–467. [309]

Schlichting, H. (1968), *Boundary Layer Theory*, sixth edition, McGraw-Hill, New York. [215, 233]

Scorer, R. S. (1958), *Natural Aerodynamics*, Pergamon Press, London. [379]

Sherman, F. S. (1955), *A Low-Density Wind-Tunnel Study of Shock Wave Structure and Relaxation Phenomena in Gases*, NACA TN 3298. [221, 222, 226, 227]

Snyder, L. J., Spriggs, T. W., and Stewart, W. E. (1964), "Solution of the Equations of Change by Galerkin's Method," *A.I.Ch.E. Jour.*, Vol. 10, 535–540. [309]

Stoker, J. J. (1957), *Water Waves*, Wiley-Interscience Publishers, New York. [214]

Strawbridge, D. R., and Hooper, G. T. J. (1968), "Numerical Solution of the Navier–Stokes Equations for Axisymmetric Flows," *J. Mech. Eng. Sci.*, Vol. 10, 389–401. [377]

Taneda, S. (1956), "Studies on Wake Vortices (III). Experimental Investigation of the Wake Behind a Sphere at Low Reynolds Numbers," *Rep. Res. Inst. Appl. Mech. Kyushu Univ.*, Vol. 4, 99–105. [300]

Taylor, G. I. (1923), "Stability of a Viscous Liquid Contained Between Two Rotating Cylinders," *Phil. Trans. Roy. Soc., London, A*, Vol. 223, 289–343. [375]

Taylor, T. D., and Acrivos, A. (1964), "On the Deformation and Drag of a Falling Viscous Drop at Low Reynolds Number," *J. Fluid Mech.* Vol. 18, 466–476. [17]

Theodorsen, T. (1935), *General Theory of Aerodynamic Instability and the Mechanism of Flutter*, NACA Report 496. [30]

Thom, A., and Apelt, C. J. (1961), *Field Computations in Engineering and Physics*, Van Nostrand, New York. [333]

Thoman, D. C., and Szewczyk, A. A. (1966), *Numerical Solutions of Time Dependent Two-Dimensional Flow of a Viscous, Incompressible Fluid Over Stationary and Rotating Cylinders*, Tech. Rept. 66–14, Heat Transfer and Fluid Mechanics Lab., Dept. of Mech. Eng., University of Notre Dame, Notre Dame, Ind. [333]

Torrance, K. E. (1968), "Comparison of Finite-Difference Computations of Natural Convection," *J. of Research of National Bureau of Standards*, Vol. 72B, 281–301. [342]

Torrance, K. E., and Rockett, J. A. (1969), "Numerical Study of Natural Convection in an Enclosure with Localized Heating from Below—Creeping Flow to the Onset of Laminar Instability," *J. Fluid Mech.*, Vol. 36, 33–54. [342]

Uberoi, M. S., and Chow, C.-Y. (1977), "Large Scale Motions in Electrical Discharges," *Phys. Fluids*, Vol. 20, 1815–1820. [358]

Yih, C.-S. (1965), *Dynamics of Nonhomogeneous Fluids*, Macmillan, New York. [379]

Yih, C.-S. (1969), *Fluid Mechanics*, McGraw-Hill, New York. [122]

Zucrow, M. J., and Hoffman, J. D. (1977), *Gas Dynamics*, Vol. 2, Wiley, New York. [204]

AUTHOR INDEX

SUBJECT INDEX